'This is an extremely valuable work ... it fu
of a textbook, for it offers an interpretativi
right.' Mark Harrison, Professor of the Hi
Wellcome Unit for the History of Medicine, ~.....u ~...v.isity, ~ix

'The book provides a narrative history in relation to empire and situates that narrative within a wider understanding of the economic, political and military functions of empire; it introduces readers to the rich and varied historiography surrounding this topic; and provides a long background to the problems of contemporary medicine and international health.' David Arnold, Emeritus Professor of History, University of Warwick, UK

The history of modern medicine is inseparable from the history of imperialism. *Medicine and Empire* provides an introduction to this shared history – spanning three centuries and covering British, French and Spanish imperial histories in Africa, Asia and America.

Exploring the major developments in European medicine from the seventeenth century to the mid-twentieth century, Pratik Chakrabarti shows that they had a colonial counterpart and were closely intertwined with European activities overseas. His themes include:

- the increasing influence of natural history on medicine
- the growth of European drug markets
- the rise of surgeons in status
- ideas of race and racism
- advancements in sanitation and public health
- the expansion of the modern quarantine system
- the emergence of germ theory and global vaccination campaigns.

Drawing on recent scholarship and primary texts, this book narrates a mutually constitutive history in which medicine was both a 'tool' and a product of imperialism, and provides an original, accessible insight into the deep historical roots of the problems that plague global health today.

Pratik Chakrabarti is Reader in History at the University of Kent. He is the author of *Western Science in Modern India: Metropolitan Methods, Colonial Practices* (2004), *Materials and Medicine: Trade, Conquest and Therapeutics in the Eighteenth Century* (2010) *and Bacteriology in British India: Laboratory Medicine and the Tropics* (2012). He is also one of the editors of the journal *Social History of Medicine*.

Medicine and Empire

1600–1960

PRATIK CHAKRABARTI

First published 2014 by
PALGRAVE MACMILLAN

Palgrave Macmillan in the UK is an imprint of Macmillan Publishers Limited, registered in England, company number 785998, of Houndmills, Basingstoke, Hampshire RG21 6XS.

Palgrave Macmillan in the US is a division of St Martin's Press LLC, 175 Fifth Avenue, New York, NY 10010.

Palgrave Macmillan is the global academic imprint of the above companies and has companies and representatives throughout the world.

Palgrave® and Macmillan® are registered trademarks in the United States, the United Kingdom, Europe and other countries

ISBN: 978–0–230–27635–2 hardback
ISBN: 978–0–230–27636–9 paperback

This book is printed on paper suitable for recycling and made from fully managed and sustained forest sources. Logging, pulping and manufacturing processes are expected to conform to the environmental regulations of the country of origin.

A catalogue record for this book is available from the British Library.

A catalog record for this book is available from the Library of Congress.

Printed in China

Contents

List of illustrations and table   vi

Preface and acknowledgements   vii

Introduction   ix

 1   Medicine in the Age of Commerce, 1600–1800   1

 2   Plants, medicine and empire   20

 3   Medicine and the colonial armed forces   40

 4   Colonialism, climate and race   57

 5   Imperialism and the globalization of disease   73

 6   Western medicine in colonial India   101

 7   Medicine and the colonization of Africa   122

 8   Imperialism and tropical medicine   141

 9   Bacteriology and the civilizing mission   164

10   Colonialism and traditional medicines   182

Conclusion: The colonial legacies of global health   200

Bibliography   206

Index   233

Illustrations and Table

Figure 1.1   'The Age of Discovery, 1340–1600' (from William R. Shepherd, *Historical Atlas*, 1911). Courtesy of the University of Texas Libraries, the University of Texas at Austin   5

Figure 3.1   James Lind uses lemon to treat a patient with scurvy on HMS *Salisbury*, 1747 (by Robert A. Thom, in *A History of Medicine in Pictures*, 1960). Courtesy of the Collection of the University of Michigan Health System, gift of Pfizer, inc, UMHS. 17   44

Figure 5.1   A map showing the likely routes of transmission of cholera from India to Europe and the Americas in the nineteenth century (by Edmund Charles Wendt, 1885). Courtesy of the Wellcome Library, London   84

Figure 5.2   An engraving of a quarantine house for Egyptian soldiers suffering from the plague, Port Said, 1882 (by Kemp). Courtesy of the Wellcome Library, London   90

Figure 6.1   An engraving of the European General Hospital, Bombay, India (by J.H. Metcalfe, 1864). Courtesy of the Wellcome Library, London   108

Figure 7.1   *A Medical Missionary Attending to a Sick African* (by Harold Copping). Courtesy of the Wellcome Library, London   132

Figure 8.1   Members of the Sleeping Sickness Commission, Uganda and Nyasaland, 1908–13. Courtesy of the Wellcome Library, London   155

Table 5.1   Crude death rates (CDRs) on ocean voyages, 1497–1917   79

Preface and Acknowledgements

While teaching courses on imperialism, medicine and science I have constantly faced the challenge of discussing these subjects over the long timescale and geographical expanse that the discipline needs. Students all too often are engrossed in the particular area which specialized research articles and monographs present. Therefore they tend to find it difficult to grasp large historical trajectories. This is not merely a problem of pedagogy. It reflects the conceptual frameworks in imperial history of medicine, which tend to focus on a particular geographical area such as South America, Africa or India; or on epidemics such as malaria, cholera, plague or yellow fever; or on tropes such as tropical medicine. Although there are a vast number of excellent research monographs, articles and edited volumes on these themes, they do not engage with the breadth of topics and period that mark the histories of medicine and empire. As a result, imperialism and medicine are not addressed as comprehensively as they ought to be. Edited volumes such as *Disease, Medicine, and Empire* (edited by Roy MacLeod and Milton Lewis, 1988) and *Warm Climates and Western Medicine* (edited by David Arnold, 1996) are extremely useful in classrooms and for research consultation, and are expansive in geographical sense, but not so in thematic and temporal range. Despite their broad titles, both books cover the period from the mid-nineteenth to the early twentieth century and focus on 'tropical medicine', a period and a theme that is also otherwise well served by monographs and research articles.

This book seeks to address this pedagogical and conceptual gap in understanding medicine and imperialism as a theme in itself. It has two main aims. The first is to make the historical and analytical links between medicine and European imperialism from the seventeenth century to the mid-twentieth century and highlight the various research questions and intellectual traditions that have emerged in this field. The second, in this process, is to mark the history of medicine and imperialism as a specialization in its own right. In various parts of the book I have tried to indicate the shared ideas, concerns, debates and intellectual frameworks across different colonial contexts. At the same time, this is not a work of essentialism; the history of medicine and empire here narrates the history of modern medicine itself. In précis, *Medicine and Empire* helps us to appreciate the history of medicine on a global scale and understand the deep problems that plague global health. I have not attempted to cover each of the imperial contexts over 350 years because that would have been impossible in a single volume. Most of the discussion revolves around British, Spanish and French colonial history, although there is some analysis of Dutch and Portuguese colonialism as well. The book broadly covers Asia, the Americas (including the West Indies) and Africa.

I have also taken a personal journey through this history, reading (some books and articles for the first time) as closely and widely as possible. That has been a most enlightening experience. While writing this book I have drawn from my own research but I have also depended heavily on other works, particularly in areas where I have little expertise. I am deeply indebted to scholars who have enriched the field with their research, rigour and insight over the years. Throughout, I have tried to refer to and reflect the viewpoints of the prominent writings on the subject. If this book can inspire and encourage students to read further and scholars to reflect on the succinct and the broader narratives then I will consider that the task of writing it to have been accomplished.

Nandini, as ever, has been the reason why I was able to write this book. Most of the chapters were written in close consultation with her. I have borrowed extensively from her teaching and research experiences. She has selflessly given her time, comments and commitment to this book.

Sarah Hodges, with her comments on the Introduction, helped me to explore and identify the interactions between imperialism and medicine more closely. I am grateful to my colleague Giacomo Macola for turning me into, in his own words, 'a closet Africanist'. I also wish to thank Mark Harrison for his detailed comments on Chapter 5. Jenna Steventon (head of humanities at Palgrave Macmillan) read parts of the manuscript with great care and made invaluable suggestions.

My greatest debt is to my students whom I have taught in India and in the UK. This book is a product of the questions they have raised, the startlingly incisive comments they have made and the occasional looks of incomprehension that they have cast in my direction. They have inspired my teaching and indeed this book.

Introduction

The history of modern medicine cannot be narrated without the history of imperialism. European medicine underwent fundamental epistemological and structural changes as European empires expanded worldwide. From the sixteenth century, small countries in Western Europe started to build global empires. The Genoa-born navigator from Spain, Christopher Columbus, reached the Americas in 1492 after crossing the Atlantic Ocean. A few years later, in 1498, Portuguese traveller Vasco da Gama reached India through a new route around the Cape of Africa. These routes to the Americas and Asia led to novel commercial and cultural contacts between Western Europe and the Atlantic and Indian oceans. Between the seventeenth and the twentieth century, major parts of these regions became colonies of European nations. As Europeans explored and exploited the natural resources of the colonies, European medicine broke out of the old Galenic (a medical tradition of Greek heritage in medieval Europe pertaining to Galen, a celebrated physician of the 2nd century AD) practices and acquired new ingredients, such as cinchona, jalap, tobacco and ipecacuanha, along with new medical insights into their uses from the colonies. European surgeons acquired vital medical skills and experience during the long colonial voyages across oceans and in the hard services of colonial outposts and battlefields. European medical experiences in warm climates, with tropical fevers, vermin and vectors led to the incorporation of environmental, climatic and epidemiological factors within modern medicine, which in turn led to what is known as the 'holistic turn' in modern medicine. European encounters with other races established ideas of race and human evolution within modern medical thinking. At the same time, modern medicine helped in the colonization of the Americas, Asia and Africa by improving the mortality rates of European troops and settlers. European physicians, travellers and missionaries offered their medicine as lifesaving drugs or as tokens of their benevolence and superiority to the colonized races. This book will explore these convergences of the histories of imperialism and medicine. It will identify the intellectual and material connections between the rise of European empires and the making of modern medicine.

Beyond this exploration of the history of medicine and empire, the book serves two further purposes. It helps us to appreciate the history of medicine on a global scale. It also provides the historical context of the deep problems that plague global health today.

The book divides this long history into four broad historical periods: the Age of Commerce (1600–1800), the Age of Empire (roughly 1800–80), the Age of New

Imperialism (1880–1914) and the Era of New Imperialism and Decolonization (1920–60). Each era had a distinct orientation in the history of medicine and imperialism but they were also marked with continuities and overlaps.

European imperialism and modern medicine

How did Europeans build global empires from the sixteenth century? These were built through long and complicated historical processes. It is helpful to approach this history by studying it through the different phases. The first was the Age of Commerce, a period in which, following the discoveries of new trade routes, Europeans (particularly the Spanish and the Portuguese) began to build maritime empires in the Atlantic and Indian oceans. The Spanish colonized the so-called New World and the Portuguese obtained territories in parts of Asia and Africa, and they exercised varying degrees of political and economic control over their subject populations. Other European nations, such as the Dutch, French and British, entered the maritime expansion from the seventeenth century, leading to major colonial warfare in the eighteenth century. Trade and commerce were the critical components of power and prosperity in this period. Historians have often described it as the first age of globalization in modern history.[1]

The Age of Empire followed in the nineteenth century. In this period, European nations, particularly France and Britain, established vast territorial empires in Asia and Africa. As Europeans now ruled and governed over large populations, they devised colonial administrative services, developed new agrarian policies, instituted laws, started universities, and established their own medical ideas and practices in the colonies as the mainstay of their rule. This was also the period of industrialization in Europe and the colonies gradually became the suppliers of raw materials for European industries. This led to a greater integration of the colonies into global economy. These economic changes also led to the migration of a large number of people as settlers and indentured labourers within the empires.

In the last part of the nineteenth century there was a greater rush to extend imperial possessions, particularly in Africa, and the period is often described as the Age of New Imperialism. Global economic competition between industrial nations to acquire natural resources and territories, the pursuit of imperial prestige and possessions, and the urge to spread European civilization led to the 'scramble for Africa' from the 1880s. Until the First World War, this was the high point of imperialism, with European colonial powers expanding across all continents of the earth. This was the time when European medicine specialized to serve colonial purposes and interests, particularly in the birth of tropical medicine.

The period following the First World War marked the phase of nationalist struggles for independence in several of the colonies. The Second World War was followed by a greater phase of such anti-imperial movements, particularly in Africa, as well as the period known as 'decolonization', during which several countries in Africa and Asia gained independence from colonial rule and began the exciting and difficult task of building their nations. This was also a period in which these countries made greater cultural and nationalist assertions in their practices of medicine.

It is important to note that imperialism did not always follow a clear or linear pattern. There were notable overlaps and parallels in imperial history in different periods. For example, the term 'civilizing mission' is associated with the late-nineteenth century European colonization of Africa (which we will study in chapters 7 and 8), when Europeans believed that by colonizing Africa they were introducing modern civilization and Christianity to the continent. The term can also be applied to the Spanish colonization of the Americas in the sixteenth and seventeenth centuries. The Spanish justified their colonization of the Americas in terms of a religious mission to bring Christianity and civilization to the native peoples of the Americas, whom they regarded as savages.[2] Similarly, the European Age of Discovery is commonly associated with the period between the seventeenth and eighteenth centuries when Europeans travelled to different parts of the world, discovered, and investigated the natural history of Asia, America and the Pacific region (see chapters 1 and 2). In most parts of Africa the European Age of Discovery only featured in the second half of the nineteenth century, when European geographers and naturalists, particularly following David Livingstone's Zambezi Expedition, started collecting specimens of central African flora and fauna to study and display in museums in Europe. These overlaps are important in our history of medicine as they help us to note and compare the connections and similarities of historical events across continents and periods.

Each of these phases of imperialism was marked by almost corresponding phases of change in the history of medicine. Not only was European medicine an important component of European imperialism from the sixteenth century but medicine itself evolved along with the history of imperialism. We will see in Chapter 1 that European *materia medica* (the various substances and preparations used in medical practice and treatment) expanded and diversified rapidly from the seventeenth century, in the Age of Commerce. The importation of exotic drugs into European markets transformed the European pharmacopoeia and medical theories. As Europeans travelled to different parts of the world in the eighteenth century and encountered different climates, European theories of diseases changed. Physicians reworked their traditional medical theories to explain the diseases, particularly the various 'fevers', that they experienced in hot climates. British physicians in the eighteenth century dwelt extensively on the so-called putrid or pestilential fevers that ravaged the health of European sailors and naval personnel on long colonial voyages. To cope with the problems of overcrowding in the ships and in colonial stations, European physicians such as James Lind and John Pringle developed theories of sanitation and hygiene. They advocated the need for adopting hygienic practices, including the disposal of waste, maintaining cleanliness and ensuring ventilation within military and naval quarters and ships. These ideas of cleanliness and hygiene gradually became part of general European preventive medicine and state policy in the nineteenth century (see Chapters 3 & 5). The several outbreaks of cholera, which appeared in Europe from Asia in the nineteenth century, paved the way for the most important public health and sanitarian measures in Europe and America.

On the other hand, in the nineteenth century, in the Age of Empire, the development of laboratory-based medicine transformed European medicine. The growth of laboratories and industrialization in Europe became important in producing modern drugs and pharmaceuticals. These also contributed to the emergence of the modern pharmaceutical industry, particularly in France and Germany. We will see in Chapter 2 how this turn towards modern pharmaceuticals from plant-based medicines was shaped by colonialism. Modern laboratory experiments also led to the emergence of the germ theory of disease from the 1880s, predominantly in France and Germany. French chemist and microbiologist Louis Pasteur developed vaccines by partial attenuation of viruses. The famous breakthrough came in 1885 when he developed the rabies vaccine. Soon French Pasteur Institutes spread to the French colonies in Africa and South East Asia, and germ theory and vaccines became part of global and imperial medicine. In the colonies, particularly during the New Imperialism of the 1890s, the Pasteur institutes became part of the French imperial 'civilizing mission'. These new developments brought a much greater assertiveness in European medicine in the colonies. The importation and spread of modern European drugs and vaccines became an important part of colonial medical policies in the nineteenth century. One key difference in the practice of medicine in this period from that of the previous era was that colonial medicine now was not only for the European sailors, soldiers and settlers but for the indigenous people. Modern pharmaceuticals and vaccines were not only vital in preserving European health in the tropics; they were also presented in the colonies as symbols of European modernity and superiority.

The emergence of germ theory in the tropics took place in the era of New Imperialism, at a time when Europeans viewed the tropical climate and environment as unhealthy regions that were ridden with diseases. The combination of germ theory with concerns of tropical climate gave rise to the new medical tradition of tropical medicine by the end of the nineteenth century. Historians have shown that tropical medicine promoted the late nineteenth-century ideas of 'constructive imperialism', particularly during the intense imperial expansion, particularly in Africa, which suggested that imperialism was ultimately for the benefit of the colonized people and nations.

On the other hand, as we will see in Chapter 10, non-Western countries and societies were not passive recipients of these various changes in modern medicine. Indigenous physicians and medical professionals in Asia and Africa engaged creatively with modern medicine, often defining their application in unique ways, but also by modernizing their indigenous medicines. This was particularly evident at the time of the rise of nationalist consciousness and decolonization in the twentieth century. Local practitioners in Asia and Africa responded to the dominance of Western medicine by codifying and standardizing their own medicines, selecting certain drugs or practices, which corresponded with modern ideas and requirement, introducing new medical substances and modern laboratory techniques, and producing indigenous pharmacopoeias. The emergence of what is known as 'traditional' or 'alternative' medicine is thus linked with the history of colonialism.

The twentieth century was also a period of international collaborations in healthcare. The First World War and the subsequent epidemics, such as the Spanish flu led to the establishment of the League of Nations Health Organization (LNHO) in 1921. The Second World War led to the formation of the World Health Organization (WHO) in 1948 in Geneva, Switzerland. The WHO marked a new era in global health and epidemic disease control. In this period, isolated colonial public health measures were linked with international policies and undertakings, which came to be known as 'global health'. The main activities of the WHO in the 1950s and 1960s was in global vaccination campaigns against measles, polio and smallpox, in anti-malarial surveys, in addressing questions of poverty and health, and in ensuring basic medical infrastructure in different parts of the world. Global health initiatives and policies in the twentieth century developed in collaboration with colonial health measures and retained a strong colonial legacy. We will see why an understanding of the history of colonial medicine is important in appreciating the contemporary challenges of global health.

What is colonial medicine?

Let us start exploring these complex issues and conceptualizing the nature of the relationship between medicine and imperialism by analysing a key term that is often used by historians of medicine to describe this history: 'colonial medicine'. What is meant by this phrase and is it useful in understanding this history?

The term 'colonial medicine' is derived from another, 'colonial science', which was used in the 1960s by historian of science and technology George Basalla to describe the second phase of his three-phase model of transmission of Western science to the so-called periphery.[3] In this phase, scientific activities in the colonies were closely linked with the interests of the metropolis, making scientific activities in the colonies dependant on those of the metropolitan institutions. Basalla's model has been criticized by historians for being too linear, simplistic and rigid.[4] It is not necessary here to get into a discussion of his model of colonial science, as I will use the phrase in a different way. I will instead analyse whether and if there are other important and useful reasons to use the phrase 'colonial medicine'.

Let us start by asking two basic questions. What is medicine and why do we need to categorize it into different types, forms or 'frames'? At an essential level, medicine is the art of healing and of preventing diseases. This art took different shapes in different social, cultural and historical contexts. Historians, philosopher and sociologists have used different categories to indicate these different circumstances.[5] Without adopting such categories and by referring to just 'medicine' it would be almost impossible to understand the various contexts within which medicine developed, and physicians and scientists operated. This in turn may also lead to accepting and reverting to the nineteenth-century positivist definitions of modern medicine and science as singular, universal and progressive.

We then need to analyse why and how medical/scientific traditions are distinguished and named. Medical traditions are named in two broad ways. The first is of ethnological or civilizational lineage, such as Arabic, Greek, Chinese, Galenical, Ayurvedic and Unani. Such practices of naming have been relatively straightforward, although, as shown in Chapter 10, can be products, of and lead to, cultural essentialism.[6]

The other is the more problematic and complicated practice of the naming of medical specializations, which I refer to as 'historical'. By this I mean that these traditions are seen to have developed from within particular historical processes and contexts. This includes categories such as 'Western', 'modern' and 'colonial'.

There have been important debates about such categories among historians, which reflect the discussions about these historical processes and contexts themselves. Historians have similarly questioned whether 'colonial medicine' is a distinct or useful category. In what senses is it distinct from Western medicine or just medicine?[7] Despite the debates about these categories and their historical processes, historians have nevertheless identified clear historical processes and episodes when medicine and science became 'Western', 'European' or even 'modern'.[8] There were two key phases in this history. The first began at the end of the thirteen century when classical Greek medical and scientific traditions were gradually severed from their Arabic or Islamic lineage and ensconced within European and Christian thought and tradition.[9] The Latin translation of Greek texts and the introduction of Hippocratic (the Greek medical tradition derived from the teachings of Hippocrates, a celebrated physician born around 460 BC) and Galenic medicine in Europe from the fourteenth century is an example of this.

The second key episode started in the late seventeenth century when European natural historians sought to develop objective views of nature.[10] European physicians similarly incorporated natural history as a key aspect of their medical knowledge, along with their study of classical Greek texts.[11] Alongside that were the changing views of the human constitution and its relationship with the environment (which we will study in Chapter 5). Although these pursuits were richly informed by colonial experiences, these developments took place within a clearly European intellectual and social problematic – that of a search for Europe's antiquity and an objective understanding of nature. These took place within European institutions (and in the process led to their pre-eminence) such as the Royal Society and the Royal College of Physicians (London), the Academie de sciences (Paris) and the Collegium Medicum (Amsterdam). The birth of modern European medicine took place in the course of these explorations of antiquity and the investigations of nature.

There is a historical divergence here, which is marked by imperialism. As we will see in this book, from the seventeenth century, precisely when medicine was thus becoming modern or European, it was becoming colonial as well. Each major development in European medicine had a colonial counterpart: the growing influence of natural history in early modern medicine; the growth and expansion of European drug markets from the seventeenth century; the rise of surgeons in status and influence within the European medical profession; developments in ideas of sanitation, hygiene and public health, particularly in the wake of cholera epidemics in the nineteenth century; growth of modern quarantine systems; the search for 'active ingredients' of cinchona, opium and tobacco, leading to the birth of modern pharmaceuticals; and finally the emergence of germ theory and prophylactic vaccination, and ideas of global health. Each of these episodes had simultaneous European and colonial histories, conjoined and yet distinct, and one cannot be understood without the other. This book seeks to narrate this mutually constitutive history of modern medicine and European imperialism.

One point needs clarification before we return to the analysis of 'colonial medicine' as a useful category. Different names for or descriptions of medicine do not need to indicate distinct research methodologies or epistemologies. The social history of medicine and science, which we will learn more about in the next few pages, enables us to appreciate that medicine or science do not only need to be analysed or defined by 'internalist' (where the focus is on the mode of cognition) parameters or research methodologies. In other words, medicine and science need not be only understood within the laboratories, in scientific formulas and theories, or within esoteric texts and mathematical calculations. The social and historical contexts of these intellectual traditions are equally, if not more, instructive.

For example, this contextual logic is used to name the medicine practised in the tropical colonies from the late nineteenth century, which is known as tropical medicine. This, as we shall see in Chapter 8, was an ambiguous category in terms of research methodology and lineage. It refers to the various medical practices which developed in the tropics from the eighteenth century and which served colonial interests in the Age of Empire. At the same time, its deep-seated links with European laboratory research or biomedicine is also evident. Thus the phrase is based more on its specific social, historical and political context rather than epistemological distinctions. To that extent, tropical medicine is a useful category in the history of medicine.

By the same rationale, 'colonial medicine' appears as a useful category because it refers to distinct contexts of medical practices, different from, although closely related to, the European or metropolitan ones. What are those contexts? This book highlights several circumstances when medicine became 'colonial'.

Some of the ways in which medicine and disease were embedded in the history of colonialism include their effect on local populations by spreading diseases, and as tools of empire in reducing the mortality of colonial troops, in the modernization of colonial armed forces or in the hegemonic roles it played in establishing European dominance in the colonies. This takes us to exploring the history of medicine and ideas of difference in the colonies. Historians such as Shula Marks and Londa Schiebinger have shown that colonial medicine was deeply implicated in promoting ideas of difference, in terms of race, gender and class.[12] We will study how ideas of difference became critical in modern medicine and in colonial policies. The other area of focus is the integration of medicine within colonial economy and governance. Here we will seek to understand that medicine was intrinsic to colonial economy and administration, in the sugar plantations of the West Indies or the rubber plantations of Africa or Malaya, in the diamond mines of southern Africa, in defining and running the colonial protectorate system in Africa, or in urban administration in Bombay (now Mumbai) or Calcutta (now Kolkata). The book will highlight the problems of a straightforward cause–effect relationship (that medicine helped in colonization) and suggest the need to understand the deeper links between colonialism and medicine.

It revolves around these themes and shows that colonial medicine had its distinct historical trajectory and characteristics, while at the same time being closely linked to the history of European or modern medicine. 'Colonial medicine' can be a useful phrase to describe and understand this distinctive history.

However, contexts are not given or fixed. These have been identified by historians. Historians have also debated intensely on these, as it is here that the essential nature and characteristics of the history of imperialism and medicine lie. An analysis of these debates through a survey of the literature on imperialism and medicine provides us with the means to understand the links between medicine and empire.

The historiography of colonial medicine

Historians have debated the most critical questions in the history of imperialism; the nature of European empires and their impact on the modern world. They have explored how imperialism led to the exploitation of resources, the imposition of alien ideas and cultures, and the disruption of existing social and ecological structures and values in the colonies. Historians have also enquired whether imperialism played a more complex and positive role by introducing ideas such as humanitarianism and Enlightenment, establishing institutions such as universities and hospitals, and economically and culturally revitalizing the colonies by connecting them with the wider world.

Similar questions have been raised in colonial medicine as well. Did Western medicine create its own hegemonic influence alongside political colonialism, by which Western medicine, hospitals and doctors became the proponents of welfare and modernity in the colonies? Did medicine play a more disruptive role by damaging indigenous social systems and medical infrastructure, leading to more diseases and greater mortality, as well as creating a dependency on Western medicine? Alternatively, did it also promote welfare, reducing sickness, epidemics and mortality rates in different parts of the world? Moreover, if medicine emerged through a complex process of exchanges and interactions, then is it possible to define it as Western or Eastern?

History writing on imperialism and medicine had three broad phases. The early historical writings on imperialism started in the nineteenth century. These generally portrayed imperialism as beneficial for the colonized countries, bringing civilization, modernity and welfare to those parts. This was followed by a second phase in the early twentieth century with a diametrically opposite view of seeing imperialism as essentially a negative influence in the colonies. The third phase began in the 1960s, when independence and postcolonial experiences of nation-building and the search for cultural identities prompted a new historical account of imperialism to emerge. This phase was often marked with a deep sense of loss of identity, culture, voices and ecology caused by imperialism and attempts at recovering and understanding the precolonial past. This also led to a composite understanding of imperialism and medicine in the late twentieth century, which took into account not only the roles of the European imperialists but also indigenous people, societies, institutions and ideas. Historians have shown that imperialism has followed a complex process where it is difficult to trace a unilateral tale of loss and suffering.

Early histories of European imperialism

Nineteenth-century writings generally represented imperialism as a triumph of European cultural superiority and military power over other races and regions in the Americas, Asia and Africa. These accounts, mostly written by Europeans, described their colonial dominance as generally beneficial as it spread ideas of progress, rationality, humanitarianism and Christianity to those who were largely seen to be backward or less civilized. These writings were by a range of people, such as European explorers, missionaries, military personnel and imperial governors. Since these people saw themselves as responsible for ushering civilization and modernity in to other parts of the world, their writing about the empire was marked by both romanticism (about how different and distinct these societies were from Europe, as much as they were untainted by modernity) and paternalism (about their responsibility to usher in modernity and progress there). It is important to remember that when these histories were written – that is, the nineteenth century – it was also a period when European imperialism was its peak and at its most aggressive. Consequently, Europeans looked at the history of their imperialism as a glorious and noble struggle against the forces of savagery, prejudice and 'darkness'.

Later historians have analysed these early imperial histories as a way to write new histories of imperialism. Alfred Crosby, a historian of early North American colonialism, defined the predominant nineteenth-century style of writing of the

European conquest of America as a 'Bardic interpretation' that romanticized the European fascination with and subsequent conquest of the New World.[13] Nineteenth-century American historians, such as George Bancroft, William Prescott and Francis Parkman described how the 'courageous' Spanish conquerors and settlers from the sixteenth century occupied the lands from the 'savage' and 'barbaric' Amerindian races. The general tone was that the European conquest of the Americas was a tale first of European 'discovery', then of glorious European adventurism and struggle, and finally of the introduction of European civilization and modernity to the Americas. The story of the indigenous Amerindians, who suffered gravely as a result of this colonization, remained largely untold. The Africans who had settled along with the Europeans as slaves were also hardly mentioned in this history.

In Asia, European travellers and statesmen wrote similar glorious histories of imperialism. George Otto Trevelyan (1838–1928), who served as a British civil servant in India, wrote about the history of colonial civil service and the Anglo-Indian society in India in *The Competition Wallah* (1863). He wrote this book to remind the British of their moral responsibility to educate, modernize and Christianize Indians.[14] Imperialism to him was thus beneficial and necessary. Valentine Ignatius Chirol (1852–1929), a British author and journalist, was deeply interested in Indian history and culture and travelled to India in the 1880s. In his history of India entitled *India Old and New* (1921), he wrote about the great ancient traditions of India, its decline during the medieval period of 'Mahomedan domination' and its revival under British rule, which introduced modern education, science and parliamentary democracy.[15]

Missionaries and explorers, such as David Livingstone and Henry Morton Stanley, wrote most of the early history of Africa in the late nineteenth century.[16] These works narrated the glorious and heroic adventures and discoveries of European explorers rather than those of the local inhabitants. Stanley wrote about his travels in Africa in *Through the Dark Continent* (1878), promising Europeans the material riches of Africa and informing them of their moral obligations in ushering civilization in to that continent in return. Joseph Conrad's *Heart of Darkness* (1903), although a novel, helped to establish the notion that Africa was the 'Dark Continent' and thus portrayed European imperialism as a history of domination, enlightenment and emancipation for Africans.[17] As we shall study in Chapter 7, the ethos of transforming Africa to a modern economy was central to its colonization by Europeans.

When Scottish missionary and explorer Livingstone travelled through the plains of Zambezi, he dreamt of cotton fields across the region as a way of transforming the African wilderness.[18] Charles Eliot, commissioner of the East Africa protectorate at the turn of the nineteenth century, believed that the problem of Africa was in its untamed nature and wilderness:

Nations and races derive their characteristics largely from their surroundings, but on the other hand, man reclaims, disciplines and trains nature. The surface of Europe, Asia and north America has been submitted to this influence and discipline, but it has still to be applied to large parts of South America and Africa. Marshes must be drained, forests skilfully thinned, rivers be taught to run in ordered course and not to afflict the land with drought or flood at their caprice...[19]

These writers portrayed Western science, technology and medicine as a gift to the colonies, which would remove prejudices, ignorance and disease. They believed that European medicine reduced the threat of diseases and epidemics of cholera, malaria, plague, leprosy and sleeping sickness, which ravaged the tropical regions. They also described how Europeans played a benevolent role in the empire by introducing modern hospitals, dispensaries and laboratories. As we will see in Chapter 7, both Livingstone and Stanley wrote at length about the positive impact of European medicine in Africa. In this period, both medicine and imperialism were seen as human [European] triumphs; of reason against diseases, ignorance and human suffering.

The critique of colonialism

The change to this tradition of imperial history writing came in the early twentieth century when a new generation of historians adopted a different approach – a structural one, focusing more on economy, society and demographic data rather than on individual stories of triumph. Such an approach exposed the negative effects of imperialism. The first writings of this genre came from the Marxist explanations of economic imperialism. Originally proposed in Karl Marx's writings, the link between European capitalism and late nineteenth-century imperialism was developed as an important thesis in Vladimir I. Lenin's *Imperialism, the Highest Stage of Capitalism* (1916). Around the same time, British liberal thinker and economic historian J.A. Hobson wrote *Imperialism: A Study*, which reflected the similar economic trends in European imperialism. Later Eric Hobsbawm, in *Industry and Empire*, linked these questions of imperialism with industrial capitalism in Britain. He made the point that late nineteenth-century imperialism was generated by pressures produced from within Western capitalism to export capital to colonial dependencies. These texts were the first body of historical writing which portrayed imperialism in a negative light by demonstrating how it essentially benefited the colonial nations and drained the colonies of their resources. These focused particularly on the period from the 1880s, when a great deal of European territorial expansion took place in Asia and Africa, a period now referred to as the Age of New Imperialism.[20]

These ideas of economic imperialism have remained an important means by which historians explained the rise of European colonialism. Scholars such as Immanuel Wallerstein later extended this thesis both historically and geographically, and viewed the entire history of imperialism from the sixteenth century as a 'world system'. According to Wallerstein, with the spread of European colonialism from the sixteenth century, a globally connected network of economic exchange relationships was created. In this system, Europe was at the centre and the colonies were at the periphery, with a unilateral flow of capital and wealth from the periphery to the centre.[21]

This trend influenced writings on the history of medicine that showed that imperialism played a negative role, particularly among the indigenous communities, by spreading epidemics, destroying local medical institutions and transforming the colonies into markets of expensive European drugs and vaccines. Historians now asked whether, rather than being a gift to the colonies, medicine was a 'tool' of empire. Did European medicine, by protecting the health of only the European troops and civilians during the phase of colonial conquest, actually facilitate colonization? Moreover, did new diseases brought and spread by colonial migration in fact aid in colonization by devastating local populations?

Roy Porter, a leading historian of medicine, summed up this approach when he wrote that 'Wherever Europeans went, they brought ghastly epidemics – smallpox, typhus, tuberculosis – to virgin populations entirely lacking resistance.'[22] From this approach the colonization of the Americas appeared to be a very different story. Historians (as elaborated in Chapter 5) showed that the Spanish colonization led to the spread of new epidemics in the Americas, devastating indigenous populations, which helped to expand European colonial territories.[23] Historians also studied how colonialism spread diseases such as malaria, sleeping sickness and smallpox in Africa, challenging the earlier trend of portraying Western medicine in Africa as beneficial and 'heroic'.[24]

This understanding of the negative impact of medicine in the colonies came from another development in the writing of history. The first change came in the historical understanding of medicine and science. In the nineteenth century, physicians rather than historians wrote almost all of the histories of medicine, and these tended to depict the history of medicine as narratives of progress and stories of the achievements of great men. The history of science and medicine in this period was based on positivist and Whiggish ideas, which viewed science and medicine as a continuous and progressive unfolding of objective thought, or as stories of great men: 'the top-down accounts of doctors, by doctors, for doctors'.[25] The history of colonial medicine in this period was similarly written as stories of heroic white doctors, waging war against colonial diseases, epidemics and prejudices.

A new trend in the writing of history of medicine and science emerged from the 1960s. The history of medicine was now written more by historians and sociologists who depicted medicine more as a 'social epistemology'[26], socially and culturally conditioned in the same way as all other forms of human endeavour.[27] Historians gradually developed a greater understanding of the social context of medical discoveries and practice.[28] Historians such as Porter moved the focus of the history of medicine from great doctors and scientists to everyday lives and stories of people. Society emerged as an important area of studying the role of medicine. It gave voice and visibility to a completely new set of actors, such as patients, nurses, drug-peddlers and middlemen; led to a new way of looking at medical artefacts and institutions, such as bedsides, laboratories, hospitals and drugs; and offered a new way of understanding medicine as a social process rather than an unfolding of a universal knowledge. This new focus helped in the appreciation of the negative role that medicine played in the colonies. Historians now undertook a more critical study of the impact of European colonial hospitals, drugs and research.

Colonial medicine and the postcolonial writing of history

Two other developments, specific to the historiography of imperial history and imperial medicine, contributed to this change in the writing of the history of colonial medicine from the 1970s. First was decolonization and the writing of the history of imperialism in the postcolonial period. Second was the attempts at recovering the lost voices of imperial history. In the postcolonial period there was also the gradual democratization of colonial societies, the rise of political and social rights movements, and the general political mobilization of different sections of the population. Thus, from the 1970s, voices of a much wider section of the erstwhile marginalized populace, such as women, lower castes, tribes, peasants and labourers, were heard and recognized, who in turn demanded their own histories.

With the independence of several African nations in the 1960s, African history writing started to incorporate a greater understanding and appreciation of African agencies, voices and, most importantly, stories of resistances. In the 1960s, postcolonial African historiography focused on two interrelated areas: precolonial African polities and African resistance during the colonial period. The classic work was Terence Ranger's article 'Connexions between "Primary Resistance"'.[29] An understanding of precolonial African social life and political structures helped in the analysis of colonial African resistance and agency. African agency thus became a major focus of colonial African history writing. Thus a sense of loss and the need to recover the lost voices of the past became an important preoccupation of imperial history writing from the 1970s.

Edward Said played an important role in the emergence of this new imperial historiography. In his seminal book *Orientalism* (1978), he argued that imperialism led to cultural hegemony.[30] He identified the roots of this cultural imperialism in the European imaginary divide of the world into the Orient (the East, originally those countries East of the Roman Empire) and the Occident (the West, originally with reference to Western Christendom or the Western Roman Empire). In common terms, Europe was the Occident while Asia was the Orient. These divides became prominent during the Crusades (between 1095 and 1291), when Europeans viewed the Arab world and the followers of Islam as backward and treacherous. According to Said, these divisions acquired new significance during colonialism when Europe colonized large parts of Asia. Colonialism confirmed to Europeans that the Orient was backward and the Occident was a superior civilization. In the nineteenth century, Europeans could travel to different parts of Asia, which were now their colonies, to collect books, texts and artefacts, and helped them to acquire a systemic, but prejudicial, knowledge about the Orient.[31] Gradually over the nineteenth century, European universities and museums through their colonial connections developed large collections of Asian texts and artefacts, and Europe became the site of learning about the Orient. According to Said, this combination of political power and cultural dominance created modern Orientalism and cultural imperialism.[32] It had two main characteristics: first, a European stereotype of the Asiatic world and people; and second, a combination of political and cultural power of Europe over its colonies. Although Said's work dealt primarily with West Asia, it has encouraged a range of research along similar lines in South Asia.[33]

An important feature of Said's analysis of cultural imperialism was a sense of loss – of identity, culture and history in the process of colonialism. Said and other historians who wrote about cultural hegemony of empire viewed imperialism as a dislocation, the disruption of indigenous cultures and knowledge traditions, and a loss of agency of the indigenous people in writing their own histories and in defining their own destinies. These feelings of loss became important in impe-rial history writing at a time when several of the colonial countries gained inde-pendence and set out on the journey to reconstruct their nations, cultures and identities, which were perceived to have been lost during colonialism. Therefore postcolonial nationalism was an important reason as to why the writings of Said became so influential.

The second major turning point in the writing of imperial history came from the 1980s, with the emergence of subaltern studies.[34] This was formed by a group of scholars who came from various disciplines and wrote about the colonial and postcolonial experiences of ordinary people. They stressed the need to identify the 'autonomous' subaltern domain (a complex collective of people defined against the 'elite' domain). This comprised marginalized groups and the lower classes who had been rendered without agency and voices through elitist and imperial policies and history writings. The predominant category of identifying a group as subaltern is by the loss of agency.[35] In its efforts to recover these histories, voices and agencies of the marginalized groups, subaltern studies was also premised on a sense of loss. It had a major impact on South Asian and African historiography. It appeared as a particularly useful concept for writing the histories of non-Western countries because it enabled a range of voices and narratives from the past, which were hidden from official documents, literature and elite texts, to be heard and chronicled as history. Subaltern studies enriched the search for African agency in African history.[36]

Even histories of the early colonization of the Americas, although they did not fall within the scope of subaltern studies, reflected this search for agencies, voices and everyday experiences of a diverse group of people affected by colonialism. Experiences of African slaves in the plantations of the West Indies and southern America or in the notorious 'middle passage' across the Atlantic have captured the attention of historians and developed as a major historical field.[37] Historians also studied the agency of slaves in the Atlantic world and in the plantations.[38] In the process, deeper links have also been created between the histories of colonial and precolonial Africa and that of the Atlantic slave trade.[39]

The historiography of imperial medicine was similarly transformed by these two changes in imperial historiography. Along with a greater understanding of the social history of imperial medicine, historians now asked new questions. Did European medicine help to establish European cultural hegemony in the colonies? Did Western medicine play a disruptive role in the colonies by destroying indigenous institutions, co-opting indigenous medical ingredients and methods into modern medicine? What roles did indigenous and subaltern groups play in accepting and shaping medicine in the colonies?

Historians such as Alfred Crosby in his *Ecological Imperialism* (1986) have gone back in time to show that this process of ecological change and destruction has taken place from the early periods of colonialism, which led to the spread of diseases in the Americas from the sixteenth century.[40] Historians have also linked medicine in the colonies with late nineteenth-century economic imperialism and the ecological changes that took place with the introduction of capitalism in colonialism. Helge Kjekshus in his *Ecology, Control and Economic Development in East African History* (1977) showed that colonialism in East Africa from the 1890s led to a series of environmental and medical disasters. New diseases, such as rinderpest, affecting both cattle and wild animals, were introduced. The clearing of forests led to droughts, and modern agriculture and plantations led to the destruction of old pastoral systems and lifestyles.[41] Other historians have argued that major disease and famines resulted from the combined influence of capitalist agriculture, industrialization and colonial administration.[42]

The consequences of ecological destruction in the spread of diseases have been demonstrated particularly in the spread of malaria in Asia and Africa during colonialism. Sheldon Watts and Ira Klein linked the increase and spread of malaria in colonial India to the rapid deforestation, expansion of the railways and consequent ecological changes.[43] For example, the British introduced large-scale irrigation networks in India: by the 1890s, nearly 44,000 miles of canals and distributaries irrigated a quarter of India's total crop area, increasing agricultural output. However, this had negative effects, including waterlogging, flooding and salination of the canals, and greater prevalence of malaria with more mosquito-breeding areas. In Africa, Randall Packard has shown that the spread of malaria was linked to general changes in land and agricultural policies, which led to rural impoverishment and general disruption of African societies.[44]

Along with these, the cultural assertions of Western medicine in the colonies became important. As we will see in chapters 6, 7 and 8, historians of colonial medicine have shown that Western medicine, particularly from the late nineteenth century, assumed a moralistic tone in Asia and Africa. At the end of the nineteenth century, the tropics and tropical bodies were seen as the natural home for the habitation of deadly pathogens. European germ theory and laboratory medicine were represented as a crusade not merely against disease, germs and social/cultural prejudices but also against the tropics themselves.[45] Other historians have studied the role played by European doctors, medical missionaries and nurses in the European civilizing missions in Africa (see chapters 6 and 7). Historians of colonial Africa have studied the hegemonic roles played by missionaries in introducing medicine to the empire.[46]

In his history of British medicine in nineteenth-century India, David Arnold identifies colonial medicine and colonialism as hegemonic, a vehicle 'not just for the transmission of Western ideas and practices to India but also for the generation and propagation of Western ideas about India and, ultimately, of Indians' ideas about themselves'.[47] Through the study of three major epidemic diseases in India – smallpox, cholera and plague – he showed how colonial therapeutics and medicine played both benevolent and hegemonic roles.

These works have been significant in changing the historiographical focus of imperial medicine, beyond great men or ideas of progress and located colonial diseases within social, ecological, cultural and economic histories. This provided a long-term account of the history of medicine, tracing the history of diseases and medicine through long-term changes, enabling historians to connect histories of the Amerindian depopulation in the sixteenth century to the sleeping-sickness epidemics in Africa in the twentieth century.[48]

Colonial medicine and history from below

At the same time, within these narratives of grand transformations, historians also sought to locate the agency of the colonized people. With the question of agency came the issue of resistance. Historians have asked whether the indigenous systems of medicine resist Western political dominance or the intrusions of Western medicine, and how and why they changed themselves in response to it. As we will see in Chapter 10, particularly in the history of traditional medicines, historians have stressed the role of native agency. They have shown that non-Western countries and societies were not passive recipients of these various changes to modern medicine. Medical professionals, healers and patients in Asia and Africa engaged creatively with modern medicine, often defining their application in unique ways, but also by modernizing of their own indigenous medicines. The volume *Western Medicine as Contested Knowledge* showed that while laboratory-based medicine appeared in the colonies as hegemonic, these interventions were also mediated by important instances of native resistances.[49]

Historians have attempted to retrieve the lost voices and agencies of the Amerindians who suffered from the Spanish colonization in the sixteenth century. Starting with the early histories of Amerindians, in 1964 Charles Gibson, for the first time, provided an extensive social history of Aztec life during the Spanish Conquest.[50] Miguel Portilla compiled the Aztec accounts of the Spanish Conquest of Mexico in *The Broken Spears*.[51] Others have studied the Spanish and the Amerindian perspectives on that invasion.[52] Dedra S. McDonald studied the history of the cultural and social interactions between Africans and Amerindians.[53] Robert Voeks wrote how these contacts between Africans and Amerindians gave rise to hybrid healing traditions in southern America combining the medicinal plants, rituals and spiritualities of these two groups.[54]

In India, with the emergence of subaltern studies, the history of medicine gained new momentum, and identified new cases of resistance and formation of collective identities. A prominent example of this form of history of medicine is David Hardiman's study of smallpox and the cult of Devi in tribal Gujarat.[55] Hardiman showed that the adivasis (aboriginal populations) of western India in the early twentieth century evoked the cult of the goddess Devi and its accompanying commandments on cleanliness and diet, to counter the threat of smallpox epidemics as well as a mode of collective assertion against the locally dominant elites, such as merchants and landlords.

The history of medicine in Africa also reflected a greater appreciation of African agency and resistance. Following Ranger's work, 'resistance', in both overt and subvert forms, became a key theme in African history. Megan Vaughan in her book *Curing their Ills: Colonial Power and African Illness* showed the complex physical and psychological impact of Western medicine in colonial Africa, which helped to establish colonial hegemony. At the same time she argued that Africans constantly devised their own modes of accepting and mediating Western medicine within their social and cultural idioms.[56]

Such mediated histories have been particularly relevant in histories of colonial hospitals, which incorporated a range of people from different backgrounds, often working alongside each other, sharing knowledge and experiences. The influence of subaltern studies is evident in the work of Walima T. Kalusa in his analysis of African agency. Kalusa studies the roles of lower class (subaltern) medical workers such as ward attendants, orderlies and nurses at the hospital run by the Christian Missions to Many Lands in Mwinilunga, Zambia, and challenges the depiction of their roles as mere 'agents' of imperialism. He shows how the medical subalterns in their everyday practices and in their translations used African languages and 'heathen' concepts to interpret Western medical terms and technologies, and 'drained Christian medicine of its scientific connotations and simultaneously invested in it "pagan" meanings'.[57] Others have questioned the monolithic structure of imperial medicine and have demonstrated the fractured and layered nature of colonial medical institutions.[58] Historians have similarly shown that colonialism in Africa led not only to ecological and cultural destruction but also to the incorporation of ideas of African ecology and environment within European biomedicine, making it hybrid and complex.[59]

In Indian history, historians have similarly shown how Indians of different sections of life redefined Western medicine and science. From within the subaltern approach, scholars have explored how Western science became both a tool of colonial hegemony and the medium of resistance and challenge to that hegemony by Indians. In the process, Western science was dislocated from its European problematic and became an Indian cultural and intellectual experience.[60] Others have shown that the intervention of Indian intellectuals and scientists led to the cultural transformation of Western science and medicine in nineteenth-century India.[61] In the histories of the introduction of colonial public health in India, Western medicine was not seen as a monolithic structure imposed from above by the colonial state. There were tensions and a lack of consensus among the colonial officials and physicians regarding the medical policies, and Indian elites, politicians and taxpayers played a vital role in defining the trajectory of Indian public health policies.[62] Arnold too recognized the subversion and negotiation in his history of state interventions of medicine.

Together these works have contributed to the emergence of a more complex picture of colonial medicine. Through these, we now know much more about the multifaceted nature of the physical and psychological impact of colonialism and colonial medicine. We also now have a deeper understanding of the roles played by not just European scientists and doctors but a diverse group of people who used and adopted Western medicine in their everyday lives. While this detailed and complex image of medical exchanges and collaborations during the colonial period has been important, it has also raised fresh questions. Two problems emerge from these relatively recent writings on colonial medicine. First is the problem of romanticization of the past and second is the question of agency with respect to colonial power and authority.

Romanticism, agency and power in colonial medicine

As mentioned above, one of the key elements in the writings of cultural imperialism has been the sense of loss. In ecological history too, this sense of the destruction of existing ecology in the colonies has been dominant, which John M. MacKenzie has described as 'Empire and the Ecological Apocalypse'.[63] However, this sense of loss can also lead to a sense of nostalgia and romanticism about the precolonial past. The critique of colonialism, seeing it as disrupting colonial societies, cultures and ecologies, raises the question of what it was like before colonialism. Were parts of Africa and Asia culturally and ecologically pristine or did these too suffer from destructions, cultural dominances and hegemonies?

There can be little doubt that colonialism created a major break and disruption in modern history. However, to understand the nature of that break, historians need also to have a critical understanding of the history of the precolonial period. This is an important undertaking because on this also depend a question about the future of postcolonial nations. What are the ideal courses for decolonization to undertake and which destinies should these nations look towards?

Two examples will illustrate why this is such a critical question. We shall see in Chapter 10 that the belief that colonialism disrupted a coherent and well-defined tradition of medicine in India led to the emergence of modern Ayurveda.

In search of a glorious and coherent precolonial medical tradition, which was both alternative and equivalent to modern medicine, traditional practitioners in the twentieth century revived a new medical tradition of Ayurveda from ancient Sanskrit texts. This meant that the complex and in some respects amorphous practices of medicine that developed in India over several centuries among different communities and within diverse socioreligious groups, beyond these classical texts, were overlooked. A new medical tradition was created. The search for a lost past can thus generate an artificial sense of tradition.

The other example is that of the forced villagization programmes in postcolonial Tanzania started by Julius K. Nyerere (1922–99) the first president of that country. A charismatic and progressive leader, he wished to connect modern socialism with African communal living. He also wanted to liberate African economies from the controls of global trade and market. He thus wanted Africans to go back to what he believed to be their precolonial self-sufficient village life. He and other leaders believed in the romantic idea that in these traditional villages the state of *ujamaa* (derived from the Swahili word meaning 'familyhood', which connoted the state of self-sufficiency in precolonial African village life) had existed for several centuries before the arrival of imperialists. He therefore moved massive numbers of people back to villages in the 1970s. Initially voluntary, the process met with increasing resistance, and in 1975 Nyerere introduced forced villagization. Unfortunately for Nyerere and for Tanzania, this caused agricultural output to plummet as people moved away from modern agricultural practices. Loans and grants from the World Bank and the International Monetary Fund in 1975 prevented Tanzania from going bankrupt from the severe deficit in cereal grains. Thus, tragically, the pursuit of precolonial self-sufficiency led to postcolonial dependency.[64] This helps us to realize that while we recognize the role of colonialism it is also important to write a critical history of the precolonial period.

This problem of romanticism, although of a different kind, is also evident in the other key element of imperial historiography: native agencies. While it is vital to study the roles and voices of different sections of indigenous and subaltern people, it is also critical to understand these with respect to social and economic contexts. In this attempt at recovering subaltern agency, the extent and nature of such agencies, and indeed why they were or became subaltern, also need to be analysed. While we seek to retrieve lost voices, we also have to understand why the voices were lost. To do so it is important to study agency with respect to the power and status of the actors under study. Otherwise the recovery of agency can lead to the problems of glorification of the roles of the people who suffered marginalization.

Agency is a critical concept in the writing of history, particularly in the history of imperialism, as it circumscribes the history of power. It is therefore necessary to understand critically its meanings and significance. Historians have demonstrated that understanding agency within the colonial context is a particularly complicated issue. Walter Johnson has written about the difficult question of agency in slavery, particularly in slave rebellion.[65] To what extent can rebellions by slaves in plantations be seen as an expression of their agencies? Johnson has suggested that there is often confusion about agency in such instances. At a basic level, agency could be about the very basic aspects of being human – that is, having the ability to feel hungry, cold, tired, angry, resentful, joy and sad, and also to rebel. At that level, slaves or indeed other marginalized communities had agency. At another level, though, agency, particularly in contexts of resistance, could mean the ability of the slaves to mobilize themselves, form collective groups and transform, to an extent, the conditions of their existence – what Johnson calls the ability of 'remapping of [their] everyday life'. At that level, Johnson suggests, it is difficult to establish or determine the nature and extent of agency that slaves enjoyed. What we need to appreciate is that individuals and groups, even in the most repressive regimes and systems, can be seen to have certain degrees of agency – that is, the ability to act. At the same time, these agencies are mediated through structures of power, of which imperialism was one. The question therefore is not just whether people or groups had agency but what the nature of that agency was. This could be understood only with reference to power.

It is for these reasons that historians of colonial medicine have also urged for the need to revisit some of the structural themes of economic and social contexts of colonialism that were developed in the 1970s. They have stressed the need to understand the material conditions that caused disease and death along with that of human agency. In the colonies in the past and in the poor countries of today, the question of agency and actors is complicated by the problem of access. People may have the agency to reformulate and redefine the medicine that they use but they often may not have the access to drugs, hospitals and general public healthcare facilities.[66] In vital sectors such as primary health and education in the impoverished countries, the question is not the agency of the poor, as there is little doubt that the poor seek or recognize the significance of health and education, but the structures of oppression and marginalization that inhibit their agency in securing these. Shula Marks in her article 'What is Colonial about Colonial Medicine' suggests that the more recent emphasis on the discursive and hegemonic nature of medicine could distract from the political economic aspects of diseases and epidemics, which concerned historians in the 1970s and 1980s.[67] She stresses the need to recognize that the greatest killers in South Africa were malnutrition and poverty, which were a product of social and economic inequalities caused by colonial capitalism, urbanization, commercialization of agriculture and mining. Historians have recently argued for the need to recognize the material contexts of health and disease: 'In shying away from questions of material interest ..."structural power" goes unexamined'.[68]

Broadly, therefore, to get a comprehensive understanding of colonial medicine, it is important to understand the multiplicity and plurality of powers that operate within colonialism.[69] It is important to understand how, in this diverse and complex history of imperialism, which comprised the lives and actions of such different groups of people, the empire was actually built and run.[70] Historians have therefore stressed the need to adhere to the broader structures (such as those on which the initial critique of colonialism was based in the early twentieth century) in our understanding of the complex interplay of human agencies in colonialism. Historian Patrick Joyce recently pointed out that 'What matters is how metropole and colony, centre and local – the multiple actors and networks of empire – were brought into a qualified degree of stability.'[71]

By studying the history of medicine and empire, we can thus understand the interplay between broader social and economic structures and human agency. Such a history provides the unique opportunity to link human histories, everyday experiences and agencies with the broad historical and structural changes and processes. This is a history of the structures of imperial power, economic systems, dominance and hegemony, and of human lives, compassion, agencies, discoveries and interactions.

Notes

1 C.A. Bayly, '"Archaic" and "Modern" Globalization in the Eurasian and African Arena, 1750–1850', in A.G. Hopkins (ed) *Globalization in World History* (London, 2002), pp. 47–73.

2 Margaret Kohn, 'Colonialism', Edward N. Zalta (ed.), *The Stanford Encyclopedia of Philosophy* (Summer 2012 Edition), http://plato.stanford.edu/archives/sum2011/entries/colonialism/.

3 George Basalla, 'The Spread of Western Science', *Science*, 156 (1967), 611–22.

4 Zaheer Babar, *The Science of Empire: Scientific Knowledge, Civilization, and Colonial Rule in India* (Albany, NY, 1996) p. 10; Ian Inkster, 'Scientific Enterprise and the Colonial "Model": Observations on Australian Experience in Historical Context', *Social Studies of Science*, 15 (1985), 677–704; Mark Harrison, 'Science and the British Empire, *Isis*, 96 (2005), 56–63.

5 For a study of how and why diseases are 'framed' in different categories, see Charles E. Rosenberg, 'Framing Disease: Illness, Society and History', *Explaining Epidemics and Other Studies in the History of Medicine* (Cambridge, 1992) pp. 305–18.

6 For a study of the problems of naming scientific traditions in the Arabic-speaking world which is relevant for other traditions of science as well, see Sonja Brentjes, 'Between Doubts and Certainties: On the Place of Science in Islamic Societies within the Field of History of Science', *NTM*, 11 (2003), 65–79, particularly pp. 67–70.

7 Harrison, 'Science and the British Empire; Margaret Jones, *Health Policy in Britain's Model Colony: Ceylon (1900–1948)* (Hyderabad, 2004), pp. 1–22..

8 For an analysis of when science became 'Western', see Chakrabarti, *Western Science in Modern India*, pp. 4–9. Mark Harrison identifies the sixteenth century

as the period when medicine, through its association with the state, humanitarian ideas and global commerce, gradually took its modern form. See his *Disease and the Modern World: 1500 to the Present Day* (Cambridge, 2004), pp. 2–5. Harold J. Cook has identified the rise of empiricism in seventeenth-century Europe as the factor that shaped modern medicine. See his 'Victories for Empiricism, Failures for Theory: Medicine and Science in the Seventeenth Century', in Charles T. Wolfe and Ofer Gal (eds), *The Body as Object and Instrument of Knowledge. Embodied Empiricism in Early Modern Science* (Dordrecht, 2010), pp. 9–32.

9 Umberto Eco, 'In Praise of St. Thomas', *Travels in Hyperreality: Essays* (San Diego, 1987) pp. 257–68. Scott L. Montgomery, 'Naming the Heavens: A Brief History of Earthly Projection', Part II: Nativising Arab Science', *Science as Culture*, 6 (1996), 73–129.

10 Michel Foucault, *Order of Things; An Archaeology of the Human Sciences* (New York, 1994/1970) pp. 128–32.

11 Harold J. Cook, 'Physicians and Natural History', in N. Jardine, J.A. Secord and E.C. Spary (eds) *Cultures of Natural History* (Cambridge, 1996) pp. 91–105.

12 Shula Marks, 'What is Colonial about Colonial Medicine?; And What has Happened to Imperialism and Health?', *Social History of Medicine*, 10 (1997), 205–19, pp. 210–11; Londa Schiebinger, 'The Anatomy of Difference: Race and Sex in Eighteenth-Century Science', *Eighteenth-Century Studies*, 23 (1990), 387–405; Schiebinger, *Plants and Empire: Colonial Bioprospecting in the Atlantic World* (Cambridge, Mass & London, 2004) pp. 105–93.

13 Alfred W. Crosby, *The Columbian Voyages, the Columbian Exchange, and their Historians* (Washington DC, 1987).

14 G.O. Trevelyan, *The Competition Wallah* (London & New York, 1863).

15 V.I. Chirol, *India Old and New* (London, 1921).

16 David Livingstone and Frederick Stanley Arnot, *Missionary Travels and Researches in South Africa.* (London, 1899); Henry Morton Stanley, *Through the Dark Continent* (London, 1880); Stanley, *How I Found Livingstone; Adventures, and Discoveries in Central Africa; Including Four Months' Residence with Dr. Livingstone* (London, 1874).

17 F. Driver, 'Geography's Empire: Histories of Geographical Knowledge', *Society and Space*, 10 (1992), 23–40.

18 P.T. Terry, 'African Agriculture in Nyasaland 1858 to 1894', *The Nyasaland Journal*, 14 (1961), 27–35.

19 Charles Eliot, *The East Africa Protectorate* (London, 1966/1905) p. 4.

20 Anthony Brewer, *Marxist Theories of Imperialism: A Critical Survey*, 2nd edition (New York, 1990), pp. 48–56.

21 Immanuel Wallerstein, *The Modern World-System: Capitalist Agriculture and the Origins of the European World-Economy in the Sixteenth Century* (New York, 1976).

22 Roy Porter, 'The Imperial Slaughterhouse', review of *Romanticism and Colonial Disease* by Alan Bewell, *Nature*, 404 (2000), 331–2, p. 331.

23 John Duffy, 'Smallpox and the Indians in the American Colonies', *Bulletin of the History of Medicine*, 25 (1951), 324–41. W. George Lovell, '"Heavy Shadows and Black Night": Disease and Depopulation in Colonial Spanish America', *Annals of the Association of American Geographers*, 82 (1992), 426–43. Noble David Cook, 'Sickness, Starvation, and Death in Early Hispaniola', *Journal of Interdisciplinary History*, 32 (2002), 349–86.

24 Raymond E. Dumett, 'The Campaign Against Malaria and the Expansion of Scientific Medical and Sanitary Services in British West Africa, 1898–1910', *African Historical Studies*, 2 (1968), 153–97; John Ford, *The Role of Trypanosomiases in African Ecology: A Study of the Tsetse Fly Problem* (Oxford, 1971); Helge Kjekshus, *Ecology Control and Economic Development in East African History: The Case of Tanganyika 1850–1950* (Berkeley, 1977); Marc Dawson, 'Disease and Population Decline of the Kikuyu of Kenya, 1890–1925', in Christopher Fyfe and David McMaster (eds) *African Historical Demography; Proceedings of a Seminar Held in the Centre of African Studies, University of Edinburgh*, vol. 2 (Edinburgh, 1981) pp. 121–38.

25 Roy Porter, 'The Historiography of Medicine in the United Kingdom' in F. Huisman and J.H. Warner (eds), *Locating Medical History: The Stories and their Meanings* (Baltimore & London, 2004) pp. 194–208, p. 196.

26 For an explanation of 'social epistemology', see Alvin Goldman, 'Social Epistemology', *The Stanford Encyclopedia of Philosophy* (Summer 2010 edition), Edward N. Zalta (ed.), http://plato.stanford.edu/archives/sum2010/entries/epistemology-social/.

28 Further reading: 'Introduction' in Roy Porter and Adrew Wear (eds), *Problems and Methods in the History of Medicine* (New York, 1987) pp. 1–12; Edwin Clarke (ed), *Modern Methods in the History of Medicine* (London, 1971). Dorothy Porter, 'The Mission of Social History of Medicine: An Historical View', *Social History of Medicine*, 8 (1995), 345–59.

29 Terence Ranger, 'Connexions between "Primary Resistance" Movements and Modern Mass Nationalism in East and Central Africa', *The Journal of African History*, 9 (1968), 631–41. See also John Iliffe, 'The Organization of the Maji Maji Rebellion', *The Journal of African History*, 8 (1967), 495–512.

30 Edward Said, *Orientalism* (New York, 1978).

31 Ibid, pp. 39–40.

32 Ibid, p. 38.

33 Javed Majeed, *Ungoverned Imaginings, James Mill's 'The History of British India' and Orientalism* (Oxford, 1992); Raymond Schawb, *The Oriental Renaissance: Europe's Rediscovery of India and the East, 1680–1880* (New York, 1984); Gauri Viswanathan, *Masks of Conquest; Literary Study and British Rule in India* (London, 1989).

34 Ranajit Guha, 'On Some Aspects of the Historiography of Colonial India', pp. 1–8.

35 Robert Young, *Postcolonialism: A Very Short Introduction* (New York, 2003).

36 For the convergences between subaltern studies and African history, see Christopher J. Lee, 'Subaltern Studies and African Studies', *History Compass* 3 (2005) doi: 10.1111/j.1478-0542.2005.00162.x; Ranger, 'Power, Religion, and Community: The Matobo Case', in Partha Chatterjee and Gyanadra Pandey (eds) *Subaltern Studies*, vol. 7 (Delhi, 1993), pp. 221–46.

37 Richard S. Dunn, *Sugar and Slaves; The Rise of the Planter Class in the English West Indies, 1624–1713* (Chappell Hill, 1972); Jane Webster, 'Looking for the Material Culture of the Middle Passage', *Journal for Maritime Research*, 7 (2005), 245–58; Eric S. Mackie, 'Welcome the Outlaw: Pirates, Maroons, and Caribbean Countercultures', *Cultural Critiques*, 59 (2005), 24–62.

38 John Thornton, *Africa and Africans in the Making of the Atlantic World 1400–1800* (Cambridge, 1998).

39 David Eltis, *The Rise of African Slavery in the Americas* (Cambridge, 2000); Stephanie Smallwood, *Saltwater Slavery: A Middle Passage from Africa to American Diaspora* (Cambridge, Mass, 2007).

40 Crosby, *Ecological Imperialism: The Biological Expansion of Europe (900–1900)* (Cambridge, 1986); and Crosby, *The Columbian Exchange: Biological and Cultural Consequences of 1492* (Westport, 1972).

41 Kjekshus *Ecology Control and Economic Development.*

42 Jill R. Dias, 'Famine and Disease in the History of Angola, 1830–1930', *Journal of African History*, 22 (1981), 349–78.

43 Sheldon Watts, 'British Development Policies and Malaria in India 1897–c.1929', *Past & Present* 165 (1999), 141–81; Ira Klein, 'Death in India: 1871–1921', *Journal of Asian Studies*, 32 (1973), 639–59.

44 Randall M. Packard, 'Maize, Cattle and Mosquitoes: the Political Economy of Malaria Epidemics in Colonial Swaziland', *The Journal of African History*, 25 (1984), 189–212.

45 Nancy Stepan, *Picturing Tropical Nature* (Ithaca, 2001) 149–79; Warwick P. Anderson, 'Immunities of Empire: Race, Disease and the New Tropical Medicine, 1900–1920', *Bulletin of the History of Medicine*, 70 (1996), 94–118; Rod Edmond, 'Returning Fears: Tropical Disease and the Metropolis', in Driver and Luciana Martins (eds), *Tropical Visions in an Age of Empire* (Chicago, 2005), pp. 175–94.

46 Ranger, 'Godly Medicine: The Ambiguities of Medical Mission in Southeast Tanzania', *Social Science and Medicine*, 15b (1981), 261–77, See also Megan Vaughan, Chapter 3, 'The Great Dispensary in the Sky: Mission Medicine', *Curing their Ills; Colonial Power and African Illness* (Stanford, 1991), pp. 55–76.

47 David Arnold, *Colonizing the Body: State Medicine and Epidemic Disease in Nineteenth-Century India* (Berkeley & Los Angeles, 1993), p. 291.

48 Further reading: John M. MacKenzie (ed), *Imperialism and the Natural World* (Manchester, 1990); J.R. McNeill, 'Observations on the Nature and Culture of Environmental History', *History and Theory*, 42 (2003), 5–43.

49 See Andrew Cunningham and Bridie Andrews (eds), 'Introduction', *Western Medicine as Contested Knowledge* (Manchester, 1997), pp. 1–23.

50 Charles Gibson, *The Aztecs Under Spanish Rule: A History of the Indians of the Valley of Mexico, 1519–1810* (Stanford, 1964).

51 Miguel León Portilla, *The Broken Spears: The Aztec Account of the Conquest of Mexico* (Boston, 1992/1962).

52 Stuart Schwartz (ed) *Victors and Vanquished: Spanish and Nahua Views of the Conquest of Mexico* (Bedford, 2000).

53 Dedra S. McDonald, 'Intimacy and Empire; Indian-African Interaction in Spanish Colonial New Mexico, 1500–1800', *American Indian Quarterly*, 22 (1998), 134–56.

54 Robert Voeks, 'African Medicine and Magic in the Americas', *Geographical Review*, 83 (1993), 66–78.

55 David Hardiman, *The Coming of the Devi: Adivasi Assertion in Western India* (Delhi, 1995).

56 Vaughan, *Curing their Ills*.

57 Walima T. Kalusa, 'Language, Medical Auxiliaries, and the Re-interpretation of Missionary Medicine in Colonial Mwinilunga, Zambia, 1922–51', *Journal of Eastern African Studies*, 1 (2007), 57–78.

58 Sanjoy Bhattacharya, Mark Harrison and Michael Worboys, *Fractured States: Smallpox, Public Health and Vaccination Policy in British India, 1800–1947* (Hyderabad, 2005); Anna Crozier, 'What Was Tropical about Tropical Neurasthenia? The Utility of the Diagnosis in the Management of British East Africa', *Journal of the History of Medicine and Allied Sciences*, 64 (2009), 518–48.

59 Helen Tilley, 'Ecologies of Complexity: Tropical Environments, African Trypanosomiasis, and the Science of Disease Control Strategies in British Colonial Africa, 1900–1940', *Osiris*, 19 (2004), 21–38.

60 Gyan Prakash, 'Science between the Lines', in Shahid Amin and Dipesh Chakrabarty (eds.), *Subaltern Studies,* vol. 9 (Delhi, 1996), pp. 59–82.

61 Dhruv Raina and Irfan S. Habib, 'Bhadralok Perceptions of Science, Technology and Cultural Nationalism', *Indian Economic and Social History Review*, 32 (1995), 95–117.

62 Mark Harrison, *Public Health in British India: Anglo-Indian Preventive Medicine 1859–1914* (Cambridge, 1994).

63 John M. MacKenzie, 'Empire and the Ecological Apocalypse: The Historiography of the Imperial Environment', in Tom Griffiths and Libby Robin (ed) *Ecology and Empire: Environmental History of Settler Societies* (Melbourne, 1997), pp. 215–28.

64 James C. Scott, *Seeing Like a State: How Certain Schemes to Improve the Human Condition Have Failed* (New Haven and London, 1998) pp. 223–61.

65 Walter Johnson, 'On Agency', *Journal of Social History*, 37 (2003), 113–24.

66 Paul Farmer had shown that the poorest communities in different parts of the world suffer from diseases as they suffer from a chronic lack of medical facilities, *Infections and Inequalities: The Modern Plagues* (Berkeley, 2001).

67 Marks, 'What is Colonial about Colonial Medicine?', pp. 215–19.
68 Harold J. Cook, 'Markets and Cultures: Medical Specifics and the Reconfiguration of the Body in Early Modern Europe', *Transactions of the Royal Historical Society*, 21 (2011), 123–45.
69 Patrick Joyce, 'What is the Social in Social History', *Past & Present*, 206 (2010), 213–48, p. 240.
70 See Pratik Chakrabarti, *Materials and Medicine; Trade, Conquest and Therapeutics in the Eighteenth Century* (Manchester, 2010), pp. 4–15.
71 Joyce, 'What is the Social in Social History', p. 240.

1

Medicine in the Age of Commerce, 1600–1800

The discovery of the two new trade routes to Asia and to the Americas marked the beginning of the Age of Commerce. Christopher Columbus and Vasco da Gama had ventured out at the end of the fifteenth century in search of an alternative route to the spice-growing regions of Asia. Why did Western Europeans seek to find new routes to the East? The answer lies in the history of late medieval trade between Europe and Asia, which was dominated by the Mediterranean trading world. Until the sixteenth century, European and Asian trade were connected by the Caspian and the Mediterranean seas, and the Mediterranean world was the point of contact between Asia and Europe. Italian and Arab merchants controlled this trade. Asian goods were brought by Arab merchants through the Red Sea or the Gulf of Persia to the eastern Mediterranean ports, such as Tyre, Constantinople and Alexandria. From there they were taken to the Italian port cities of Amalfi, Naples, Genoa and Venice by Italian merchants who then supplied them to the rest of Europe. Spices were a vital and lucrative commodity in Europe and were used for various purposes: for meat preservation, and as condiments, perfumes and medicines. Merchants and traders in Western Europe, particularly in the then ascendant states of Spain and Portugal, were eager to establish their own direct trading links with Asia to avoid the Arab and Italian traders who consumed much of the profit.

Between 1500 and 1800, Western European commerce expanded globally, so historians have termed the period the 'Age of Commerce'. It is important to note that the term has a predominantly Western European connotation, as Western Europe gained most from this new maritime expansion. For the Arab and Mediterranean traders it was the beginning of a period of commercial and economic decline, as most of the Asian trade passed on to the Portuguese, and then to the Dutch and English, through the new maritime routes. As for Asian traders, there was a general increase in their trade, but they now traded predominantly with the Portuguese, Dutch, French and English, rather than the Arabs. For the indigenous populations in the Americas, the Age of Commerce and the Spanish Conquest brought new diseases, death, subjugation by the Spanish and enslavement on the new plantations.

In the sixteenth century, Spain expanded its colonies in the Americas. It occupied the Caribbean islands and the conquistadors (the Spanish soldiers

1

and explorers who undertook the initial conquest of the Americas) toppled native empires, such as the Aztecs and Incas, on the mainland of North and South America. The small Caribbean islands occupied an important place in the Atlantic trade between the Americas and Europe. These were important entrepôts of supplies and trading commodities, and European powers struggled against each other to gain control over the islands. Initially dominated by the Spanish, the Caribbean Sea was referred to as a 'Spanish lake'.[1] Between 1519 and 1522, the Spanish achieved an important circumnavigation when Ferdinand Magellan and Juan Sebastian Elcano sailed around South America to reach Asia though the western maritime route, where the Spanish established their colonies in the Philippines. In the seventeenth century, Spain controlled an empire that covered the Americas and Asia.

Meanwhile the Portuguese initiated European trading networks in the Indian Ocean. Following da Gama's expedition, they opened up rich areas of maritime trade for Europeans. Alphonse de Albuquerque, a naval officer, established the Portuguese Empire in the Indian Ocean, starting with Goa in India in 1510. Soon the Portuguese assumed control over Mocha in the Red Sea; Hormuz in the Persian Gulf; Goa, Diu and Daman in western India; the island of Ceylon (now Sri Lanka); and Malacca in southeast Asia. They also succeeded in cutting off much of the Arab trade and managed to have nearly all spices headed for Europe loaded on Portuguese ships.

The new trading routes that the voyages of Columbus and Vasco da Gama established marked important changes in the commercial history of Europe and other parts of the world. There were three main consequences: first, the decline of the older trading networks of Italian and Arab merchants through the Mediterranean and the Red Seas and the subsequent rise of Western European commerce; second, the establishment of first direct contact between Western Europeans and distant populations, cultures, flora and fauna; and third, these events marked the beginning of Western European maritime empires in Asia and the Americas (see Figure 1.1).

This chapter will explore the ways in which the Age of Commerce transformed the basic premises of modern medicine. Commercial and colonial expansion from the seventeenth century gave rise to a new world of interaction between Europeans, Asians, Africans and Amerindians, the emergence of a hybrid medical culture, the availability of new drugs in the markets of Asia and Europe, and the emergence of surgeons and apothecaries as the new agents of medicine. In Chapter 2 I will focus more specifically on plants as essential elements in this emerging history of medicine and colonialism.

Commerce and empire

A journey intended to find a new trade route to the spice islands of the East took Europeans to a 'new' world. Columbus landed in the Caribbean islands, believing that he had reached India. Over time, these Atlantic islands and the great continents lying beyond them provided Europe with several new resources. In South America, the Spanish found something even more precious than spices – gold and silver – a discovery that led to the growth of European bullionism and the rise of mercantilism in the seventeenth century.[2] American gold and silver provided currency to facilitate the increase in intra-European trade, but, more significantly, it also provided the bullion that the Europeans could exchange for the products of Asia, where European goods were not in reciprocal demand. The influx of American bullion in Asian trade also led to the decline of Venice and the Hanseatic cities, and the rise of new trading centres along the west coast of Europe: Cadiz and Seville in Spain; later, as the centre of trade shifted, Bordeaux, Saint-Malo, Nantes and Calais, in France, and then Antwerp and Amsterdam in Holland; and, finally, London and Bristol in England. All of these centres also had trade connections with the Americas.

In Europe, this led to the Commercial Revolution of the sixteenth century. The inflow of money, bullion and new items of trade in Europe led to the breakdown of old feudal economies, the expansion of markets, new ports, towns, a rise in prices and the formation of a new merchant class. Bullionism increased monetarism (the belief that the supply of money is a major determining factor in the behaviour of an economy). It also paved the way for the rise of mercantilism (the economic theory that supposes that the prosperity of a state depends on its ability to amass bullion by maintaining a favourable balance of trade). With American bullion, Europeans from the late sixteenth century could obtain substantial quantities of commodities such as spices, textiles, plants and herbs, and manufactured goods from different parts of the world. How did this Commercial Revolution lead to colonies?

The answer is in the commercial rivalries that started among different European nations. With this rivalry came the need to secure trading monopolies – both to secure the sources of exotic goods and products, and to reduce the outflow of bullion to rival nations. Monopoly over commerce was crucial to mercantile economies. Therefore European merchants sought to gain direct political and economic control of the regions that produced the goods of their commercial interest. This search for trading monopolies marked the beginning of European colonial establishments in Asia and the Americas. European commercial countries, starting with the Portuguese and the Spanish, and then the French, the Dutch, the Danes and the British, established colonies in their attempt to secure commercial monopolies.

By the seventeenth century, the Dutch had rapidly demolished the Portuguese power in the Indian Ocean and then displaced the Arabs and Italians to gain complete monopoly over the East Indies spice trade. Throughout the seventeenth century, the Dutch gained decisive shipping, commercial and financial dominance in the East Indies as well as in the European markets. Due to the Dutch control of the East Asian Spice Islands, the British settled for a relatively less important source for spices: India.

The Dutch and the British introduced European joint stock private companies in Asian and Atlantic trade. In 1600 the English East India Company (EEIC) was established, followed by the Dutch East India Company (Oost Vereenigde Indische Kompagnie) in 1602. The Dutch West India Company (Geoctroyeerde Westindische Compagnie) was granted a charter in 1621 for a trade monopoly in the West Indies as well as in Brazil and North America. The South Sea Company, founded in 1711, was a British joint stock company that traded in the West Indies and South America during the eighteenth century. Through most of the seventeenth, eighteenth and even nineteenth centuries, these companies traded in Asia and the Americas, accumulating wealth and establishing colonies. They also employed large armies, which required the services of surgeons. The surgeons were recruited from European medical colleges and sent to the colonies in Asia and the Americas.[3] As we shall see later in this chapter, colonial surgeons became the agents of change in European and colonial medicine in the eighteenth century.

The English made important colonial gains in North America in the seventeenth century. They established colonies on the east coast in Massachusetts and New England, and then in Pennsylvania. In the south, the Spanish retained colonies for much longer than in the north. In the Caribbean, the British earned a decisive strategic victory when they gained the colony of Jamaica from the Spanish in 1655. Jamaica became the hub of British trade in the Caribbean and, later, a rich sugar plantation colony in the eighteenth century. Because of these colonial settlements and expansion in Atlantic trade, British overseas trade during this period became 'Americanized'. By the end of the eighteenth century, North America and the West Indies received 57 per cent of British exports and supplied 32 per cent of its imports. Between 1660 and 1775, commodities such as sugar, tobacco and coffee, which were procured from the Americas or grown in the plantations, became vital components of British imports. These had also become items of mass consumption in Britain.[4]

The French and the Dutch secured colonies on the east coast of North America from the middle of the seventeenth century. France established a colony in Quebec in North America by the middle of the seventeenth century. The Dutch established settlements near what is now New York. In the West Indies, the French established colonies in Guyana and the islands of Martinique and Guadeloupe. They entered into a prolonged conflict with the Spanish over the major island of Hispaniola, and in 1697 the island was divided into two parts. The French secured the western part, which they named Saint-Dominique. This soon became a French plantation colony for sugarcane and coffee. The main motive for British and French expansion in Atlantic trade in the seventeenth and eighteenth centuries was to retain the profit for themselves by securing control over the sources of items of commercial import and to avoid intermediary trading groups.[5] In the Indian Ocean throughout the eighteenth century, the French and British competed with each other for colonial expansion. India was the main base for the English trading company, where it established important trading ports in Calcutta (now Kolkata), Bombay (now Mumbai) and Madras (now Chennai). The French established a colony in Pondicherry, south of Madras on the Coromandel Coast.

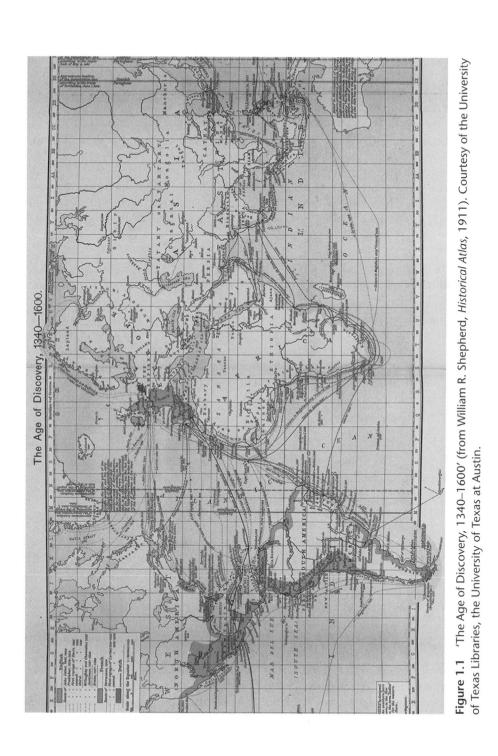

Figure 1.1 'The Age of Discovery, 1340–1600' (from William R. Shepherd, *Historical Atlas*, 1911). Courtesy of the University of Texas Libraries, the University of Texas at Austin.

The commercial and colonial rivalries led to mercantilist warfare among the British, the Spanish and the French. Large numbers of troops were mobilized and sent to fight battles in North America, the West Indies and Asia. This resulted in large-scale European mortality, which posed new medical challenges, which I will explore in Chapter 4.

The end of the eighteenth century marked the first two phases of European imperialism. The first was under the Spanish and the Portuguese who from the sixteenth century built maritime empires in Asia and the Americas, which had become large territorial colonies by the seventeenth century. From the middle of the seventeenth century there was a new phase of commercial and colonial expansion when other Northern European nations, such as the British and the Dutch, gradually established their own empires in India, South East Asia, the West Indies and in the Pacific Ocean.

Medical exchanges in the Age of Commerce

Commerce was the main conduit of European contact and encounter with other parts of the world in the seventeenth and eighteenth century. It was through trade and commercial exchanges that the rich vegetation, foliage, unfamiliar plants and animals, as well as different cultures of the tropical regions of Asia and the Americas, came to be linked with European experience. For the Europeans in trading settlements away from Europe, in the West Indies, Asia and the Americas, exchange was key to survival and making profits. Due to the commercial motif, the early colonial history has often been seen as a history of 'exchanges'. In the Americas, this has been described as 'the Columbian Exchange'.[6] In Asia there has been no specific phrase to describe it, but exchanges and interactions have been seen as the key feature of early European colonialism here.

When Christopher Columbus reached the Caribbean islands, he was struck by their beauty and the variety of the plants that he saw:

> so green and so beautiful, like all the other things and lands in these islands, that I do not know where to go first nor do my eyes tire of seeing such beautiful greenery and so different from our own. I believe that there are on the islands many plants and trees which would be of great value in Spain as dyes and medicinal spices, but I do not recognise them, which I much regret.[7]

This combined sense of wonder, the disappointment of not knowing enough about these plants and the promise of the 'value' of these plants drove Europeans to explore and collect plants, seeds, bark and fruits, export them to European markets and include these in their own medicine. The supply of medicaments in Europe increased massively between the sixteenth and nineteenth centuries. This European exploration of exotic plants in the tropical colonies has been described as 'colonial bioprospection'.[8]

Historians have studied in detail the nature of the changes that were introduced in European natural history and medicine by this contact with the wider world from the sixteenth century. On the one hand the Age of Commerce introduced items of curiosity in Europe, stimulating intellectual discussions about exotic natural objects and medicinal plants and providing European thinkers with a global view of the natural world.[9]

On the other hand, collecting plants, drugs and seeds also had their commercial benefits. Eighteenth-century political economists believed that a knowledge of nature was essential in amassing national wealth. They understood the potential of coffee, cacao, ipecacuanha, jalap and Peruvian bark as profitable items. The collection and study of these plant materials was also associated with commercial expansion, the acquisition of colonies and the establishment of colonial gardens and plantations.[10] This combination of intellectual and commercial benefits is useful to understand the history of medicine during this period.

Although this appears to be the dominant narrative, there are other aspects besides the enrichment of the European pharmacopoeia to this history of colonial medicine in the Age of Commerce. Alongside the European encounters with the tropical world and nature, intimate interactions about nature, medicines and healing practices took place between the Amerindians and the Africans. African-American plant pharmacopoeias formed a very important part of the cultural history of the Caribbean islands. The colonization of the Americas by Europeans was accompanied by a diverse blending of diseases, medical systems and plant-based pharmacopoeias between the Old World and New World.[11] In the Caribbean islands and in South America, the plantations placed Africans and Amerindians in close contact throughout the seventeenth and eighteenth centuries. As they worked together, they exchanged their knowledge about plants and therapeutics. For example, in Brazil, African slaves adopted the use of indigenous medicinal plants from the Tupinamba Amerindians.[12]

At the same time, Africans had carried many West African plants and weeds with them, which entered the medical practices of the indigenous populations and African healing systems often flourished in the Americas. In the West Indian plantations, slaves easily outnumbered the whites. Therefore the West Indian slaves could preserve much more of their African roots, languages and rituals compared with those in North America.[13] In that process, African-based ethnomedicine travelled to, and survived in, various parts of the Caribbean islands.

These exchanges were not just about plants and therapeutics; they were about a wider sharing of spiritual and religious beliefs and worldviews between Africans and Amerindians.[14] Within this system, cures were achieved through both pharmacological and magical powers.[15] Thus spirituality was an important component of this hybrid healing tradition that developed in South America and the West Indies in the Age of Commerce. Even when converted to Christianity, slaves and Amerindians retained their faith in the power of magical healing. In Jamaica, for example, African healers treated their patients with herbal baths and infusions of African origin in combination with Christian prayers.

Voodoo and obeah were the two products of such diverse exchanges in plants, culture and spirituality across the Atlantic. Voodoo is a religious practice that originated from the Caribbean island of Haiti. It is based upon a merging of beliefs and practices of West African origin with Roman Catholic Christianity, from the time when African slaves were brought to Haiti in the sixteenth century and converted to Christianity, while they still followed their traditional African beliefs. Voodoo was also part of slave rebellions, as during the Haitian Revolution of 1790, voodoo priests used ritualistic chants to inspire slaves to rise against their white masters. Throughout the seventeenth and eighteenth centuries, voodoo continued as a medicoreligious practice among the slaves and peasants in the West Indies and the southern United States.

Obeah was another complex healing tradition of African origin that survived in the West Indies. It became a spiritual and intellectual institution that unified slaves from different regions and cultures of Africa. Obeahmen (those who practiced obeah) used medicinal herbs of African and American origin to treat slaves. They contributed to the assertion of cultural identity among slaves throughout the eighteenth century.[16] Obeah was also an important part of slave rebellions from the late seventeenth century.[17] In 1736 an obeahman was among the rebels executed in Antigua.[18] In Montego Bay, Jamaica, a rebellious obeah practitioner was captured by the British authorities and burnt to death.[19] Due to this link with slave rebellions, the white planters perceived obeah and voodoo as threats to their authority.[20] The authorities sought to suppress them by legislation and by the end of the eighteenth century, African magic and medicine were outlawed in most parts of the West Indies, to be punished by death or deportation.

Although the spiritual and political motifs of these Afro-Caribbean practices were repressed, European naturalists and physicians often found their knowledge of local medicinal plants useful. British physician Hans Sloane, who visited Jamaica at the end of the seventeenth century, searched among the obeahs for curative herbs and plants.[21] European physicians gained from the therapeutic knowledge of these practitioners and adopted this knowledge by sifting their spiritual and ritual content. They believed that the influence of Christianity would eventually 'liberate' these practices and their practitioners. One English surgeon wrote that he hoped that Christian faith would 'emancipate [the slaves] from the mental thraldom of ignorance that made them so susceptible to the malady [obeah].'[22] The spread of the evangelical movement among the slaves did seek (and it succeeded to an extent) to abolish practices of African witchcraft and magic among slaves.[23] However, these practices survived, sometimes combining African herbal practices and Christian rituals. The medical practices were passed on through generations by the slaves and survived even in the nineteenth century in the plantations in the two Americas. Some of the herbal traditions were even adopted by white physicians and slave owners.[24]

Another distinct and important aspect of medical exchanges in the Commercial Age was in the revival of the use of minerals in medicine. This marked a shift from the dependence on herbal medicine. Minerals were the most important materials of early European Atlantic commerce. The Spanish, French and British searched for mines in South America and the West Indies. As William Robertson wrote in his *History of America*, 'precious metals were conceived to be the peculiar and only valuable productions of the New World, when every mountain was supposed to contain a treasure, and every rivulet was searched for its gold sands'.[25]

The discovery of Peruvian silver also led to the exploration of mines in Europe in the seventeenth century.[26] This was not a simple quest for great wealth in the form of rare metals. Mercantilism ushered in a new awareness in Europe of minerals in general, not just gold and silver. Europeans from Spain to Norway found a new fulfilment in minerals – these came to signify their source of wealth and wellbeing. This led to a growing interest in mineralogy among physicians in Europe and the revival of Paracelsian medicine in the seventeenth century.[27] Paracelsus was an early sixteenth-century German physician and alchemist who believed in the virtue of chemicals and minerals as medicine. Physicians in the seventeenth century advocated the need to drink mineral water and explored the chemical properties of certain springs. They also stressed the use of stones and minerals in medicines. The increasing chemical analysis of the virtues of bath waters in Europe stimulated interest in mineral, metallic and chemical medicines.[28] From the early seventeenth century, physicians in London showed a greater interest in mineral cures.[29]

The Commercial Revolution and the revival of interest in the medicinal virtues of minerals led to the growth of mineral spas in Europe in the seventeenth century as important health institutions. The French established thermal spas in Vichy and Saint Galmier.[30] The British established or revived the older mineral baths in Bath, Buxton, Tunbridge Wells and Epsom.[31] Even in South America and the West Indies, mineral springs and baths became popular. For several years during the early period of colonization of the West Indies, the British were convinced that these islands were rich in silver and gold mines. With the help of slaves, British surgeons and prospectors searched for minerals in the mountains and forests of Jamaica and St Kitts. They did not find any silver, but they discovered springs rich in minerals. Believing in their therapeutic virtues, the settlers established baths on these islands, which became favourite convalescent places in the eighteenth century.[32]

A useful way to understand how these cultural, social and geographical experiences in the seventeenth and eighteenth centuries shaped European medicine is by studying the practices and writings of European surgeons. The surgeons, who worked for the trading companies or in the plantations, or served the colonial armies, became the mediators of cultural, medical and ecological knowledge between different cultures and societies. Their colonial experiences and knowledge proved vital to European medicine, as they often distilled, incorporated and disseminated hybrid medical materials and traditions. This process also contributed to a rise in the status of surgeons in the medical fraternity in general, and their position was now very different from that of the earlier barber surgeons in Europe.

The changing world of European surgeons

Historians have shown that colonial surgeons transformed European medicine, by finding new cures and modes of preventive medicine moving away from the older practices of bleeding and expulsive drugs.[33] European surgeons played a dual role. As surgeons, they were the principal bearers of European medicine to the colonies, but they also investigated new ingredients and medical insights among the diverse groups of people they met. At the same time, throughout the eighteenth century, the status of surgeons within European medical practice itself was changing. Surgeons rose to prominence from lowly placed barber surgeons within European medicine; came to be respected as theoreticians of diseases and cures, particularly those of exotic regions; and were invested with the care of the navy and army by the state. Physicians of the Royal College of Physicians or naturalists of the Royal Society during this period depended on the correspondences sent by surgeons from distant continents for the enrichment of their medical practice.

Surgeons who served in the West Indian islands worked in a unique geographical and social site which enabled them to study the botanical resources of the islands as well as to record the therapeutic traditions of the slaves and the Amerindians, and then to incorporate them into their own therapeutic practices. In Asia they visited local markets in search of spices and medicinal herbs, interacted with local practitioners, even learnt local languages to read and translate the medicinal texts.

The surgeons worked in diverse circumstances, which enriched their medical and cultural knowledge. In the Caribbean they even served the pirates and sometimes participated in their buccaneering activities. Richard Sheridan has provided an interesting account of the practice of medicine by British and French surgeons who were also occasionally involved with piracy and privateering in the Americas and in the Caribbean in the seventeenth and eighteenth centuries. The buccaneers who suffered from fevers, scurvy and wounds preferred to be in close contact with the surgeons, who were also 'given special consideration in the sharing of the booty'.[34] The buccaneer surgeons prescribed medicines drawing upon pharmacological remedies of Europe, but also used the tropical medicinal plants and folk remedies they learnt from Amerindian and African slaves.[35] Some of the Atlantic surgeons, such as Thomas Dover and Lionel Wafer, participated in privateering piracy activities.[36] Alexander Olivier Esquemeling, a French surgeon who served the pirate Captain Morgan, also studied the medicinal plants in Tortugas near Haiti. He wrote the book *The Buccaneers of America*. In 1666 he served the French West India Company and went to Tortugas. There he joined the buccaneers, probably as a barber-surgeon, and remained with them until 1674. After a brief stay in Europe, he returned to the Caribbean and served as a surgeon during the attack on Cartagena in 1697.

Many buccaneer surgeons amassed substantial wealth from their association with these lucrative activities. Dover, a surgeon trained under English physician Thomas Sydenham, travelled in a Bristol slaver in 1708 to the West Indies where he learnt new cures for smallpox from the slaves.[37] He participated in piracy activities and even organized an expedition to the Spanish Main. He invested the wealth acquired from this trade and his successful medical practice in the Bristol syndicate in backing Woodes Rogers' privateering venture to the South Seas in 1708.

The private fortunes they amassed from the colonies also helped the social status of the surgeons back home. Hans Sloane, who went to Jamaica as a surgeon from England, married Elizabeth Langley, widow of Fulk Rose, a rich Jamaican planter, and inherited part of the income from of her former husband's plantation. This wealth allowed him to set up his private practice in Bloomsbury in London. He also purchased the Manor of Chelsea in 1712, which made him the landlord of the physic garden. He later became president of the Royal Society.

In the Indian Ocean, European surgeons assimilated Asian spices and medicinal plants into their medical practice. Samuel Browne, an English surgeon who was stationed at Madras hospital in the late seventeenth century, spent most of his time collecting medical plants from the neighbouring forests and interacting with the locals, finding out about their curative properties and tracing their South East Asian, Dutch and Portuguese origins. He gathered a large collection of spices and aromatic and medicinal plants. His notes on Asian medicinal plants, which were published in the *Philosophical Transactions of the Royal Society*, provide a detailed account of the cultural practices associated with these plants.[38] They described the early European, particularly Portuguese, connections in bringing aromatic and medicinal plants from places such as Batavia to India. These were then incorporated into British and even local medicine. He sent his specimens and notes to the famous London-based apothecary James Petiver, who was involved in a broad scheme of comparing these with similar other accounts sent from the West Indies, Guinea, East Asia and elsewhere in India. His aim was to compile a comprehensive treatise on the medical botany of different parts of the world.[39]

Edward Buckley, chief surgeon and a contemporary of Browne in Madras, sent a 'China Cabinet' full of instruments and samples used by the Chinese surgeons to Hans Sloane at the Royal Society.[40] Along with collecting plants and medicinal recipes, European surgeons also engaged in private trade in spice in Asia. They even ran their own commercial establishments, such as preparing and selling arrack, spices and medicines.

One episode highlights the cultural and medical experience that European surgeons had in the eighteenth century. Brown, a surgeon of the EEIC, met Mr Morad, a wealthy Armenian merchant in Masulipatnam (an old spice trade port on the Coromandel coast of India), whose wife was in labour. When Brown went to see her, he found an interesting medicinal collection at the merchant's home. There were several French and British cordials and some other local drugs that he was not familiar with. Among them he found a bottle containing a 'mysterious' liquid, which Morad had collected from a nearby village. This was supposed to be a useful remedy for diseases of the bile. Brown learnt that the liquid was in fact bottled dew or fog, which fell heavily in that region in the months of September and October and was collected by spreading a fine muslin cloth in the evening. The following morning the dew was wrenched from the cloth and poured into a bottle. When Brown narrated his experience to a Portuguese surgeon at the English hospital in Masulipatnam, the latter confirmed that he too had heard about that drug and a Portuguese trader had in fact once tried to collect it himself. Brown added that while the British had very little intercourse with the Armenians in Masulipatnam, the Portuguese, due to their long presence in Asia and familiarity with their language, were much better acquainted with them and their medicines.[41]

The surgeons wrote about their myriad experiences of novel therapies to physicians and surgeons in Europe. In the eighteenth century, journals such as the *Philosophical Transactions* or the *Medical Commentaries*, the latter published in Edinburgh, became important repositories for the diverse forms of experiences that surgeons wrote about in this period. Surgeons also collected these drugs and sent them to Europe through their private trading connections to be sold in the European markets. Because of this, various medical items entered Europe's medical practice and pharmacopoeia. These transformed the European medical trade and its apothecary shops.

Markets and medicines in the Age of Commerce

Markets were the sites around which European economic, political and cultural history revolved in the seventeenth and eighteenth centuries. In Europe in the early modern period, markets changed from the older provincial and local institutions, selling predominantly agriculture-based products, to more cosmopolitan and urban ones, which sold items collected from different parts of the world and catered to a greater number and variety of consumers. In Asia, the markets that traded locally within the Indian Ocean were from the seventeenth century integrated with the global market system. In the Americas and on the coasts of Africa, markets exchanged bullion, sugar, curiosities, animal and plant products, and slaves.

Annales historian Fernand Braudel has shown that the emergence of 'national markets' (a network of markets, which connected the cities and the provinces) in Europe in the eighteenth century was critical to the creation of nation states in Britain, France and the Netherlands. This provided a move away from the agriculture-based regional economics to a more unified fiscal structure and polity. He also argued that this fiscal and commercial unification was preceded by the expansion in foreign trade by these nations.[42]

Apart from fiscal unification, markets also represented the new consumer culture that developed in Europe due to the arrival of colonial goods and exotics. Fuelled by the wealth earned from the bullion trade and stimulated by exotic products from the Orient and the Americas, the European elite thronged to the markets. Over the seventeenth century, major European cities, such as London and Amsterdam, became the centres of a new consumer culture.[43] Colonialism transformed the consumer and manufacturing culture in London, giving rise to new forms of markets selling exotic goods.[44] Markets in Amsterdam too became the centres of Dutch colonial and national trade.[45] The East India companies successfully developed luxury and semiluxury markets in Europe for the middle class and gentry for Asian textiles, porcelain, teaware, dinner services and armorial ware, associating these items with taste and fashion.[46] In this section we will see how the history of modern medicine in this period was intertwined with this emergence of markets both in Europe and in the colonies to produce a new medical practice and culture.

The Commercial Revolution led to the emergence of what is known as the 'medical marketplace' in Europe. Harold J. Cook coined the phrase in his book *The Decline of the Old Medical Regime* to show the declining control over the medical practice and market of the Royal College of Physicians in England and the diversity of sources from which medicines could be bought or acquired in the seventeenth century.[47] In Britain the growth of the new medical marketplace reflected the new 'commercial capitalist, spectacle-loving, consumer oriented society' of the country.[48] By the end of the seventeenth century, new drugs appeared from the Orient and the Americas in European markets, and new medical experiences that surgeons accumulated in the colonies became known to English medicine. European drug markets became harbingers of the changing times. Historians have used the concept of the 'medical marketplace' to analyse these changes in the history of medicine in early modern Europe. They have highlighted various aspects of the social and economic organization of medicine in early modern Europe, including the commercialization of medical practice, professional competition and restructuring of the professional hierarchies. The apothecaries who sold the drugs in the markets became the sources of information about the new medicines. Prominent apothecaries, such as Petiver in London and Jan Jacobsz in Amsterdam, ran their global medical networks collecting botanical and medical specimens from distant lands. Petiver also wrote to the Royal Society regularly about the new varieties of medicines that he collected.

Sloane worked closely with Petiver in London. He invited Petiver to the Royal Society's meetings. Petiver in turn used his global network to secure medicine and plants for the Royal Society. He collected medicinal plants and drugs from all over the world, pushing the need to find medicinal alternatives globally. In 1699, Petiver published a piece in *Philosophical Transactions* encouraging botanists to search for alternative species of medicinal plants in areas under British colonial control.[49]

Between 1615 and 1640, 40 per cent of drugs on the English market came from the East Indies. By the second half of the seventeenth century, this had gone up to 60 per cent.[50] New apothecary shops opened for the sale of these 'exotic' drugs in London and elsewhere. Due to the sheer number of new drugs available, the apothecary shops also became highly specialized. Containers and jars were now systematically arranged, and ordered onto separate shelves. Vendors adopted innovative methods to advertise and display their exotic products. These apothecary shops, their innovative methods of sale and their exotic merchandize popularized the use of exotic drugs in European medicine.[51]

This history of markets in Europe was connected to the history of markets and ports in the colonies. The interest in Oriental drugs took European surgeons to the local markets of Asia, which were the hub of seventeenth- and eighteenth-century trade in spices.[52] Markets in Asia became the nodal points of commodity exchange.[53] Asian markets (or bazaars) in the eighteenth century were of various kinds – contiguous lanes of shops, situated within large roofed buildings, arranged over a large square in the village or town centre, or scattered around the ports.[54] The Asian bazaars of the seventeenth and eighteenth centuries were the meeting places of diverse worlds; visited by French, British, Portuguese, Armenian and Indian merchants, local rulers, petty traders and buyers; goods arriving from across the sea and the hinterlands; and European dogs being sold next to Arabian horses.[55] European surgeons and merchants explored these markets in search of exotic drugs to export to European markets or to use in their colonial settlements.

In the colonial establishments of Asia in the eighteenth century, European surgeons working for the trading companies often used what they called 'bazaar medicines' (medicines that were bought from the local markets) along with European ones, at times combining the two. By the middle of the century, bazaar medicines became a regular component of the supply of drugs for the British military hospitals in India. These medicines were themselves of eclectic origins. They were brought to the large marketplaces in Asia from distant and relatively obscure locations. For example, an English surgeon based in Madras, Whitelaw Ainslie, found that aloes, which was commonly found in the bazaars of India, was brought from China as well as Borneo; cardamom, commonly used as a medicine by British surgeons, grew on the Malabar Coast but was also brought from Cambodia;[56] tabasheer, a popular medicine in Asia, was rarely found in India and was brought as an article of trade from West Asia;[57] mercury, the vital medicine of the Coromandel Coast used for salivation, was procured from remote Tibet and brought to the Indian ports through China by country trade sloops[58] (supplied mostly by the Dutch).[59] Other items available in Indian markets were benzoin (brought from Sumatra and Java), *Calamus aromaticus* (extracted from plants found in the Malabar region), camphor (brought to Indian markets from Sumatra and Java), China root (from south China and the East Indies), 'Dragons Blood' (from a kind of cane found in Java), Galingale (from South East Asia), gamboge (from Cambodia and Thailand) and myrrh (from Arabian and African coasts).[60]

These drugs were then taken to Europe to be sold in European apothecary shops. European medical text and catalogues too incorporated names and used of plants and drugs herbs from different parts of the world, which I will study in detail in Chapter 2. In these diverse ways, European medicine, similar to European mercantile economies, was enriched in the Age of Commerce through colonial connections.

Before I end this story of the enrichment of European medicine, the stimulating experiences of the surgeons, the growing influence of apothecaries and the general history of prosperity of the Age of Commerce, there is a need to take note of the darker sides of this new market economy as well. The growth of slave markets was an integral part of the Age of Commerce. The increasing commercialization and monetarism, rise of colonial power, and growth of capitalism and plantations created a paradigm in which the bodies of slaves were viewed more easily as objects or products for the market.[61] The same is true of the sailors who were captured from the streets of European cities by pressgangs and thrown into the unknown and hard colonial voyages. The sailors in crowded ships were treated almost no better than the slaves. They were often flogged for punishment and died due to a lack of food and medicine.[62]

Selling and buying slaves became a major commercial enterprise and slave markets developed in West Africa, in the West Indies and in South and North America from the seventeenth century. European traders earned great fortunes from this trade. Trevor Burnard and Kenneth Morgan have estimated that in one year, 1774, the total value of the slave trade in the island of Jamaica was £25 million, nearly totalling the wealth of the whole island.[63]

Medicine played its role in these slave markets. As slaves were considered to be valuable commodities, the traders and their surgeons used ingenious ways to keep them, or make them look, healthy. The trading contract between slave merchants and slave traders obliged the captains of ships to certify that there were no contagious diseases on the ships before they were admitted to the ports.[64] The ships involved in the slave trade employed surgeons to look after the crew and slaves. In order to deceive buyers, surgeons sometimes blocked the anuses of slaves suffering from dysentery with oakum, causing them excruciating pain.[65] When a ship docked, slaves were greased with palm oil to make them look 'sleek and fine', and the prospective buyer looked into their mouths and tested their joints before buying them.[66] Slaves and slavery were an integral part of the market economy, of the prosperity, global scale of exchange and the often brutal pursuit of profit of the Age of Commerce.

Conclusion

The history of the Age of Commerce is one of mixed shades. The Commercial Revolution ushered in a period of great economic growth in several parts of Western Europe and also in parts of Asia and the Americas. Encounters with diverse cultures, plants and markets enriched European medicine, introduced new drugs and stimulated European medicine and medical marketplaces. Commercial activities enriched Europe's knowledge of medicine, environment and health. Europeans discovered new medical herbs, and interacted with different cultures and practices of medicine. New drugs entered medicine in Asia and the Americas as well, as Africans carried their medicines and medicinal plants to the Americas, sharing these with the Amerindians. In the Indian

Ocean, European demands and the growth in the spice trade brought new drugs and plants to the markets. European surgeons and missionaries incorporated these in their own medicine.

However, there were also negative consequences to the history of commerce. The commercial expansion brought new diseases to the Americas, which led to the extermination of large sections of the Amerindian population (See Chapter 5). Commercialization also increased in the enslavement and buying and selling of humans. The history of modern medicine is testimony to both of these aspects of the Age of Commerce. It should also be added here that seeing this history predominately in terms of exchanges could also obfuscate the military expansions and conquests that took place simultaneously in the Age of Commerce.

Notes

1 Phyllis Allen, 'The Royal Society and Latin America as Reflected in the *Philosophical Transactions* 1665–1730', *Isis*, 37 (1947), 132–8, p. 132.
2 Peter Bakewell (ed.), *Mines of Silver and Gold in the Americas* (Aldershot, 1997).
3 Iris Bruijn, *Ship's Surgeons of the Dutch East India Company; Commerce and the Progress of Medicine in the Eighteenth Century* (Leiden, 2009).
4 Jacob M. Price, 'What did Merchants do? Reflections on British Overseas Trade, 1660–1790', *Journal of Economic History*, 49 (1989), 267–84, pp. 271–2.
5 Robert Brenner, *Merchants and Revolution: Commercial Change, Political Conflict, and London's Overseas traders, 1550–1650* (Cambridge, 1993).
6 Crosby, *The Columbian Exchange*.
7 Christopher Columbus (edited and translated with an Introduction and notes by B.W. Ife), *Journal of the First Voyage (diario Del Premier Viaje) 1492* (Warminster, 1990), pp. 47–9.
8 Schiebinger, *Plants and Empire*.
9 Pamela H. Smith and Paula Findlen (eds), *Merchants & Marvels; Commerce, Science, and Art in Early Modern Europe* (New York & London, 2002). See also Kay Dian Kriz, 'Curiosities, Commodities, and Transplanted Bodies in Hans Sloane's "Natural History of Jamaica"', *The William and Mary Quarterly*, 57 (2000), 35–78.
10 Lucille Brockway, *Science and the Colonial Expansion: The Role of British Royal Botanic Gardens* (New York, 1979); Emma C. Spary, '"Peaches Which the Patriarchs Lacked": Natural History, Natural Resources, and the Natural Economy in France', *History of Political Economy*, 35 (2003), 14–41.
11 Robert Voeks, 'African Medicine and Magic in the Americas', *Geographical Review*, 83 (1993), 66–78, p. 67.
12 Ibid, p. 72.
13 Richard S. Dunn, *Sugar and Slaves; The Rise of the Planter Class in the English West Indies, 1624–1713* (Chappell Hill, 2000/1972), p. 250.

14 Ibid, pp. 66–78.
15 George Brandon, 'The Uses of Plants in Healing in Afro-Cuban Religion, Santería', *Journal of Black Studies*, 22 (1991), 55–76.
16 See Jerome S. Handler, 'Slave Medicine and Obeah in Barbados, Circa 1650 to 1834', *New West Indian Guide*, 74 (2000), 57–90, pp. 82–3.
17 Sharla M. Fett, *Working Cures: Health, Healing and Power on the Southern Slave Plantations* (Chapel Hill, NC, 2002), p. 134
18 Frank Wesley Pitman, 'Fetishism, Witchcraft, Christianity Among the Slaves', *The Journal of Negro History*, 11 (1926), 650–68, pp. 652–3.
19 Ibid, p. 653. See also Fett, *Working Cures*, p. 134.
20 Christiane Bougerol, 'Medical Practices in the French West Indies: Master and Slave in the 17th and 18th Centuries', *History and Anthropology*, 2 (1985), 125–43, p. 136.
21 Hans Sloane, *A Voyage to the Islands of Madera, Barbados, Nieves, S. Christophers and Jamaica, with the Natural History*, vol. 1 (London, 1707), Preface.
22 As quoted in J.S. Haller, Jr., 'The Negro and the Southern Physician: A Study of Medical and Racial Attitudes 1800–1860', *Medical History*, 16 (1972), 238–53, p. 240.
23 Pitman, 'Fetishism, Witchcraft and Christianity Among the Slaves', pp. 664–8. Obeah practices became less prevalent as the proportion of new slaves from Africa declined by the end of the eighteenth century: Handler, 'Slave Medicine and Obeah in Barbados', p. 69.
24 Todd L. Savitt, *Medicine and Slavery: The Disease and Healthcare of Black in Antebellum Virginia* (Urbana, London, 1978) p. 173.
25 William Robertson, *The History of America*, vol. 3 (London, 1800–1), p. 238.
26 Alix Cooper, *Inventing the Indigenous: Local Knowledge and Natural History in Early Modern Europe* (Cambridge, 2007), pp. 87–115.
27 Ferdinando Abbri, 'Alchemy and Chemistry: Chemical Discourses in the Seventeenth Century', *Early Science and Medicine*, 5 (2000), 214–26.
28 Antonio Clericuzio, 'From van Helmont to Boyle; A Study of the Transmission of Helmontian Chemical and Medical Theories in Seventeenth-Century England', *The British Journal for the History of Science*, 26 (1993), 303–34.
29 Roy Porter and Dorothy Porter, 'The Rise of the English Drugs industry: The Role of Thomas Corbyn', *Medical History*, 33 (1989), 277–95.
30 Alain Clément, 'The Influence of Medicine on Political Economy in the Seventeenth Century', *History of Economic Review*, 38 (2003), 1–22, p. 14.
31 Nigel G. Coley, '"Cures Without Care" "Chymical Physicians" and Mineral Waters in Seventeenth-Century English Medicine', *Medical History*, 23 (1979), 191–213.
32 Chakrabarti, *Materials and Medicine*, pp. 175–82.
33 Paul E. Kopperman, 'The British Army in North America and the West Indies, 1755–83: A Medical Perspective', in Geoffrey L. Hudson (ed.) *British Military and Naval Medicine 1600–1830* (Amsterdam, 2007), pp. 51–86.

34 Richard B. Sheridan, 'The Doctor and Buccaneer: Sir Hans Sloane's Case History of Sir Henry Morgan, Jamaica, 1688', *JHMAS*, 41 (1986), 76–87, p. 79.

35 Ibid.

36 James William Kelly, 'Wafer, Lionel (*d.* 1705)', *Oxford Dictionary of National Biography* [Hereafter *Oxford DNB*], www.oxforddnb.com/view/article/28392, accessed 9 September 2011.

37 Kenneth Dewhurst, *The Quicksilver Doctor, the Life and Times of Thomas Dover, Physician and Adventurer* (Bristol, 1957), p. 54.

38 James Petiver and Samuel Brown, 'Mr Sam. Brown His Seventh Book of East India Plants, with an Account of Their Names, Vertues, Description, etc', *Philosophical Transactions of the Royal Society* [hereafter, *Philosophical Transactions*], 23 (1702–3), 1252–3.

39 James Petiver, 'An Account of Mr Sam. Brown, his Third Book of East India Plants, with Their Names, Vertues, Description', *Philosophical Transactions*, 22 (1700–1), 843–64.

40 Sloane, 'An Account of a China Cabinet, Filled with Several Instruments, Fruits, &c. Used in China: Sent to the Royal Society by Mr. Buckly, Chief Surgeon at Fort St George', *Philosophical Transactions*, 20 (1698), 390–2.

41 'Medical News', *Medical Commentaries*, 1 (1786), 385–9.

42 Fernand Braudel, *Civilization and Capitalism, 15th–18th Century: The Perspective of the World* (Berkeley, 1992), pp. 277–89.

43 Linda L. Peck, *Consuming Splendor: Society and Culture in Seventeenth-Century England* (Cambridge, 2005).

44 Nuala Zahedieh, 'London and the Colonial Consumer in the Late Seventeenth Century', *The Economic History Review*, 47 (1994), 239–61.

45 Woodruff D. Smith, 'The Function of Commercial Centers in the Modernization of European Capitalism: Amsterdam as an Information Exchange in the Seventeenth Century', *The Journal of Economic History*, 44 (1984), 985–1005.

46 Maxine Berg, 'In Pursuit of Luxury: Global History and British Consumer Goods in the Eighteenth Century', *Past & Present*, 182 (2004), 85–142.

47 Harold J. Cook, *The Decline of the Old Medical Regime in Stuart London* (Ithaca, 1986).

48 See, Introduction, in Patrick Wallis and Mark Jenner (eds), *Medicine and the Market in Early Modern England and its Colonies, c. 1450–c. 1850* (Basingstoke, 2007), pp. 1–23, p. 5

49 Petiver, 'Some Attempts Made to Prove That Herbs of the Same Make or Class for the Generallity, have the Like Vertue and Tendency to Work the Same Effects', *Philosophical Transactions*, 21 (1699), 289–94.

50 Wallis, 'Exotic Drugs and English Medicine: England's Drug Trade, c. 1550–c. 1800', *Social History of Medicine*, 25 (2012): 20–46.

51 Wallis, 'Consumption, Retailing, and Medicine in Early-Modern London', *The Economic History Review*, 61 (2008), 26–53.

52 For a more detailed account of the link between medicine and Indian markets in the eighteenth century, see Chakrabarti, 'Medical Marketplaces Beyond the West: Bazaar Medicine, Trade and the English Establishment in Eighteenth Century India', in Wallis and Jenner (eds) *Medicine and the Market*, pp. 196–215.

53 For an account of the East India Company and markets on the Coromandel Coast, see Arasaratnam, *Merchants, Companies and Commerce on the Coromandel Coast 1650–1740* (Delhi, 1986), pp. 213–73.

54 S. Hajeebu, 'Emporia and Bazaars', in J. Mokyr (ed.) *Oxford Encyclopaedia of Economic History*, vol. 2 (Oxford, 2003), p. 258.

55 H. Furber, 'Asia and the West as Partners before "Empire" and after', *Journal of Asian Studies*, 28 (1969), 711–21, p. 712.

56 Whitelaw Ainslie, *Materia Medica of Hindoostan, and Artisan's and Agriculturalist's Nomenclature* (Madras, 1813) pp. 8–9.

57 Ibid, p. 46.

58 Ibid, p. 57.

59 See Niklas Thode Jensen, 'The Medical Skills of the Malabar Doctors in Tranquebar, India, as Recorded by Surgeon T.L.F. Folly, 1798', *Medical History*, 49 (2005), 489–515, p. 506.

60 Chakrabarti, *Materials and Medicine*, pp. 42–4.

61 For a historiographical survey of this topic, see Lesley A. Sharp, 'The Commodification of the Body and its Parts', *Annual Review of Anthropology*, 29 (2000), 287–328.

62 Isaac Land, 'Customs of the Sea: Flogging, Empire, and the "True British Seaman" 1770 to 1870', *Interventions: International Journal of Postcolonial Studies*, 3 (2001), 169–85.

63 Trevor Burnard and Kenneth Morgan, 'The Dynamics of the Slave Market and Slave Purchasing Patterns in Jamaica, 1655–1788', *The William and Mary Quarterly* (2001), 205–28, p. 209.

64 'The trade granted to the South-Sea-Company: considered with relation to Jamaica. In a letter to one of the directors of the South-Sea-Company by a Gentleman who has resided several Years in Jamaica' (London, 1714), p. 10.

65 Alexander Falconbridge, *An Account of the Slave Trade on the Coast of Africa* (London, 1788), pp. 46–7.

66 Emma Christopher, *Slave Ship Sailors and Their Captive Cargoes, 1730–1807* (New York, 2006), pp. 198–9; Dunn, *Sugar and Slaves*, p. 248.

2

Plants, medicine and empire

Plants were vital to European colonialism. Modern colonialism had begun in search of exotic spices and tropical plants. It later thrived on deriving profits from growing these plants in colonial gardens and plantations. However, there was more to this European engagement with exotic tropical flora than the pursuit of commercial profit. Exotic plants from the colonies pervaded European natural history, science and medicine from the seventeenth century. Exotic medicinal plants transformed European medicine. During the seventeenth century the import of drugs from the Orient and the New World increased 25-fold.[1] By the late eighteenth century, medical drug imports in England were valued at around £100,000 a year, 50 times greater than the £1,000 to £2,000 a year, two centuries earlier.[2] In the eighteenth century, European medical texts, catalogues and dispensatories regularly referred at length to the plants and herbs from different parts of the world.[3] Many of these items, such as ipecacuanha, cinchona, sarsaparilla and opium became effective and highly popular in European medical practice.[4] These changes in European medicine took place due to the exploration, observation and exploitation of the plants and herbs of Asia and the Americas by European naturalists, missionaries, travellers and surgeons. Through them a new relationship between the natural world of the tropics and the medical world of Europe was forged.

Colonial prospection and medical botany

Historians have described this European search for valuable plants in the colonies as 'colonial bioprospecting'.[5] From the seventeenth century, botanists and naturalists travelled in search of new and exotic plants and carried them to their colonies. This was part of the mercantilist enterprise; European nations sought to grow exotic plants in the gardens and plantations of their own colonies to check the drain of bullion that would have occurred if they had been acquired through trade. This mercantilist interest led to the transportation and transplantation of plants such as cotton, tobacco, coffee, pepper and sugarcane between Malacca, Virginia, India and the West Indies. This also led to the expansion of colonial gardens and plantations. By the end of the eighteenth century, European nations possessed around 1,600 botanical gardens around the world.

Although several colonial plants were grown in the prominent European botanical gardens, such as Kew Gardens in London, Jardin du Roi in Paris and the Hortus Botanicus in Leiden, these were mainly for intellectual and aesthetic purposes. The actual commercial cultivation of vital tropical plants, such as coffee, sugar and cinchona, which could not grow in the temperate European climate and needed large acres of land, took place in the colonial gardens and plantations. The profits from the sale of these plants and medicines then went to Europe. The process of searching for exotic medicinal plants and cultivating them for European drug markets, apothecaries and pharmaceutical companies started in the seventeenth century and continued until the twentieth century when several African medical plants, such as *Stropanthus kombe*, were commercially exploited by European pharmaceutical companies for their medicinal qualities.[6]

European study of Asian medicinal plants began with an initial interest in Oriental spices. Spices were items of commercial interest but had also generated great botanical and medical curiosity in Europe. By the eighteenth century, spices formed an important part of the apothecary's recipe books in Europe.[7] Spices and spiced foods were supposed to have valuable medicinal functions, as digestives, stimulants and cures for fevers, headaches and colic. Therefore European merchants searched the spice markets of Bantam and Masulipatnam, and European naturalists and physicians, such as Caspar Bauhine (an Italian physician and botanist), Prosper Alpinus and Jacobus Breynius, explored the plants and herbs of India for their medicinal value.[8] European physicians invested in networks of traders and surgeons to collect Asian spices. Henry Draper Steel (1756–1818), a prominent physician from London, issued detailed instructions for travellers to Asia for identifying, purchasing and carrying spices and drugs back to Europe.[9] London-based apothecary Petiver developed networks in the Asian spice trade to collect medical specimens.[10] European naturalists started to explore plants in the alluring coasts, forests, ports and markets of the Orient. The medicinal plants of the Spice Islands and of the Malabar and the Coromandel coasts of India became well known in Europe through the explorations by the Portuguese and Dutch.

The Portuguese were the first Europeans to commence botanical and medical studies in Asia. Garcia D'Orta (1479–1572), a physician and professor in Lisbon, travelled to India in 1534 and stayed there until his death. The Portuguese authorities leased the island of Bombay to D'Orta in perpetuity. From there he studied Indian plants and drugs, and in 1563 he wrote a book called *Dialogues on Samples and Drugs* (*Colóquios dos simples e drogas he cousas medicinais da Índia*). This became popular across Europe and was translated into Latin and other European languages. This was the first modern European text which provided a first-hand knowledge of the diseases and drugs of India.

Another Portuguese traveller and physician who travelled to Asia in the sixteenth century was Cristobal Acosta (1515–94). His interests were in natural history, pharmacy and botany, and he was another European pioneer in studying Indian medicinal plants for their pharmaceutical uses. Acosta was one of the early explorers. He went to Asia around 1550 as a soldier and visited Persia, India, Malaya and probably China. He was appointed physician to the royal hospital in the Portuguese colony of Cochin in India in 1569. In India he also met D'Orta. Over the next few years he collected botanical specimens in various parts of India. When his term of service ended in 1572, he returned to Lisbon. Back in Portugal he practised medicine between 1576 and 1587, when he was the appointed as the physician to the city of Burgas. In his medical practice he introduced the various medicinal herbs and drugs that he had collected from India and elsewhere in Asia. He documented his entire knowledge of Asian medicinal plants in his *Tractado de las drogas y medicinas de las Indias Orientales* (*Treatise of the drugs and medicines of the East Indies*, 1578), which contained the first European systematic observations of Oriental drugs. This text too was widely translated into different European languages.[11]

By the seventeenth century, as colonial possessions expanded, the study of natural history diversified in Europe and became part of the European university curriculum. Scholars from not just Spain and Portugal, but the Netherlands, England and France participated in the study of Oriental spices and medical plants.[12] The Dutch East India Company was established in 1602 and it rapidly demolished Portuguese power in the Indian Ocean to gain a monopoly on the East Indies spice trade. Throughout the seventeenth century the Dutch economy flourished from the East Indies spice trade. This growing prosperity was reflected in the numerous works of natural history of Asia that were produced by the Dutch during this period. After the Portuguese, the Dutch took a very active interest in Asian botany and medicine.

The engagement in the spice trade and commerce generally enabled the Dutch naturalists and scientists to develop their tools of observation, objectivity, accumulation and description, which were the hallmarks of late seventeenth-century natural history. With commercial expansion, natural history became an important topic at Dutch universities, botanic gardens flourished in the Netherlands, and collections of exotic objects multiplied in apothecary shops and private collections. This gave rise to a new philosophy regarding the objective view of nature in the Netherlands.[13] In this convergence of commerce and natural history, spices and medicinal plants of Asia were highly coveted by Dutch physicians, naturalists and apothecaries. Thus, in the eighteenth century, spices and medicinal plants formed the foundation of the apothecary's recipe books in the Dutch Republic.[14] Dutch physicians believed that spices and spiced foods served valuable medicinal functions, as digestive agents, as stimulants, in the cure of fevers and colic, and as 'carminatives' (cures for flatulence). Throughout the seventeenth and early eighteenth centuries, trade in Asian plants remained the cornerstone of Dutch medicine.

In 1678, Henry van Rheede, the Dutch governor of the Malabar Coast of India, compiled a systematic and massive 12-volume text on Indian botany. It was entitled *Hortus Indicus Malabaricus* and was based on his own observations, with advice and assistance from local experts.[15] Rheede listed the Latin, Malabar, Arabic and Sanskrit names of the plants with translations of testimonies of local practitioners. His work became the most important reference for botanists working in India for several centuries.

Another important Dutch work was on the medical botany of Ceylon by Paul Hermann (1646–95). He spent seven years (1670–7) exploring the island at the expense of the Dutch East India Company. His notes reached English botanist William Sherard (1659–1728), who edited them to produce a catalogue published as *Musaeum Zeylanicum* (1717). Hermann's collections of Ceylonese plants were disseminated all over Europe. Some of them were collected by the Danish Apothecary-Royal August Günther. Günther loaned them to Swedish biologist Carl Linnaeus (1707–78), who studied them at length and placed them in his new taxonomical system. Hermann's collections contained many plants that were new to Linnaeus, and he felt the need to compile a separate volume on plants of Ceylon in his *Flora Zeylanica* (1747). After those of Rheede and Hermann, the most significant Dutch work on South Asian plants was by Nicolai Laurentii Burmanni, who published his *Flora Indica* in 1768.[16] Nicolai was inspired by Linnaeus to classify the Indian plants according to the new taxonomy. His studies were based on the plants collected by his father, John Burmanni, professor of botany in Amsterdam. John too had gathered some of Hermann's collections. Thus the Portuguese and the Dutch initiated the exploration and study of Asian flora and medicinal plants. By the middle of the eighteenth century, these had also been well entrenched in European natural history traditions, situated within the new classificatory schemes as well as established in European medical training, practices and recipes.

With the relative decline of the spice trade around the middle of the eighteenth century, the Dutch hegemony began to wane. In 1799 the Dutch East India Company became bankrupt. The British, who gained colonial supremacy in India from the middle of the eighteenth century, were a late entrant, in trade as well as in botanical studies. William Roxburgh's *Flora Indica* (published posthumously in 1832) was the first significant work by the British in this field. Roxburgh's work was on the Coromandel Coast in the east, which also had its attraction with regard to medicines and spices.

On the other side of the world, the Spanish opened up vast natural resources in the Americas. From the time of Columbus' arrival in the Americas, the Spanish had interacted with the local *Tainos* (an extinct Arawakan people who inhabited the Greater Antilles and the Bahamas) and gained knowledge from them of plants, medicines and their healing practices. However, rapid depopulation meant that by the end of the sixteenth century the Spanish became the authority in American plants and drugs. The Spanish also dominated the Atlantic drug trade until the early seventeenth century. One of the early Spanish works on American plants was by Spanish physician and botanist Nicolas Monardes (1493–1588). His book (*Joyfull Newes out of the Newe Founde Worlde*) published in 1577, became a standard text on American diseases and their cures for European physicians and naturalists in the seventeenth and eighteenth centuries. Although the Spanish physicians interacted with Amerindian physicians to learn about the plants of the Americas, they mostly rejected the magicoreligious context in which the Amerindians used their plants as medicine. They instead favoured their own Galenic humoural tradition, within which they located the newly acquired American plants and medicines.[17]

The Spanish, by virtue of their presence in the Indian and Atlantic oceans following Magellan's voyages, also introduced Asian medicinal plants, such as the *Cassia fistula*, into the West Indies.[18] Other plants too travelled between different regions through Spanish and Portuguese networks. The ipecacuanha of South American origin had an interesting history in the East Indies. Originally a Tupi-Guarani name for a plant of the Rubiaceae family found in Brazil, ipecacuanha in the eighteenth century came to specify several plants whose roots showed emetic properties. Several varieties of the ipecacuanha were discovered in India and Ceylon, where they were used as emetics (e.g., *Cynanchum loevigatum* produced the white ipecacuanha of Bengal, and *Cynanchum tomentosum* and *Polygala glandulose* were the white and black ipecacuanhas of Ceylon, respectively).[19]

French naturalist Esquemeling, about whom we have learnt previously, was initially employed by the French West India Company and sent to Tortugas. He then joined the buccaneers and left the French company to work privately. During his various explorations in the Caribbean islands, Esquemeling learnt about plants such as the *Lignum sanctum* or the guaiacum, *Cassia lignea aloes* and *Lignum aloes*, which were native to the Caribbean and South America.

Through these networks, plants and knowledge of therapeutic practices were transferred across the world, often from the West Indies to the East Indies and vice versa. The Cape of Good Hope in South Africa, which was located strategically between the trading routes of the East and the West, acquired great significance in such exchanges. The botanical interest in the Cape developed with Dutch involvement from the 1690s. The Dutch were keen to maintain the Cape as a provisional base for their trade in the Indian Ocean, and invested in the development of a botanical garden where they grew plants from Ceylon and South East Asia.[20] By the end of the eighteenth century, the Cape and its garden had gained a reputation among English botanists for its rich botanical collection.[21]

Through these explorations and networks across the Atlantic and Indian oceans, the Spanish merchants and apothecaries developed a thriving export industry of drugs from the New World to the Old World. Monardes once declared: 'not only Spain is provided with this product [*Cassia fistula*] but all of Europe and nearly the whole world'.[22]

The European discovery of the cinchona tree is an essential part of this history of colonial botany and the emergence of global trade in medicinal plants. It also reflects the character of colonial bioprospecting. The bark of the cinchona tree, which grew wild in several parts of South America, particularly Peru, was one of the most vital drugs in the seventeenth and eighteenth century as it was the main medicine used against different kinds of intermittent and remittent fevers. It is debatable to what extent the Spanish actually 'discovered' it. According to some sources, the name 'cinchona' came from the countess of Chinchon, the wife of a Peruvian viceroy in Peru, who was cured of a fever by using the bark of the cinchona tree in 1638 by a local physician. The countess of Chinchon then supposedly introduced it to European medicine in 1640, even before botanists had identified and named the species. According to others, in the seventeenth century a monk of the Augustinian Order noticed its medicinal qualities. He too seemed to have been informed by the locals about its medicinal effects as he wrote: 'A tree grows which *they* call "the fever tree" in the country of Loxa, whose bark...cures the fevers and tertiana; it has produced miraculous results in Lima'.[23] Since the bark was popularized in Spain by the Jesuits who brought it from South America, it came to be known as the 'Jesuit's bark'. Cinchona bark was first advertised for sale in England in 1658, and entered the London Pharmacopoeia in 1677.

However, English botanist Clements R. Markham, who took cinchona saplings from Peru to India in the nineteenth century, claimed that the medicinal virtues of the bark were not well known to the Amerindians and they attached 'little importance to them'.[24] Debates have also continued about whether malaria, in the treatment of which cinchona was found to be particularly useful, existed in the New World before the advent of the Spanish.[25] Whatever the nature of this European discovery of cinchona, it soon became a vital and popular drug against various kinds of fever. The Spanish acquired a monopoly over the tree and the bark, and developed a lucrative trade of supplying cinchona to the apothecaries of Europe in the seventeenth century.

Europeans were not content with collecting and trading tropical plants. They wanted to grow them on a large scale for greater commercial profit. Although the Spanish and the Portuguese had started the first plantations of tobacco, coffee and sugar in the Americas in the seventeenth century, the British, Dutch and French colonization of the Americas and the West Indies from the second half of the seventeenth century led to the cultivation of tropical plants in huge plantations. The plantation system was a new and unique agricultural system, which required detailed knowledge about the climatic conditions, the soil, different varieties of the plants and a well ordered labour force. In Europe the plantation experience introduced a new link between natural history and commercial agriculture. From the early eighteenth century, French naturalists and medical botanists associated themselves with global trading networks and plantation colonies, and explored the different species, varieties and productivities of plants which they considered to be useful.[26]

In the eighteenth century, as the French established sugar and coffee plantations in the West Indies, Nicolas-Louis Bourgeois, secretary of the Chamber of Agriculture, while posted on the newly acquired island of Saint-Dominique, studied the Caribbean medical traditions, which, he believed, reflected a fusion between Amerindian and West African traditions. Other French botanists studied and collected medicinal plants in the West Indies. Botanist Michel

Etienne Descourtilz travelled to Saint-Dominique and compiled a compendium of local medicinal plants, some of which he suggested could act as substitutes for cinchona, such as Rhizophore bark. He also recorded that Caribbean slave healers administered herbal teas to reduce pain and fever, especially a decoction of 'stinking peas' (*Cassia occidentalis*). Descourtilz recorded several of those anti-febrile herbal teas, such as those made from the flowers of *Poincillade* or of *Quassia amara*.[27]

In the English sugar plantation colony of Jamaica, the same ethos of utility was present in the investigations of medical botany. Early English naturalists there studied plants that could be grown in the plantations, cultivated as sources of food for slaves and introduced as medicine in Europe. Soon after its occupation in 1655, Thomas Tothill, the collector and receiver-general of the island, conducted a survey to find out the useful commodities produced in the island. Among others he found China roots, *Cassia fistula* (introduced from India) and tamarind growing wild, and believed that these could be good sources of medicine for the plantations. Hans Sloane (1660–1753), a physician and most prominent English collector of natural history specimens of the early eighteenth century, arrived in Jamaica in 1687. This was the time when the newly occupied island was being transformed from a sporadic maritime Spanish settlement to an organized plantation for growing sugar. Sloane thus experienced the coexistence of Spanish, Amerindian and African knowledge of plants and healing traditions. He found that Jamaica contained 'most of the Natural Productions of Espanolia, Barbados and the other hot American Isles, but also many of those of Guinea and the East-Indies'.[28] He referred to several Spanish sources in his research, such as the works of John de Barrios and Franciscus Uria while describing the pimento or the Jamaican Pepper-tree.[29] He noted that Spanish historian Francisco Hernández adopted the Amerindian names for these plants.[30] Sloane also collected medical recipes from the Africans who lived freely after the Spanish deserted the island, all of which was published in his two-volume *Voyages*.[31] He observed how different races who lived or had settled on the island, such as the Caribs, the Spanish and the Africans, had brought different medicinal plants to the island. He also identified various medicinal plants within the plantations. Another British surgeon, James Thomson, produced a list of medicinal plants which were available in Jamaica and recommended these to be cultivated on every estate and used in the medical practices there. These included the aloes, the cabbage tree, bitterwood, capsicum peppers, contrayerva and arrowroot.[32]

Missionaries and medical botany

Missionaries were the other group of Europeans who worked along with the botanists and surgeons in exploring and collecting medicinal plants in the colonies. Plants were a vital part of early missionary life there, not only for their medical potential but for their spiritual purposes as well. Christian missionaries viewed nature as the work of God, thus the study of the natural world and the laws of nature in all its diversity, which was now made evident through colonial explorations, widened the scope of this exploration of divine creation. Spanish and Portuguese Jesuit missionaries were driven by the urge to discover the 'rare and wonderful' productions of nature in Asia and the Americas. The Jesuit order was created in response to the colonial explorations in 1534, and formed important intellectual and religious links between Europe, Asia, Africa and the Americas from the sixteenth century. Jesuits travelled to various parts of the two worlds to find out about the plants, people and places. Natural history – the collection, improvement, cultivation and classification of nature – had a strong evangelical connotation.[33] Jesuits made significant contributions to botany and *materia medica*, along with astronomy, cartography, geography and natural history.

Following closely on the heels of the explorers and colonialists, Spanish and Portuguese missionaries settled in the Americas and Asia. Alongside observing nature, they studied local medical practices and the uses of the local vegetation by indigenous people, with whom they often formed close links.[34] Jesuits were the first missionaries to study tropical nature from the sixteenth century.[35]

In the Americas and the West Indies, the missionaries combined preaching the gospel with studying local plants. Missionaries there were involved in a range of medical activities: collecting medicinal plants from the forests in the southern United States, using medicines to attract slaves and Amerindians to Christianity, and even selling European drugs. The missionaries established long-distance networks of exchange of botanical specimens across the Atlantic. They sent carefully prepared botanical specimens across the ocean to Joseph Banks in London and to others in England, as well as to their patrons in Northern Europe, particularly Halle in Germany in the eighteenth and nineteenth centuries.[36] Christian Oldendorp, a Moravian missionary, carried several plants from the Caribbean islands of St Thomas and St Croix to Europe. In North America, on the other hand, Pietist missionaries were involved in selling European pharmaceuticals among the white settler communities.[37]

At the same time, Christianity and missionary medicine underwent transformations when in contact with Amerindian and African religions and healing traditions.[38] Christian missionaries viewed Amerindian 'medicine men' as the main obstacle in promoting their religion. In Lima in Peru, missionaries came into conflict with local Nahua physicians, who they believed were the sources of old religious beliefs which they were seeking to uproot.[39] At the same time they actively sought to learn about their medicinal practices, which had strong spiritual connotations within Amerindian religions and healing traditions. The communities which were converted retained their older rituals. In order to cure the local communities with local drugs and medicine, missionaries fused their practices with local rituals and healing traditions. In the process, Christianity and colonial medicine in South America incorporated Amerindian customs and rituals. The Spanish adoption of Peruvian bezoar stones in the sixteenth and seventeenth centuries took place through a convergence of Andean occult practices and Christian values.[40] The word 'bezoar', on the other hand, came to Spanish vocabulary through Arabic contact – it was derived from the Arabic word *badzahr*, which means antidote or counterpoison.

In the plantations of the West Indies, Protestant missionaries worked among the slaves and they treated the sick as part of their medical and evangelical duties. Following the Reformation, Protestant churches gradually began to send physicians to serve as medical missionaries in the colonies.[41] From 1760 to 1835, the missionaries operated a mission station at the Mesopotamia sugar estate in Western Jamaica. Joseph Foster Barham, the owner there, had urged them to settle in his plantation.[42] Protestant missionaries also settled on the Danish West Indian island of St Thomas, on Antigua and on Barbados. Augustus Gottlieb Spangenberg (1704–92) of the Moravian Church wrote in detail about the methods and ideals to be followed in converting the slaves on the plantations to Christianity.[43] Preaching Christianity and salvation among slaves and working within the steeply hierarchical structure of the plantation system often met with difficulties. A great hindrance was the fear among the planters and merchants that the slaves would ultimately demand greater freedom, which would inter-fere with their commercial interests. Spangenberg thus urged a distinct code of principles to be followed while preaching among the slaves. The missionaries were instructed not to interfere with commercial or plantation matters; they were to teach the slaves that 'it is not by chance, but it is of God, that one man is a master and another slave' and thus to follow the path of God for their inner salvation.[44]

In the botanical gardens established by the Moravian missionaries in Greenland and in the West Indies, plants and medicines served a variety of missionary purposes; to heal and feed the slaves and to preach the messages of the Bible to them.[45] Feeding and healing the slaves who carried out hard labour on the plan-tations was central to preaching Christianity. The missionaries also frequently referred to the plants in their gardens or surrounding their settlements in their discourses on God and his creations.

In the plantations, in a world driven by profit, ravaged by human suffering, reli-gion, more specifically Christian charity and medicine, was often the only source of salvation, care and cure for the slaves. On the other hand, medicine became one of the most important modes through which Christianity established itself firmly among Amerindians and Africans in the West Indies and the Americas.

Portuguese Jesuits were the first European missionaries to arrive in India following the Portuguese colonial settlements.[46] Father Mathew of the Order of St Joseph's Carmelite helped Van Rheede in his compilation of *Hortus Indicus Malabaricus*, from his own observations and with the advice and assistance of local physicians.[47] Father Papin of the French Jesuit mission in India wrote to Paris about the drugs, plants and the medical practices of the local communities.[48]

Protestant missionaries, such as the Pietists and Moravians, followed the Jesuits in India in the eighteenth century. The Danish brought the first Protestant mission to India, namely the Danish-Halle Mission in Tranquebar on the Coromandel Coast. In 1706, Bartholomäus Ziegenbalg (1683–1719) and Heinrich Plütschau (1678–1747), both German Lutheran missionaries from the Pietist tradition at the University of Halle, landed at the Danish settlement of Tranquebar. They were formally known as the Lutheran missionaries of the Royal Danish-Halle Mission, as well as the Tranquebar missionaries.

In India the Pietists were joined by the Moravians, who arrived in 1759. They were granted revenue-free land by the Danish king in Tranquebar and the Nicobar islands to cultivate and set up their own botanical gardens.[49] The Moravians purchased and cultivated lands for their subsistence.[50] This improved their knowledge of local flora and also brought them closer to the local communities.

The early Protestant missionaries such as Ziegenbalg and Plütschau developed associations with the local population as part of their religious duties and out of their curiosity about local customs. They studied Indian cultures, practices, languages and texts, and also explored and grew local medicinal herbs in their mission gardens. The missionaries maintained diaries, which they sent to Europe regularly. These diaries, along with their letters and private communications, narrated Indian festivals, temples, arts and crafts, music, legends, rituals, the diseases that people commonly suffered from and the medicines that they used to cure them. This resulted in the collection of detailed information about local medical practices. Johann Ernst Gründler of the Tranquebar mission spent his life studying Tamil medical texts and culture. To understand these more closely, in 1710 he left Tranquebar and settled in Poreiar, a nearby village. There he started eating and dressing like the local people. He acquired many medical palm leave bundles from them, which had information about various diseases, medicines and herbs, all of which was compiled in his *Malabar Medicus*.

Plütschau, another Protestant missionary in India, wrote about Indian medical practitioners, their own 'search into the secrets of nature' and how their medical skills could surprise physicians in Europe: 'our Physicians in *Europe* would wonder at the Performance of our *Malabar* Doctors here'.[51] While translating Tamil medical texts, missionary Benjamin Heyne appreciated Indian medical traditions: 'The medical works of the Hindoos are neither to be regarded as miraculous productions of wisdom, nor as repositories of nonsense.'[52] The textual and botanical interests that the missionaries took in Indian medicinal plants were part of their broader engagement with local languages and cultural traits. The Danish missionaries in Tranquebar learnt Tamil in order to get closer to the people among whom they were working. In the process they also studied and translated local medicinal texts into German.

They shared these with British surgeons and botanists who were based at the nearby English settlement of Madras in their search for medicinal plants in India. John Peter Rottler, a missionary with the Danish settlement in Tranquebar, was responsible for looking after the garden there. Between 1799 and 1800 he made several trips from Tranquebar and Madras, collecting botanical specimens. He sent these to the EEIC surgeons in Madras as well as to London.[53] From there his collections were distributed at Kew and in Liverpool. A 12-volume herbarium of plants sent from India entitled *Plantae Malabaricae* was found in the Department of Systematic Botany at the University of Göttingen. The university acquired this as part of the large herbarium from the collection of August Johann Von Hugo (1686–1760) in 1764. The missionaries in Tranquebar had in fact collected this herbarium, with names in Tamil with German commentaries, on the request of by Dr Hugo around 1732–3.[54] In this way the different missionary collections of medical botany were integrated with European medicine and natural history.

Missionaries added a new dimension to this history of colonialism, plants and medicine. At a time in Europe when the study of natural history and the practice of medicine was becoming an increasingly secular pursuit (i.e., based on observation and empiricism, distanced from spirituality, occultism and religion, and coming into conflict with the medicoreligious practices of other societies), missionaries retained and even revitalized Christian spirituality in the European exploration of nature and in the art of healing. This enabled them to have closer contact with and an insight into those communities with which they worked. This illustrates an aspect of colonialism that is different from colonial warfare and the exploitation of natural resources for commercial purposes. On the other hand, this helped in the spread of Christianity and European medicine in the colonies, particularly in the Americas. This continued beyond the eighteenth century. As we will see in Chapter 7, missionaries were at the forefront of introducing Western medicine into colonial Africa as well as collecting information about African plants and their local medical uses in the late nineteenth century. While doing so they often drew inspiration from the activities and experiences of the earlier missionaries in the colonies during the seventeenth and eighteenth centuries. In terms of the history of medicine, the missionaries paved another path through which medicinal plants and medical practices from the colonies entered European medicine and medical texts.

There is another final episode of this narrative to explore: how the diverse cultural, medical and botanical ideas were absorbed within modern medicine. How did these plants of distant places enter the European pharmacopoeia and European medicine?

From colonial plants to modern drugs

Modern medicine emerged through these various practices of collecting, recording and using plants in the Western and Eastern colonies. Two processes were involved in the transformation of colonial medical botany to modern medicine. The first was the new classification and order in which these plants were placed, which led to the emergence of the modern *materia medica*, the knowledge of various substances used as medicine. The other was the emergence of modern drugs and pharmaceuticals. In this final section we will briefly see how the history of colonial botany is linked to the history of modern pharmaceuticals.

As apothecary shops in Europe became the centres of exchange and knowledge of exotic drugs in the seventeenth and eighteenth century, physicians of the Royal College and similar other orders in Europe were faced with the possibility of losing their authority as experts in medical knowledge, particularly of the variety of drugs then appearing in Europe from the Orient and the New World. In response they started an extensive enterprise to collect, study and classify exotic medicinal plants and items. As explained earlier, the practice of collecting and observing plants became the hallmark of eighteenth-century European natural history.[55] From observation and collecting came the need for ordering and cataloguing botanical specimens. As European naturalists studied plants sent from different parts of the world, they also felt the need to assign a new form of classification or order to these, one that was useful and familiar to them.

The ordering and classification of the natural world was a complex historical process. This took place simultaneously in Europe and in the colonies. In the colonies, this codification of plants within European systems involved the naturalists, missionaries and surgeons who studied plants, medical practices and texts. They introduced modern European botanical ideas and principles to this local knowledge as they codified them in their own texts. These explorations of colonial nature and collecting them in Europe marked the beginning of the Age of Discovery. Historians have challenged the earlier romantic view of this as one of great expansion of knowledge about the natural world of the earth. They have instead argued that cataloguing and classifying plants in the colonies was not only for intellectual interests but also helped in the incorporation of these plants within the colonial economy. Historians have also seen this European classification and codification as a general 'ordering' of tropical nature which not only helped in their exploitation as resources but also transformed the ecology and geography of the colonies.[56]

Londa Schiebinger has described the eighteenth-century European linguistic traditions that developed around plants, which was part of a single imperial project of codifying colonial plants as 'linguistic imperialism'.[57] Lucille Brockway showed that colonial botanical explorations were linked to the exploitation of plants such as coffee, cacao, ipecacuanha, jalap and Peruvian bark as commercial crops.[58] Schiebinger and Claudia Swan have shown that colonial botanical science developed in close connection with European commercial interest, territorial expansion and state politics.[59] This often led to the rejection of local names and the imposition of European forms of nomenclature and classification.[60]

However, there were, as we have seen, other motives behind the classification of colonial botany. The missionaries and surgeons investigated and classified the plants and medicinal herbs as a result of a spiritual and romantic fascination with Oriental and American nature and culture. Missionaries and botanists were some of the first to produce Spanish, German, French and English texts on tropical plants. British surgeons, such as Whitelaw Ainslie, started the systematic recording of Indian medicinal plants into English. His *Materia Medica of Hindoostan* was a compilation of various medicinal specimens that he collected from the local markets, whose names he then verified along with their uses in ancient Sanskrit and Tamil texts, finally incorporating them into a Linnaean taxonomic order in Europe.[61]

In Europe the classification and naming of plants entailed a dual journey: a study of the natural history of plants corresponded with their simultaneous authentication in ancient and classical texts.[62] By the seventeenth century, European naturalists and philosophers had gathered a vast collection of Latin texts, which had been translated from classical Greek texts, which they regarded as the foundation of their knowledge of nature. While exploring and finding new plants in the colonies, European botanists continuously returned to these texts to identify the origins of exotic or unusual flora and their uses.

Changes in philosophical reasoning and within European natural history from the seventeenth century influenced the use of colonial plants in European medicine. Natural history in this period underwent a shift from philosophical reasoning to objective investigations of nature.[63] The objective observation of nature led to empiricism, which meant that henceforth knowledge was to be acquired not only by philosophical disputation but also by the investigation and ocular observation of nature. This empiricism was of a global nature – it was driven by the belief that nature of the entire earth should be observed and classified under universal categories.

To serve the purposes of observation and empiricism, the newly formed Royal Society of London under its first secretary, Henry Oldenburg, developed a global network of correspondence. Oldenburg created an elaborate series of questionnaires, which he sent to as many respondents as he could get in touch with in as many parts of the known world to which European ships then sailed. Explorations of the natural world of the Americas, Asia and parts of Africa became vital to the expansion of European medicine, which was by then firmly based on this empiricism. European botanists and physicians became increasingly interested in plants of the New World and Asia, and they wrote texts which reflected this trend of observing and classifying plants on a global scale. Robert Morrison's *Plantarum Historiae Universalis Oxoniensis* (1680) and Nehemiah Grew's *Anatomy of Plants* (1682) are examples of the global scale of dissemination and textualization of new botanical knowledge. In 1686, English naturalist John Ray summarized the known flora of the world in a three-volume *Historia Plantarum*. Physicians and apothecaries such as Hans Sloane and James Petiver also developed global networks for new discoveries of medicinal plants, and cataloguing them and displaying them in public museums and gardens.[64] Around the middle of the eighteenth century, Swedish botanist Carl Linnaeus, developed a new taxonomic system based on these large collections of plants that had arrived in Europe. He classified plants not according to their vernacular names, or their geographical or cultural origins, or even by their uses, but according to the number and arrangement of the male and female parts of the flower.

Yet confusion about origins and true qualities of different species of plants remained, as botanists and surgeons still depended on visual observation to distinguish one species from another. Several types of plants, seed and bark arrived in Europe in a damaged state, which made visual identification difficult or impossible. By the end of the eighteenth century a new form of taxonomic and laboratory tradition emerged in Europe. These were based on identifying the chemical elements of matters and were part of a larger change in eighteenth-century science, known as the 'Chemical Revolution', initiated by Antoine Lavoisier in France. His investigations had two parts: the use of language and experiments conducted in the laboratory to define the essential composition of matter. Based on this he developed a new chemical taxonomy as well as a radically new understanding of all elements and compounds. Through a new language and laboratory experiments he sought to redefine the active composition of matter. Following his work, medical abstracts of plants came to be known by their chemical components. Lavoisier's taxonomic experiments in naming the chemical elements of matter developed in conjunction with laboratory experiments on plants and other substances in search of their 'active principles'. This led to the discovery and nomenclature of the key medicinal elements of the prominent and exotic medicinal plants of the tropics in the early 1800s, such as morphine from opium (1804), emetine from ipecacuanha (1817/18), strychnine from *Strychnos nux-vomica* (1819), quinine from cinchona (1820), caffeine from coffee (1820) and nicotine from tobacco (1828).

The history of the discovery of quinine was a significant part of this new search for the elements of matter through laboratory experiments, which transformed herb-based medicines into modern pharmaceuticals. Although by the middle of the eighteenth century cinchona bark had become a vital medical substance in Europe and was used for various kinds of fever, European physicians and botanists were often not sure of the true nature of the bark. Many of those who were based in Europe had never seen a cinchona tree. The genus *Cinchona* includes more than 30 species, which grow in scattered clumps in forests of Columbia, Ecuador, Peru and Bolivia, all varying widely in terms of their colour and the alkaloids present in them. Moreover, unscrupulous traders, seeking to make a profit from the strong demand for cinchona, often sent ordinary bark to European markets but called it cinchona.[65] This confusion existed for many other medicinal plants that entered the expanding European medical marketplace in the eighteenth century.

Chemical research based on Lavoisier's principle of chemical elements of matter sought to resolve some of these problems of ambiguity. As the use of cinchona bark increased throughout Europe, apothecaries and chemists began to investigate the specific element contained in the cinchona bark which was active against intermittent fever. Finding that, they believed, would end the trade in spurious bark. Sustained efforts were therefore made by the end of the eighteenth century to identify the 'active principle' of the bark.

This new tradition developed mainly within French and German laboratory traditions. In 1779, French scientists J.B.M. Bucquet and C.M. Cornette announced that they had successfully extracted the 'essential salt' of cinchona.[66] In 1790, Antoine François Fourcroy, another French chemist, discovered the existence of a 'colouring' matter (a resinous substance with the characteristic colour of the bark) and for some time maintained that he had isolated the active principle.[67]

In 1820, working in a small laboratory located at the back of an apothecary shop in Paris, two French chemists and pharmacists, Pierre-Joseph Pelletier and Joseph-Bienaimé Caventou, finally discovered quinine as the active ingredient of cinchona. Between 1817 and 1820 they were involved in a series of laboratory experiments to extract the active ingredients or alkaloids of several medicinal plants. They discovered active ingredients such as emetine, strychnine and quinine and searched for the alkaloids of tropical spices, such as black pepper (piperine).

The identification of active components of botanical specimens, which came to Europe from distant parts of the worlds, helped to reduce the dependency on traders and apothecaries who often sent or sold spurious drugs. Moreover, as even the real bark varied in potency, extracting the active agent also helped the standardization of the dosage and the manufacturing of pills containing the active ingredient in the laboratory. While it was botany in the seventeenth and eighteenth century, it was chemistry in the nineteenth century which held the clue to the emergence of modern medicine. This new chemical analysis also reduced the earlier dependence on empiricism and 'ocular demonstration', which was the cornerstone of seventeenth- and eighteenth-century natural history and medicine.

The discovery of the active ingredients of medicinal plants marked the beginning of the large-scale manufacturing of modern drugs in modern laboratory-based industries in France, Germany and Britain. From the mid-nineteenth century, manufacture of these active ingredients as drugs started in European laboratories and pharmaceutical factories, which were then exported to the tropical colonies.

Although the discovery of active alkaloids marked a new phase in medical history, it did not put an end to the exploration of colonial and tropical plants. In fact, the growth of pharmaceutical industries in the West started a new era of 'colonial bioprospection'. From the 1850s, colonial botanists, often supported by European pharmaceutical companies, searched for and established commercial monopolies over new and known tropical medicinal plants. These were then sent to Europe and the United States as industrial raw materials to be converted into modern drugs through the identification and extraction of their active ingredients and then supplied to the global market.

In the meantime, from the end of the eighteenth century, physicians and scientists in Europe and North America embraced this new science of taxonomy and laboratory experiments. New medical literature published at the time reflected this. Physician Andrew Duncan Jr (1773–1832) of the Edinburgh Royal College of Physicians played an important role in incorporating Lavoisier's new science into the new medical texts. In 1803 he published the *Edinburgh New Dispensatory*, in which he synthesized and classified the different medicinal plants that came to Britain from the colonies and other parts of the world, according to the new chemical nomenclature introduced by Lavoisier. He was deeply influenced by Lavoisier's taxonomy based on chemical elements, which developed a radically new understanding of all elements and compounds. He adopted the new nomenclature to replace the earlier pharmaceutical names of vegetable and animal substances used in apothecary shops.

This combination of natural history and chemistry shaped Duncan's analysis of the myriad medicobotanical information arriving from various parts of the empire. Although the new chemical nomenclature was derived from the earlier traditions of colonial bioprospection, botanical and linguistic studies, Duncan assigned it to these medicinal items derived from plants to identify their true chemical elements. Duncan Jr believed:

> If the general principle be admitted, it naturally follows, that the names of all substances employed in medicine should be the same with the names of the same substances, according to the most approved systems of Natural History and Chemistry; and that the titles of compound bodies should express as accurately as possible the nature of their composition.[68]

The *Edinburgh New Dispensatory* became the most prominent and widely circulated pharmacopoeia, and it was published in several editions and in different languages.

These new texts, drugs and names led to the emergence of a new discipline in European medical training. From the late eighteenth century, *materia medica* emerged as a distinct medical discipline in European medical colleges with separate courses in medical curriculum and university professorships. Outside the universities, it developed as a specialized pharmaceutical discipline, ultimately paving the way for pharmacology and pharmaceutics and the professional rise of chemists and druggists.[69] Medical students in Europe or in the colonies who went to study in the modern medical colleges in the nineteenth century had to learn this new *materia medica* as part of their medical training.

Conclusion

Colonialism transformed European medicine by uncovering a new world of nature in Asia, the Americas and Africa. Plants were vital commodities of European colonialism. European colonialism had started in search of exotic spices and tropical plants, and it thrived on transplanting plants of one region to areas under their control, developing gardens in the colonies and in Britain, and establishing plantations. Through the various forms of collections, observations and cataloguing that took place during the early phases of colonialism, European botanists and physicians acquired a global perspective on nature. Institutions such as the Royal Society became global in the eighteenth century.

As a result of this engagement with plants from the seventeenth century, European medicine was transformed. Studying, classifying and experimenting with plants became an important part of medical training in Europe from the eighteenth century. New medicinal plants entered European medicine, new apothecary practices emerged, and new dispensatories and medical texts were produced. In the nineteenth century through laboratory research, the 'active' ingredients of several exotic plants were discovered, which gave rise to modern pharmaceuticals. Tropical medicinal plants were from then on used to manufacture modern pharmaceuticals. At the same time the search for exotic medicinal plants in the tropical forests of Asia, Africa and South America continued.

Notes

1 Dunn, *Sugar and Slaves*, p. 279.
2 Wallis, 'Exotic Drugs and English Medicine'.
3 Porter Roy, Dorothy Porter, 'The Rise of the English Drugs Industry'.
4 Ibid, p. 279.
5 Schiebinger, *Plants and Empire*, pp. 73–104.
6 Markku Hokkanen, 'Imperial Networks, Colonial Bioprospecting and Burroughs Wellcome & Co.: The Case of *Strophanthus Kombe* from Malawi (1859–1915)', *Social History of Medicine* (2012) doi: 10.1093/shm/hkr167.
7 A.M.G. Rutten, *Dutch Transatlantic Medicine Trade in the Eighteenth Century Under the Cover of the West India Company* (Rotterdam, 2000), p. 124.
8 [An Account of Some Books] *Philosophical Transactions*, 13 (1683), 100.
9 Henry Draper Steel, *Portable Instructions for Purchasing the Drugs and Spices of Asia and the East-Indies: Pointing out the Distinguishing Characteristics of Those That are genuine, and the Arts Practised in Their Adulteration* (London, 1779).
10 Petiver, 'An Account of Mr Sam. Brown, his Third Book of East India Plants'. See also Petiver, 'The Eighth Book of East India Plants, Sent from Fort St George to Mr James Petiver Apothecary, and F.R.S. with His Remarks on Them', *Philosophical Transactions*, 23 (1702–3), 1450–1460.
11 Jacob Seidi, 'The Relationship of Garcia de Orta's and Cristobal Acosta's Botanical Works', *Actes du VIIe Congress International d'Histoire des Sciences* (Paris, 1955), 56407.
12 H.J. Cook, *Trials of an Ordinary Doctor: Joannes Groenevelt in Seventeenth-Century London* (Baltimore, 1994).
13 H.J. Cook, *Matters of Exchange, Commerce, Medicine, and Science in the Dutch Golden Age* (New Haven, 2007).
14 Rutten, *Dutch Transatlantic Medicine*, p. 124.
15 Henry van Rheede, *Hortus Indicus Malabaricus* (Amsterdam, 1678)
16 Nicolai Laurentii Burmanni, *Flora Indica: Cui Accedit Series Zoophytorum Indicorum, Necnon. Prodromus Florae Capensis* (Amsterdam, 1768).
17 Teresa Huguet-Termes, 'New World Materia Medica in Spanish Renaissance Medicine: From Scholarly Reception to Practical Impact', *Medical History*, 45 (2001), 359–76.

18 Julia F. Morton, 'Medicinal Plants – Old and New', *Bulletin of the Medical Library Association*, 56 (1968), 161–7, p. 162.
19 Andrew Duncan, *Supplement to the Edinburgh New Dispensatory* (Edinburgh, 1829), p. 57.
20 Richard Grove, *Green Imperialism: Colonial Expansion, Tropical Island Edens and the Origins of Environmentalism, 1660–1800* (Cambridge, 1995) pp. 128–30.
21 Francis Masson, 'An Account of Three Journeys from the Cape Town into the Southern Parts of Africa; Undertaken for the Discovery of New Plants, Towards the Improvement of the Royal Botanical Gardens at Kew', *Philosophical Transactions*, 66 (1776), 268–317.
22 G.M. Longfield-Jones, 'Buccaneering Doctors', *Medical History*, 36 (1992), 187–206 pp. 194–5.
23 Quoted in Marie Louise de Ayala Duran-Reynals, *The Fever Bark Tree: The Pageant of Quinine* (New York, 1946), p. 24.
24 Clements R. Markham, *Peruvian Bark: A Popular Account of the Introduction of Cinchona Cultivation Into British India, 1860–1880* (London, 1880), p. 6.
25 Markham argued that malaria was not found in the pre-Spanish Americas and that the bark was absent from the *materia medica* of the Incas, ibid, p. 5. For similar views, see Norman Taylor, *Cinchona in Java: The Story of Quinine* (New York, 1945), p. 29. For an alternative view, see Marie Louise de Ayala Duran-Reynals, *The Fever Bark Tree: The Pageant of Quinine* (NY, 1946), pp. 25–6.
26 Spary, 'Peaches which the Patriarchs Lacked', p. 15.
27 Marie-Cecile Thoral, 'Colonial Medical Encounters in the Nineteenth Century: The French Campaigns in Egypt, Saint Domingue and Algeria', *Social History of Medicine* (2012), hks020v1-hks020.
28 Sloane, *A Voyage to the Islands*, vol. i, pp. xvi–ii.
29 Sloane, 'A Description of the Pimienta or Jamaica Pepper-Tree, and of the Tree That Bears the Cortex Winteranus', *Philosophical Transactions*, 16 (1686–1692), 462–8.
30 Ibid, p. 465.
31 Sloane, *A Voyage to the Islands*, vol. i.
32 James Thomson, *Treatise on the Diseases of the Negroes, as They Occur in the Island of Jamaica with Observations on the Country Remedies, Aikman Junior, Jamaica* (Kingston, 1820) p. 167.
33 Sujit Sivasundaram, 'Natural History Spiritualized: Civilizing Islanders, Cultivating Breadfruit, and Collecting Souls', *History of Science*, 39 (2001), 417–43.
34 Ines G. Županov, *Missionary Tropics: The Catholic Frontier in India, 16th–17th Centuries* (Ann Arbor, 2005); M.N. Pearson, 'The Thin End of the Wedge. Medical Relativities as a Paradigm of Early Modern Indian-European Relations', *Modern Asian Studies*, 29 (1995), 141–70.
35 Steven J. Harris, 'Jesuit Scientific Activity in the Overseas Missions, 1540–1773',

Isis, 96 (2005), 71–9.

36 J.L. Reveal, and J.S. Pringle, 'Taxonomic Botany and Floristics', in: *Flora of North America North of Mexico*, 1 (1993), 157–92.

37 Renate Wilson, *Pious Traders in Medicine: A German Pharmaceutical Network in Eighteenth-Century North America* (Philadelphia, 2000).

38 For a detailed study of interactions between indigenous faiths and Christianity in the United States, see Nicholas Griffiths and Fernando Cervantes (eds) *Spiritual Encounters: Interactions Between Christianity and Native Religions in Colonial America* (Birmingham, 1999).

39 Osvaldo Pardo, 'Contesting the Power to Heal: Angels, Demons and Plants in Colonial Mexico', in Griffiths and Cervantes (eds) *Spiritual Encounters*, pp. 163–8.

40 Marcia Stephenson, 'From Marvelous Antidote to the Poison of Idolatry: The Transatlantic Role of Andean Bezoar Stones During the Late Sixteenth and Early Seventeenth Centuries', *Hispanic American Historical Review*, 90 (2010), 3–39.

41 H. Glenn Boyd, 'A Brief History of Medical Missions', *Gospel Advocate*, 132 (1990), 14–15.

42 Richard S. Dunn, *Moravian Missionaries at Work in a Jamaican Slave Community, 1754–1835* (Minneapolis, 1994), pp. 8–10.

43 August Gottlieb Spangenberg, *An Account of the Manner in Which the Protestant Church of the Unitas Fratrum, or United Brethren, Preach the Gospel, and Carry on their Missions Among the Heathen*. Translation, H. Trapp (London, 1788).

44 Ibid, p. 42.

45 Michael T. Bravo, 'Mission Gardens: Natural History and Global Expansion, 1720–1820', in Schiebinger and Swan (eds), *Colonial Botany: Science, Commerce, and Politics in the Early Modern World* (Philadelphia, 2005), pp. 49–65.

46 Ines G. Županov, *Missionary Tropics: The Catholic Frontier in India, 16th–17th Centuries* (Ann Arbor, 2005); M.N. Pearson, 'The Thin End of the Wedge. Medical Relativities as a Paradigm of Early Modern Indian-European Relations', *Modern Asian Studies*, 29 (1995), 141–70.

47 J. Heniger, *Hendrik Adriaan van Reede tot Drakenstein (1636–1691) and Hortus Malabaricus: A Contribution to the History of Dutch Colonial Botany* (Rotterdam, 1986), pp. 41–6.

48 'A Letter from Father Papin, to Father Le Gobïen, Containing some Observations Upon the Mechanic Arts and Physick of the Indians', *Philosophical Transactions*, 28 (1713), 225–30.

49 Johan Ferdinand Fenger, *History of the Tranquebar Mission: Worked out From Original Papers, Published in Danish and Translated in English From the German of Emil Francke* (Tranquebar, 1863), p. 265.

50 Ibid, pp. 83–4.

51 'Extract of another Letter, relating to some diseases incident to the Malabarians: Likewise of some remedies they commonly use against them', *An Account of the Religion, and Government, Learning, and Oeconomy, &c of the Malabarians:*

Sent by the Danish Missionaries to their Correspondents in Europe, Translated from the High-Dutch (London, 1717), pp. 61–2.

52 Benjamin Heyne, *Tracts, Historical and Statistical, on India with Journals of Several Tours. Also an Account of Sumatra in a Series of Letters* (London, 1814), p. 124.

53 C.S. Mohanavelu, *German Tamilology: German Contribution to Tamil Language, Literature and Culture During the Period 1706–1945* (Madras, 1993), p. 151.

54 G. Wagenitz, 'The "Plantae Malabaricae" of the Herbarium at Göttingen Collected near Tranquebar', *Taxon*, 27 (1978), 493–4.

55 Jennifer Thomas, 'Compiling "God's Great Book [of] Universal Nature", The Royal Society's collecting strategies', *Journal of the History of Collections*, 23 (2011), 1–13. See also Michael Hunter, *Establishing the New Science: The Experience of the Early Royal Society* (Woodbridge, 1989), pp. 123–55.

56 Peder Anker, *Imperial Ecology: Environmental Order in the British Empire, 1895–1945* (Cambridge, MA, 2001).

57 Schiebinger, *Plants and Empire*, pp. 194–225.

58 Brockway, *Science and the Colonial Expansion*.

59 Schiebinger and Swan (eds), *Colonial Botany*.

60 John Gascoigne, *Science in the Service of Empire: Joseph Banks, the British State and the Uses of Science in the Age of Revolution* (Cambridge, 1998); Richard Drayton, *Nature's Government: Science, Imperial Britain, and the 'Improvement' of the World* (New Haven, London, 2000); Zaheer Baber, 'Colonizing Nature: Scientific Knowledge, Colonial Power and the Incorporation of India into the Modern World-System', *British Journal of Sociology*, 52 (2001), 37–58, Schiebinger, *Plants and Empire*.

61 Ainslie, *Materia Medica of Hindoostan*, p. i.

62 H.J. Cook, 'Physicians and Natural History', pp. 92–3.

63 H. J. Cook, *Matters of Exchange*, pp. 21–3.

64 Lisa Jardine, *Ingenious Pursuits: Building the Scientific Revolution* (London, 1999).

65 For a detailed account of this difficulty with cinchona, see P. Chakrabarti, 'Empire and Alternatives: *Swietenia Febrifuga* and the Cinchona Substitutes', *Medical History*, 54 (2010), 75–94.

66 Markham, *Peruvian Bark*, pp. 30–1.

67 Ibid, pp. 31.

68 Duncan, *The Edinburgh New Dispensatory*, p. viii.

69 J.K. Crellin, 'Pharmaceutical History and its Sources in the Wellcome Collections. I. The Growth of Professionalism in Nineteenth-Century British Pharmacy', *Medical History*, 11 (1967), 215–27.

3

Medicine and the colonial armed forces

The commercial and colonial rivalries of the seventeenth and eighteenth centuries led to major military conflicts between European nations in different parts of the world. Most of the military conflicts around the middle of the eighteenth century were between the Spanish, the French and the English. The main wars were the Anglo-Spanish War of 1739, the Seven Years War (1756–63), the American War of Independence (1776–82) and the Napoleonic Wars (1799–1814). Some of the wars were fought on a global scale. France and Britain, for instance, fought the Seven Years War (1756–63), which is also known as the 'first world war' as it took place in various parts of the world, in the Caribbean, the Americas and India.[1] It is also regarded as one of the greatest military triumphs of British imperial history. Britain became the largest maritime empire in the eighteenth century by defeating the Spanish, Dutch and French in Asia and the United States. British forces were also involved in major military conflicts in South Asia against the local rulers, such as Hyder Ali and Tipu Sultan, known as the Anglo-Mysore Wars (the four Anglo-Mysore Wars), against the rulers of Bengal in the battles in Bengal (the Battle of Plassey in 1757, and in Buxar in 1764) and against the Marathas in western India.

The wars took large numbers of European troops and seamen to the West Indies, Asia and North America. At the height of the Seven Years War in 1760, the navy had increased to 70,000 personnel.[2] In most of these wars, European soldiers fought in difficult and unfamiliar terrains and climate, lived in putrid and unhealthy conditions and suffered huge causalities. This led to a major drain on the human and financial resources of European nations, which tended to counteract the profits that they earned from the colonies.

One striking fact about European military mortality in the eighteenth century is that more people died from disease than from battle injuries. Some 70 per cent of the British and American forces besieging Cartagena in 1741 died from disease, as did 40 per cent of the British troops that captured Havana. Around half of the French troops sent out to Saint-Dominique in 1792 perished and 50,000 British soldiers and seamen died of yellow fever there and on other West Indian islands between 1793 and 1798. An equal number of patients were discharged as invalids.[3] Most soldiers died from diseases such as scurvy, dysentery and yellow fever.

In the eighteenth century there was a growing sense among European physicians and surgeons that these diseases could be prevented. Gilbert Blane, a famous Scottish naval physician, claimed that most deaths from disease in the West Indies during the American War of Independence were the result not of climatic factors but of preventable causes.[4] The realization that diseases were the greatest killers of Europeans in the colonies and that these could be prevented marked a turning point in European medicine and military organization.

Colonial warfare and the European military crisis

Colonial warfare and loss of lives through diseases posed a mercantilist concern. Continuous warfare from the seventeenth century had led to mounting national debt in most European nations. The increased size and cost of the eighteenth-century armies weighed heavily on their economies. This was the main reason for the increase in taxation and the principal stimulus for administrative rationalization and centralization initiated by the British army in the eighteenth century.[5] Taxation to pay for military and naval expenditure was a growing burden on all countries, including Britain.[6] The average annual tax revenue in England doubled during the Nine Years War (1688–97). This doubled during the War of Austrian Succession (1740–8) and doubled once again by the time of the American War of Independence, which started in 1775. This amounted to a six-fold increase in the period from 1660 to 1783.[7]

The other major concern for European nations was the condition of European navies and armies in this period. Continuous warfare had increased the size of the armed forces but they remained disorganized. Until the middle of the eighteenth century, most European navies lacked clear recruitment strategies and discipline, and suffered from the difficult conditions of service. Until the mid-seventeenth century, the British navy did not have a permanent naval service. Even after that, recruitment was irregular and often forced by notorious pressgangs through the 'impressments' system. During the wars, pirates and privateers provided a lot of the naval service in the colonies. The British Crown and navy sanctioned privateering to assume control over the bullion trade as well as to ward off enemy ships in the Atlantic. The Crown depended on these privateers and pirates to provide protection and military support. For example, it relied on Henry Morgan to provide protection from the Spanish in the Caribbean Sea during the 1660s and 1670s.

Private contractors similarly ran naval victualling and supplies, as the navy was unable to supply its growing fleet. Private individuals and ships' captains made huge personal profits from this. The Lascelles family, for example, held naval contracts between 1720 and 1750 in Barbados and in the Leeward Islands, and Henry Lascelles amassed a vast personal fortune.[8]

The supplies were not regulated by the navy. Alcohol, particularly rum, was supplied in abundance, which led to the problem of drunkenness and desertion,

and added to the lack of discipline. Debauchery, drunkenness and moral trans-
gressions became an important concern for the European naval establishment,
particularly in the West Indies in the eighteenth century.[9] Ulcers were very
common among the soldiers and slaves, and they were caused by their salty
diet and excessive consumption of rum.[10] Due to the disorganized state of naval
supplies, recruitment policies and the coerced nature of service, naval personnel
suffered from low morale, disease and addiction. The physical and moral condition of the troops became a major concern for European navies and armies.

The medical service, which served the navies and armies during this period,
was also disorganized. The eighteenth-century naval medical service was predominantly provided by individual surgeons who were often appointed on ad hoc
basis. Until the end of the eighteenth century, the navy rarely assumed direct
responsibility for providing medical service to its ships or in its hospitals.[11] Most
of the naval ships and medical establishments were served by barber-surgeons
who generally enjoyed a lowly status within the medical profession. There was a
sharp divide in seventeenth-century medicine between physicians and surgeons.
Physicians thus enjoyed a higher status than surgeons, whom they considered
untrained and essentially performing manual tasks. Physicians practised intellectual forms of medicine. They were trained in medical theories of fevers and
natural history in the universities, and they were familiar with various medical
texts and the latest drugs and forms treatments.

Surgeons, on the other hand, were those who treated wounds, fractures, defor-
mities and disorders through manual and surgical means. They were mostly
responsible for practices such as amputation, surgery and bloodletting. Therefore
the predominant form of medicine that surgeons practised was crisis management or, in other words, curative and not preventive. Surgeons served in the
ships and military camps in the colonies of Asia and the Americas, with little
knowledge of drugs, therapies and the diseases that often confronted them.

Moreover, navies had very few hospitals or permanent institutions of care for
patients during this period. Until then, only religious orders or charities built
and maintained hospitals. The vast majority of sick and wounded seamen were
kept in sick quarters, which were usually rented rooms, scattered in different
parts of the colonial port cities. While this was cheap and flexible, when scattered across towns it was impossible for surgeons to supervise their patients
adequately. Unlicensed apothecaries and quacks who often supplied spurious-
quality drugs provided a major part of the supplies of medicine in these quarters in the colonies.

However, by the early nineteenth century, this crisis seems to have been largely
resolved. European navies and armies had by then become disciplined and organized fighting forces. Disease had ceased to be a major cause of mortality. The
medical care in the navy and army had become more organized, with surgeons
playing a prominent role in defining military policies. How did this transformation take place? What role did medicine play in disciplining and modernizing
European armed forces?

Historians have seen this transformation as the convergence of two historical processes: the emergence of the European 'medicine of hot climates' and the rise of preventive medicine. The work of physicians such as James Lind in finding the causes of mortality of European troops in the tropical colonies and on the ideal modes of acclimatization resulted in improving the health of European armed forces in the colonies. On the other hand, the rise of preventive medicine and military discipline introduced efficient medical care for European soldiers and seamen in Europe and in the colonies.[12] Preventive medicine, which became the foundation of nineteenth-century European public health, was first introduced within European military establishments by surgeons. These military institutions provided greater scope for observation of causes of disease and the application of various experimental methods of cure and prevention. In the army and the navy, sufferers from disease could be kept under adequate control and observation, and their accurate statistics of sickness and health could be compiled.[13]

Historians have also seen the growth of hospitals in the eighteenth century as institutions, which enforced discipline and efficient medical care, as the base of this new regime of medical care. The emergence of military and naval hospitals in Europe and the development of regulations governing hygiene and sanitation were part of a more general process whereby medicine became an important administrative process, in which the human body became subject to control and discipline by the state.[14] This was a complex process of change, which took place in various spheres of military, medical and colonial history.

Changes in eighteenth-century medicine and the navy

Although the process of change had started from the early eighteenth century, dramatic changes in theories of disease and in structures of medical care within the European military forces took place from the mid-eighteenth century. I will explore the case of scurvy to see how the cure of scurvy was linked to the larger changes within the navy and medicine. Scurvy was a major health problem for the navy in its colonial voyages in the seventeenth and eighteenth centuries. It was a disease characterized by the general debility of the body, extreme tenderness of the gums, foul breath, subcutaneous eruptions and pains in the limbs, induced by exposure and the heavy diet of salted foods. Now recognized as due to insufficient ascorbic acid (vitamin C) in the diet, in the eighteenth century it was associated with bodily putrefaction. James Lind, a Scottish-born naval surgeon and physician, is credited as having introduced the use of lemon as a preventive for scurvy on the ship HMS *Salisbury* in 1747, which resolved a major medical problem for the British navy (see Figure 3.1).

However, this depiction may give a false impression of what actually happened on the ship and more importantly about its broader historical context. A much greater story lies behind this image, which explains why and how Lind chose

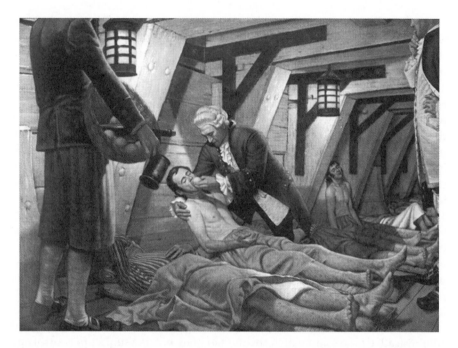

Figure 3.1 James Lind uses lemon to treat a patient with scurvy on HMS *Salisbury*, 1747 (by Robert A. Thom, in *A History of Medicine in Pictures,* 1960). Courtesy of the Collection of the University of Michigan Health System, gift of Pfizer, inc, UMHS. 17.

to use lemon to treat those who were suffering from scurvy. More importantly, it helps to explain the broader transformation, beyond scurvy and lemon juice, that took place in British navy and medicine in the eighteenth century.

Disease in the eighteenth century was understood in terms of traditional miasmatic theories, which suggested that to maintain good health, the body and the environment needed to be in harmony. Hygiene and health were associated with the harmonious circulation of the atmosphere, and the origins of disease were found in stagnation, putrefaction and bodily decay. Since European physicians saw disease and putrefaction as the physical manifestation of a lack of harmony between the body and the environment they attached moral stigma to disease. Putrefaction was seen as moral corruption and decadence. Thus putrid diseases (those that were identified as being caused by putrefaction) were also seen as the result of moral degeneration. Physical cleanliness and social and moral order were now linked. All of these became important in the cure of scurvy in the navy, which, as we have noted, was a disease associated with putrefaction and decay. The same stigma was attached to leprosy, which also represented physical and moral decay. Today, physical afflictions and moral degeneration are generally separated, but it is important not to underestimate the significance of this link in the seventeenth and eighteenth centuries. Even today, physical afflictions are associated with moral prejudice, both by the medical profession and in the popular imagination. This is the case for the new and emerging diseases such as HIV/AIDS in the 1980s and bird flu in the new millennium.

In the mid-eighteenth century, British physicians dwelt extensively on putrid or pestilential fevers that ravaged ships, jails and other confined spaces. These were also called 'crowd diseases' as these physicians believed that they were caused by overcrowding and a lack of fresh air.[15] John Pringle, British military physician and the physician general of the army, was a central figure in introducing new medicine into the British armed forces. According to him, putrid diseases constituted a distinct class, which were caused by bodily degeneration.

In 1750 he published *Observations on the Nature and Cure of Hospital and Jayl Fevers*, where he attributed outbreaks of disease in the army and jails to the air vitiated by filth, perspiration and excrement, in which circumstance the blood underwent putrid changes. According to Pringle, putrefaction was the main cause of disease in jails, ships and any confined space. Fever was a putrid change in the blood arising from stagnant air.

Scurvy and leprosy were the most obvious examples. Scurvy in particular was seen as a disease of putrefaction particularly due to the visible decay of the gums and other body parts. Scurvy became more common in ships due to the long voyages and the lack of fresh food.[16] It was seen to be caused by an unhealthy atmosphere and moral laxity.

Due to the close link that Pringle created between disease and putrefaction or degeneration, cleanliness, both physical and moral, was paramount to him – it was conducive not only to health but also to moral order and virtues. As a cure, Pringle recommended the virtues of cleanliness, discipline and proper habits of work, 'to have the soldiers early out, and exercised'[17]. Thereby, according to him, the cure for scurvy or for any putrid disease had to be likewise, physical and moral cleansing. Pringle suggested various means of 'purifying the air' among which was the use of, what he called, 'antiseptics'.[18] According to him, antiseptics were various acids, limejuice, vinegar, oil of vitriol and camphor. This is how lemon juice came to be used for scurvy as an 'antiseptic', not because it contained vitamin C, which was not then known. Before I continue with the story of lemon and antiseptics, let us ask, what antiseptics were in the eighteenth century.

Putrefaction is defined as the state of being putrid or rotten; the process or action of putrefying or rotting. In other words, putrefaction was a deviation from the natural form. Antiseptics derive their name from septic, which means putrefactive, thus antiseptics were entrusted to counteract putrefaction. Thus antiseptics had a moral connotation; they were seen as regenerative agents which removed sources of putrefaction. Antiseptics represented the new regime of hygiene and preventive medicine that transformed European medicine and society in the nineteenth century. This almost immediately had its implications and applications beyond disease. Around the middle of the eighteenth century, physicians and surgeons were arguing that order and good health in society required the use of these new medical interventions. This is how medicine became important in the modern world, since it promised both physical and moral regeneration, particularly at a time when disease and putrefaction had taken such a toll on European life and had caused a threat to its mercantilist and humanitarian ideals.

James Lind's experiment onboard HMS *Salisbury* in 1747 needs to be seen against the backdrop of these ideas of putrefaction, disease and the use of antiseptics. He was a naval surgeon onboard *Salisbury*. He conducted a clinical experiment with the use of various antiseptics for the cure of scurvy among the crew of the ship. He divided the 12 seamen into pairs, and prescribed for each pair a different antiseptic as a potential remedy. The groups were given cider, elixir of vitriol, vinegar, seawater, oranges and lemons, a purge prepared from garlic, mustard seed and other substances, all considered antiseptics by contemporary physicians. The pair who were prescribed the oranges and lemons quickly recovered. The others did not. Today this outcome may seem logical because we know that scurvy is caused by a deficiency of vitamin C, which is present in citrus fruit. However, this is a good example of why historical processes of the past cannot always be appreciated from the perspectives of the present. It is important to remember that Lind did not prescribe lemons because he knew about the vitamin C deficiency that caused scurvy. He had no conception of vitamins. His experiments with lemons were based on contemporary ideas of seeing scurvy as a putrid disease, which needed to be treated with antiseptics.

Lind's findings about the use of lemons therefore did not have a great significance for him, beyond it being an effective antiseptic. In his work he placed his *Salisbury* findings as part of a wide set of recommendations for maintaining the health of the sailors onboard ships. Among Lind's other recommendations were the need for improved atmosphere and fresh air – ships should be regularly fumigated and ventilated. Seamen should take regular baths. He still believed that scurvy was common among those who are 'lazy, indolent and less cleanly fellows' or those wearing 'old unclean cloathes'.[19] He recommended that to prevent such laxity and unclean habits, sailors should be issued with uniforms rather than having to clothe themselves, and also provided with a proper diet, with lemon and 'sufficient quantities of greens'.[20] Their diet should comprise pickled vegetables and 'rob' – an extract of oranges, and lemons – should be carried on the ships. Shallots and garlic should be included in 'the surgeon's necessaries' and ships on station should be regularly supplied by small boats bearing fresh vegetables. Lind also recommended proper recruitment: 'Such idle fellows as are picked from the streets or prisons' should not be included in ships' crews, for they brought contagious diseases and low morale onboard with them.[21] The drinking of spirits, rather than beer, was to be discouraged. Wholesome drinking water could be manufactured by distillation. Lind in general argued for better hygiene on ships, and more humane and efficient treatment of seamen. In doing so, he also adopted a paternalistic tone, which is reflective of the changes in the status of surgeons in the navy by the middle of the eighteenth century. This was because surgeons by then played a role that was more vital to the care of seamen and for the introduction of preventive medicine in the navy.

Lemon juice became commonly used in the British navy as part of these wider recommendations and changes. It is important to remember that the acceptance of lemon as a cure for scurvy in the navy was neither immediate nor uncontested. Since the medical rationale of using lemon juice as effective for scurvy was part of the broader rationale of preventive medicine recommended by Lind and Pringle, the navy attached little significance to lemons themselves. The year after Lind published his recommendations, the Sick and Hurt Board of the navy rejected a proposal to provide sailors with supplies of fruit juice. Various other methods and antiseptics for the treatment of scurvy continued to be suggested by navy surgeons.[22] Doctors suggested that elixir of vitriol was another antiseptic that was equally effective. They also believed that Lind's method of preserving health over long voyages by not giving meat preserved in salt and using lemon was not practical. Others suggested that cinchona bark was the better antiseptic medicine to prevent putrefaction of the blood.[23] It was only in the 1790s, during the Napoleonic Wars, under the leadership of the chairman, Gilbert Blane, that the Sick and Hurt Board employed Lind's prescription of issuing fresh lemons. Thus it was not as if the navy suddenly started serving lemons and oranges and the problem of scurvy was over – it was a broader shift in medical thinking and in naval discipline that led to their use.

The emergence of the Sick and Hurt Board in the British navy in the eighteenth century was associated with these transformations in naval medicine. Established in 1702, the board organized and oversaw much of the health of the naval forces of Britain and in the colonies. It had strong colonial concerns, particularly about the quality of medicine sent there. Initially the board only operated in times of war and was not primarily a medical body; it was also concerned with prisoners of war.[24] By the mid-eighteenth century, the Sick and Hurt Board changed in character and structure, and played an active role in regularizing navy supplies and the establishment of naval hospitals. It increased its medical members in 1755, when the Seven Years War was imminent, and also focused exclusively on the medical affairs of the navy. The board also gradually developed its own supplies of medicines and competed with the Company of Apothecaries, the main suppliers of medicine to the navy. It often complained to the Admiralty about the packaging and cost of medicine supplied by the Company of Apothecaries. It also complained that the latter was more concerned with profit than with serving the sick. It suggested instead that cheaper drugs could be procured at the naval establishment in Plymouth itself.[25]

The Sick and Hurt Board also took an active interest in setting up naval hospitals, replacing the temporary privately rented establishments in which sick seamen were kept. The contract system was also suffering from various malpractices and corruption. By the 1740s the Admiralty was considering the idea of the navy providing its own hospitals and medical supplies. These and the war with Spain that began in 1739 provided the impetus for wider naval medical reforms. A new Commission for Sick and Hurt Seamen was formed of three commissioners, a fourth being added in 1745 to cope with the increasing work. When the Seven Years War was imminent, the navy started establishing its own naval hospitals in Haslar near Portsmouth, Plymouth, Jamaica, Antigua, Barbados, Halifax, Nova Scotia and Gibraltar. By the 1760s the navy was sending its casualties to its own hospitals.[26]

These were part of the general changes undertaken in naval healthcare, as victualling become important and the navy took direct responsibility for providing fresh vegetables and medicines to its crews. Surgeons were also being recognized as an important component of the navy, particularly in ensuring preventive medicine, in terms of cleanliness, hygiene and regular medical check-ups. A new breed of surgeons emerged in Europe who were trained in universities and had gained substantial experience from serving in the colonies in hot climates, which many physicians did not have. They were no longer known as barber-surgeons and took up more responsibilities and positions of authority in the navy and army. Because of these changes, during the Napoleonic Wars the Royal Navy symbolized efficiency and discipline, and appeared as a modern armed force.

The 'Mortality Revolution' among nineteenth-century colonial troops

Similar to the navy, changes in the army were brought about through a combination of two concepts that emerged in the eighteenth century: humanitarian care and discipline. On the one hand, based on the military experiences of eighteenth-century warfare, army generals and surgeons felt the urge for more humane care of their soldiers. On the other, European armies felt the need to employ more regular and disciplined forces, for the sake of efficiency and reduced costs. This concern for disciplining the army developed from the mercantilist concern about the increased scale and cost of eighteenth-century military campaigns. There was a general realization that better preventive medicine for soldiers improved military efficiency and boosted the morale of the soldiers. Medicine helped in making military campaigns more efficient as it was far less expensive to treat a sick or wounded serviceman than to train a replacement.

Under the impact of continuous warfare from 1793 to 1814, the British army modernized its medical department and introduced sanitary reforms and hygiene regulations. In 1806 the *Regulations and Instructions* enforced the first significant medical rules.[27] Due to the army's campaigns in the Netherlands, the West Indies, North Africa and Iberia, British military medicine became an area of specialization.[28] The Army Medical Department was reorganized in 1793 and it concentrated on three main activities: prevention of diseases, superintendence of medical stores and supplies, and regular inspection of military hospitals. Military surgeon James McGrigor (1771–1858) who was the inspector-general of army hospitals played an important role in reorganizing military hospitals in Britain during the Napoleonic Wars. He later became the director general of the Army

Medical Department. He instituted the system of field hospitals where the less severely wounded could be treated and could be returned to service sooner, while the general hospitals had space for those who required major and long-term treatment. He also stipulated the provision of a proper diet for the wounded soldiers.

By the late eighteenth century, port sanitation became a prominent practice in military medicine as part of the new regime of preventive medicine. In 1794 the War Office established sanitary regulations at Plymouth and Portsmouth, the main ports of embarkation of the military forces. Jeremiah Fitzpatrick (1740–1810), inspector of health for land transport, enforced a new hygiene code and supervised the cleaning of ships, particularly those bound for India, the Continent and the West Indies.[29]

Henry Marshall (1775–1851) is often regarded as 'the father of military medical statistics' and 'the foremost early Victorian advocate of army reforms to benefit enlisted men'.[30] He inaugurated the practice of collecting military statistics and the general modernization of health services in the army. He started the movement for the humane care of soldiers in the British army. Marshall believed that such care would help to improve both their health and their morale. He collected detailed information about the relative insalubrity of every British military garrison in the West Indies and in the East Indies. The substantial statistics regarding the health and diseases of European forces in the colonies helped him to devise better modes of acclimatization of troops in the tropics, to identify healthier sites for stationing the troops (mostly on the higher grounds that were away from swamps and thus from mosquitoes) and to recommend a better diet. He concluded that it was necessary to improve the sanitary condition of the troops, including more frequent rotation of regiments from long service in the Caribbean to the cooler climate of the Mediterranean or British North America. He instructed the construction of larger and better-ventilated barracks at higher elevations, urged for restrictions on liquor rations and made suggestions for a more wholesome diet for soldiers.[31]

Marshall targeted the indiscriminate use of alcohol in the armed forces as part of his attempt to instil moral and physical discipline. He criticised the polices of the army of supplying alcohol to its forces without any rationing as being counterproductive to the regime of discipline: 'We instill the moral and physical poison with one hand, and hold out the lash with the other.'[32] He claimed that alcohol led to a cycle of crime, disease, indiscipline and punishment. He urged for the abolition of alcohol allotment in order to introduce better discipline and morale in the army, particularly in the colonies. He initiated measures by which divisional commanders received discretionary powers to limit the issue of spirits.

These recommendations, although they might appear ordinary and common today, were revolutionary in the eighteenth century, both in terms of giving shape to the modern army and navy as well as in situating medicine at the heart of the modernization of the armed forces. Medicine played an increasingly important part in this process of rationalization. These recommendations and their gradual application instilled a sense of order and discipline in ships and barracks, led to better healthcare of soldiers and seamen by the state and helped to improve their morale. It also fundamentally changed the status of surgeons who had adopted a more important role in naval and military health, which included both preventive as well as curative medicine.

P.D. Curtin has argued that these changes led to a 'Mortality Revolution' among British colonial troops in the East and West Indies in the nineteenth century.[33] This was principally led by military medical officers such as McGrigor and Marshall. As we have seen, Marshall's work across the British Empire enabled him to accurately document and attempt to improve the lives of British soldiers stationed across the globe. His measures included improved sanitary conditions, regular regimental rotations and a stylized design of wide, well-ventilated army barracks that dramatically enhanced living conditions in the colonies for British troops. Marshall's work in the army coincided with a period of unprecedented medical advancement at home and abroad during the late eighteenth and early nineteenth centuries, led by the naval and military physicians who had experienced and attempted to control the high mortality rates of colonial voyages. By instilling simple but rigorous practices of hygiene, cleanliness, medical discipline and a healthy diet aboard shipping vessels, the major European powers succeeded in expanding their colonies in the New World, India and Africa. Curtin showed that these simple but highly effective reforms included various new measures such as 'a heavy emphasis on barracks design, hospital design and location, ventilation, drainage, clean water ... and disciplined attention to sanitary conditions of all kinds'.[34] The new reforms were brought about by the predominant desire for colonial expansion that characterized the European armies and navies of the nineteenth century. Curtin, for instance, has shown that in India, 'in 1869, the mortality rate for all British soldiers in India dropped below 20 per thousand' and has argued that these successes were a great encouragement to attempt 'still more imperial ventures at some later time'.[35] In short, not only were medical practitioners such as Marshall able to stem the tide of high mortality rates in the colonies, but because of these developments many of these empires were able to grow and expand further. This supported the intent of many major European powers to increase their influence over native populations without the worry of contracting, or spreading themselves, any major epidemics. The development of advanced sanitary regimes therefore not only aided mortality rates on colonial voyages to and from a nation's respective settlements but, according to Curtin and other historians, also greatly aided the increasingly expansive role of empire across the globe, as many of the major European powers grew vastly in size and wealth once basic medicinal practices had been fully ingrained into military and civilian life. This was later aided by Robert Koch and Louis Pasteur's germ theory. They provided more clinical evidence of causation of diseases and popularized the use of vaccines for protection against several diseases that afflicted Europeans in the tropics.

From the late eighteenth century, scientific principles of preventive medicine were applied within the navy and the army, where sufferers from disease could be kept under adequate control in hospitals and proper statistics of sickness and health could be compiled. Curtin attributes the decline of mortality rates of European armies in the nineteenth century to preventive measures adopted by the army (greater access to clean water, improved sewage disposal, less crowded and better ventilated barracks), the relocation of troops to hill stations when epidemic disease threatened and the more widespread use of quinine as a treatment for malaria.[36]

Military medicine in the colonies in the eighteenth century

However, there were limits to this story of gradual progression of medicine and welfare, and the modernization of the armed forces through the application of new medical theories and practices. It is important to note that these changes were more evident in the armed forces which were stationed in Europe than those in the colonies. While a new military establishment emerged in Europe with the help of medicine, with new ideas of cleanliness and discipline being enforced and hospitals caring for the sick and wounded in a more efficient and humane manner in Europe, in the colonies the impact of these medical ideas, even on European military forces, was not quite so revolutionary. Even according to Curtin, in the period between 1817 and 1838, European troops stationed in Europe experienced a much greater decline in mortality than those serving in the tropical colonies in Asia, the Caribbean islands and Africa.[37] Decline in mortality rates among European troops serving in the colonies took place only in the 1840s and 1860s.[38]

The emergence of modern medical institutions such as hospitals in the colonies followed a more ad hoc and erratic pattern. Acceptance and application of the new principles of medical care were often patchy and mortality decline even among European troops in the colonies remained uneven. This I will study closely in the case of India and Africa in chapters 6–9, which will aid an understanding of some of the divergences between European and colonial medicine.

Although several naval and military hospitals were established in the colonies around the middle of the eighteenth century, they remained disorganized and suffered from chronic shortages of medical staff, equipment and drugs. Throughout the Seven Years War, British military and naval care establishments in the West Indies suffered from disorganized medical care and a lack of discipline.[39] The colonial hospitals often did not have regular surgeons, medical supplies continued to be procured from local traders and contractors, and patients often suffered from a lack of drugs.

The naval hospital in Kingston, Jamaica, was established in 1740 at a time when the Sick and Hurt Board was establishing several other naval hospitals. While the contemporary naval hospitals in England, such as those at Haslar and Plymouth, became crucial naval health stations in the years to come,[40] the one at Kingston became a site of suffering and neglect. It was deserted by the 1750s as it was considered unhealthy. In 1768 a fire destroyed it. Until the end of the eighteenth century, naval and military personnel in Jamaica had to be kept in temporary medical establishments.

The story of the Kingston hospital reflects that of other colonial hospitals in the region. In 1759 in Antigua, Commodore John Moore (1718–79) contracted a local trader, Robert Patterson, to build two hospitals in Antigua for the services of the navy. Patterson converted his sugar and rum factories there into temporary hospitals. Moore found that the hospitals lacked bedding and adequate supplies of medicine. Patterson died in the next year and no new naval hospitals were built. At the same time, complaints reached the Sick and Hurt Board about the condition of the naval hospitals in Barbados. In 1762, following the occupation of Cuba, the surgeons wrote from Havana to Admiral George Pocock about the grave medical situation there following the occupation. They had no stocks of medicine or money to procure them locally.[41] In 1764, surgeons in Havana sent another message to London 'praying' for an allowance for medicines for the sick and wounded. They added that the place had no hospitals on shore, nor were there any hospital ships.[42] Ironically, at the same time that this great financial and humanitarian crisis was being faced by the naval men, Pocock amassed a huge personal booty of £123,000 from the Havana expedition.[43] By 1764 the situation had deteriorated even further in Antigua and Barbados. The surgeons who worked there resigned and those appointed in their place were not considered sufficiently qualified by the Sick Board.[44]

The formation of the Army Medical Department had little positive impact on the military establishment in the West Indies. The Jamaican garrison in fact suffered worse mortality rates in the first three decades of the nineteenth century than in the last three of the eighteenth.[45]

The hospitals in India, such as those in Madras, Bombay and Calcutta, also suffered from a lack of discipline, proper diet and sometimes any permanent base throughout the eighteenth century. While hospitals in Britain such as those in Haslar and Plymouth became symbols of the new naval regime, those in the colonies in the West Indies and India continued to suffer from the earlier problems of indiscipline, drunkenness and desertion.[46]

In terms of military discipline, in India, even in the nineteenth century, the British army struggled to establish firm moral and physical discipline and order.[47] Venereal diseases, which were identified with moral and physical indiscipline among troops, remained one of the main scourges of the British troops in India throughout the nineteenth century, affecting between half and a quarter of them, a much higher proportion than in Europe. In 1897 the Onslow Commission found that venereal diseases affected 44 per cent of British troops in India, more than double that for those stationed in Britain.[48] Often in the eighteenth and nineteenth centuries, colonial troops secured important military victories and sustained empires, despite relatively poor medical facilities, indiscipline and irregularity of medical supplies and provisions. The successes were more due to the greater access to resources and labour and the use of more effective military strategies than medical progress.

The Mortality Revolution that Curtin refers to in European troops in the East and West Indies took effect only by the 1860s. European armies had secured their major military successes in India and the West Indies by then and it is difficult to make a direct correlation between those military triumphs and the improvements in medical facilities. This rationale is more applicable to the colonization of Africa, which took place from the 1860s, when modern medicine and discipline played a far greater role in those colonial military activities. However, as we shall see in Chapter 7, here too the explanations for European military successes were more complicated.

The military successes of European forces in the colonies, and the history of colonization in general, thus cannot be explained only in terms of medical modernization. The success of the European powers in establishing colonies in the Asia and the Americas in the eighteenth century was primarily due to the commercial dominance and power they gained in these parts in the Age of Commerce. This enabled them to amass huge wealth and large quantities of material resources to recruit many troops, particularly from indigenous populations, and to sustain warfare, which offset to a large extent the losses from the mortalities incurred in the battlefields and from disease.

Conclusion

Diseases and problems of discipline, which resulted from colonial warfare and long voyages, transformed modern medicine and the European armed forces. These transformations helped to establish preventive medicine at the heart of modern European state policies in both military and civilian life. The introduction of preventive medicine led to a European mortality decline by the end of the eighteenth century, as we shall see in detail in Chapter 5. It also made European armies better equipped to fight in tropical regions, particularly in the nineteenth century.

At the same time, it is important to note the skewed nature of these changes, which reflects the history of colonialism. The benefits of modern medicine and humanitarianism were primarily and predominantly enjoyed by European states, although many of the stimulants of such changes came from colonial experiences. The military-medical revolution and modernization was predominantly a European phenomenon. Military establishments in the colonies did not enjoy the fruits of preventive medicine equally. Throughout the eighteenth and early nineteenth centuries, colonial naval and military establishments continued to suffer from disorganized medical facilities and hospitals systems.

Notes

1 Tom Pocock, *Battle for Empire: The Very First World War, 1756–63* (London, 1998).
2 Stephen Conway, 'The Mobilization of Manpower for Britain's Mid-Eighteenth-Century Wars', *Historical Research*, 2004 (77), 377–404.
3 Benjamin Moseley, *A Treatise on Tropical Diseases*, 2nd edition (London, 1789), pp. 119–53; David Geggus, 'Yellow Fever in the 1790s: The British Army in Occupied Saint Dominique', *Medical History*, 23 (1979), 38–58.
4 Gilbert Blane, 'On the Medical Service of the Fleet in the West Indies in the Year 1782', in Blane, *Select Dissertations on Several Subjects of Medical Science* (London, 1833), pp. 65–86, 65.
5 Alan J. Guy, *Oeconomy and Discipline, Officership and Administration in the British Army 1714–63* (Manchester, 1985).
6 M.S. Anderson, *War and Society in Europe of the Old Regime 1618–1789* (London, 1988), p. 71.
7 John Brewer, *The Sinews of Power: War, Money and the English State 1688–1783* (London, 1994), pp. 89–90.
8 Douglas Hamilton, 'Private Enterprise and Public Service: Naval Contracting in the Caribbean, 1720–50', *Journal of Maritime Research*, 6 (2004), 37–64.
9 Anita Raghunath, 'The Corrupting Isles: Writing the Caribbean as the Locus of Transgression in British Literature of The 18th Century' in Vartan P. Messier and Nandita Batra (eds) *Transgression and Taboo: Critical Essays* (Puerto Rico, 2005), pp. 139–52
10 Thomas Dancer, *The Medical Assistant; Or Jamaica Practice of Physic: Designed Chiefly for the Use of Families and Plantations* (Kingston, 1801), p. 294.
11 P.K. Crimmin, 'British Naval Health, 1700–1800: Improvement over Time?', in Geoffrey L. Hudson (ed.), *British Military and Naval Medicine, 1600–1830* (Amsterdam/New York, 2007), pp. 183–200.
12 R. Harding, *Amphibious Warfare in the Eighteenth-Century: The British Expedition to the West Indies 1740–1742* (Suffolk, 1991), B. Harris, 'War, Empire, and the "National Interest" in Mid-Eighteenth-Century Britain', in J. Flavell and S. Conway (eds), *Britain and America Go to War: The Impact of War and Warfare in Anglo-America, 1754–1815* (Gainesville, 2004), pp. 13–40; J.R. McNeill, 'The Ecological Basis of Warfare in the Caribbean, 1700–1804', in M. Ultee (ed.) *Adapting to Conditions: War and Society in the 18th Century* (Alabama, 1986), pp. 26–42 ; N.A.M. Rodger, *The Command of the Ocean: A Naval History of Britain, 1649–1815* (London, 2006); R. Pares, War and Trade in the West Indies, 1739–1763 (London, 1963); C. Lawrence, 'Disciplining Disease: Scurvy, the Navy, and Imperial Expansion, 1750–1825', in D.P. Miller and P.H. Reill (eds), *Visions of Empire: Voyages, Botany, and Representations of Nature* (Cambridge, 1996), pp. 80–106; Paul E. Kopperman, 'Medical Services in the British Army, 1742–1783', *JHMAS*, 34 (1979), 428–55.

13 Charles Singer and E. Ashworth Underwood, *A Short History of Medicine* (New York & Oxford), 1962, p. 181. See also Peter Mathias, 'Swords into Ploughshares: the Armed Forces, Medicine and Public Health in the Late Eighteenth Century', in Jay Winter (ed.), *War and Economic Development: Essays in Memory of David Joslin* (Cambridge, 1975), 73–90.

14 Colin Jones, *The Charitable Imperative: Hospitals and Nursing in Ancien Regime and Revolutionary France* (London, 1989), Chapter 6, Christopher Lawrence, 'Disciplining Visions of Empire'.

15 Alain Corbin, *The Foul and the Fragrant: Odor and the French Social* Imagination (Cambridge, Mass., 1986), Chapter 6.

16 Mark Harrison, *Medicine in an Age of Commerce: Britain and its Tropical Colonies, 1660–1830* (Oxford, 2010), pp. 65–7.

17 John Pringle, *Observations on the Diseases of the Army, in Camp and Garrison. In Three Parts. With an Appendix, Containing Some Papers of Experiments* (London, 1753) p. 95

18 Ibid, pp. 134–59.

19 Lind, *Essay on the Most Effectual Means of Preserving the Health of Seamen in the Royal Navy* (London, 1762), p. 115.

20 Ibid, p. 3.

21 Ibid, p. 28.

22 ADM/F/22, 27 August 1761, Caird Library, National Maritime Museum, London.

23 ADM/F/11, Ibid.

24 Crimmin, 'The Sick and Hurt Board and the Health of Seamen c. 1700–1806', *Journal for Maritime Research*, 1 (1999), 48–65.

25 Ibid, ADM/F/13, Office of the Sick and Hurt Board, 12 March 1756, Caird Library, National Maritime Museum, London.

26 M.S. Anderson, *War and Society*, p. 107.

27 Richard L. Blanco, 'The Development of British Military Medicine, 1793–1814', *Military Affairs*, 38 (1974), 4–10, p. 6.

28 Ibid, p. 4.

29 Blanco, 'The Soldier's Friend Sir Jeremiah Fitzpatrick, Inspector of Health for Land Forces', *Medical History*, 20 (1976), 402–21.

30 Blanco, 'Henry Marshall (1775–1851) and the Health of the British Army', *Medical History*, 14 (1970), 260–276, p. 260

31 Ibid, pp. 260–76.

32 Quoted in ibid, p. 266

33 Philip D. Curtin, *Death by Migration: Europe's Encounter With the Tropical World in the Nineteenth Century* (Cambridge, 1989), pp. 1–39.

34 Curtin, 'Disease and Imperialism', p. 102.

35 Ibid.

36 Curtin, 'Disease and Imperialism', pp. 99–107.

37 Curtin, *Death by Migration*, p. 11.

38 Curtin, 'Disease and Imperialism', p. 99.

39 For a detailed account of British medical experiences during the Seven Years War, see Erica Charters, 'Disease, War, and the Imperialist State: The Health of the British Armed Forces During the Seven Years War, 1756–63', unpublished DPhil thesis, Faculty of Modern History, University of Oxford, 2006.

40 Haslar by the end of the eighteenth century could accommodate nearly 2,000 patients, while Plymouth accommodated 1,200.

41 ADM 1/237, Surgeons to George Pocock, 30 August 1762, p. 104, Caird Library, National Maritime Museum, London.

42 ADM/F/24, Report on the memorial signed by several surgeons, 17 April 1764, of His Majesty's ships employed in the reduction of the Havana, 1 May 1764, ibid.

43 Tom Pocock, 'Pocock, Sir George (1706–1792)', Oxford DNB, www.oxforddnb.com/view/article/22421, accessed 12 November 2005.

44 ADM/F/24, Sick and Hurt Board to the Admiralty, 24 September 1764, Caird Library, National Maritime Museum, London.

45 Geggus, 'Yellow Fever in the 1790s'.

46 For a detailed study of the state of colonial hospitals in the eighteenth century in South Asia and the West Indies, see Chakrabarti, Materials and Medicine, pp. 59–74, 89–100.

47 Kenneth Ballhatchet, Race, Sex and Class Under the Raj: Imperial Attitudes and Policies and Their Critics (London, 1980).

48 Eric Stokes, 'The Road to Chandrapore', London Review of Books, 2 (17 April 1980), 17–18.

4

Colonialism, climate and race

By the end of the eighteenth century, European nations had established empires in almost all parts of the world, except for the interiors of Africa. Britain in particular, at the end of the Seven Years War (1756–63), had acquired the single largest empire in the world. P.J. Marshall has argued that the expansion of the British Empire was the most significant political and economic process of the eighteenth century.[1] Although Britain lost the 13 northern colonies in the United States (due to the American War of Independence, 1775–83), it gained substantial territories in South Asia, consolidated itself in the Caribbean islands and southern Africa, and established new colonies in Australia and other Pacific regions. Apart from the territorial expansion, by the end of the Napoleonic Wars in 1815, Britain had also established itself as the foremost global trading nation with trading connections across the Atlantic, Indian and Pacific oceans. The Dutch, Portuguese and French retained colonies in Asia, the Americas and Africa. Although the French colonial power declined after the Seven Years War, the French had expanded their colonial territories in the first half of the eighteenth century over vast territories in North America, Asia and parts of Africa. In the second half the French continued to expand in the Caribbean islands, in Saint-Dominique and in Hispaniola, and in the Indian Ocean, in the Seychelles and Mauritius. Consequently, by the early nineteenth century, a large number of European troops, civilian population, traders and diplomats had migrated and settled in different parts of the world.

Such colonial expansion and migration gave rise to a concern. Can Europeans survive in hot climates? Can they live and fight battles in the tropics to maintain the empire? These were important issues. As we will see in Chapter 5, mortality rates of Europeans in colonial voyages and settlements, both military and civilian, continued to be high throughout the eighteenth century. The future of empires depended on resolving this vital question.

These concerns acquired significance in another aspect of colonialism, in the question of race. With growing colonial power and authority, Europeans came to view themselves as different from and superior to those over whom they now ruled. Therefore the question of European survival in the tropics was also connected to contemporary ideas of race. With the expansion of European colonies, concepts of climate, geography and race came to dominate European medical, political and economic ideas. Over the eighteenth and nineteenth centuries, these questions became critical in shaping the settlement patterns of Europeans across the globe, in Asia, Africa, the Americas and Asia-Pacific. By the middle of the nineteenth century two types of colonial settlement emerged as a consequence of these debates about race and climate: the 'non-settler colonies' (South Asia, South East Asia and Sub-Saharan Africa) and the 'settler colonies' (North and South America, Australia, New Zealand, Canada and South Africa). Such distinctions between the two forms of colony were not so clear in the eighteenth century, as Europeans often believed that they could indeed settle in tropical Asia and Africa. How did this change take place in the nineteenth century?

Tropical climate and racial difference

Climate, particularly tropical climate, had become an important preoccupation with European physicians, surgeons and colonial administrators during the colonial expansion of the eighteenth century. Geographically, the tropics are broadly the regions within the lines known as the Tropic of Cancer and the Tropic of Capricorn. However, more than being regions on the map, 'tropics' were widely discussed as medical, cultural and geographical concepts in European literature. These discussions conjured up contrasting ideas, describing the tropics as places of luxuriant natural bounty or of excessive heat and putrefaction. By the late eighteenth century, however, as the Europeans conquered the interiors and hinterlands of tropical regions and tried to establish settlements, the romanticization of the tropics turned darker. Experiences of death, disease and the discomforts of living in hot countries seemed not so much a decadent luxury as a sore trial, and perhaps even a death sentence. British soldiers in the colonies in the West Indies, as well as in India, suffered from high mortality rates from diseases such as malaria, various types of fever, plague and cholera.

From the late eighteenth century, therefore, in scientific and medical discourse as well, aestheticism of the tropics was gradually obliterated. At the end of the eighteenth century, the writings of James Johnson encapsulated the growing pessimism among Europeans about their ability to colonize India and to adapt to its climate.[2] This was a combination of the growth of polygenic racial theories and the effects of colonization. With most of the tropical world under their domination, Europeans now identified fundamental differences between the ruler and the ruled, believing that European climate, culture and constitutions were superior to those of Africa and India.[3] With the arrival of 'Asiatic cholera' in Europe in the 1830s, the tropical regions, particularly India, came to be viewed as unclean, unhygienic and unhealthy, and culturally and socially backward, which produced such diseases.[4]

It was during this time that the term 'tropics' also lost its ambivalence in European medical and scientific discourse. Scientists and medical men explored possibilities of understanding the tropics and its inhabitants in terms of a new medical specialization and professional eminence.[5] This scientific clarity had emerged in conjunction with broader cultural, moral and climatic ideas, which were used to describe colonized populations in large parts of the globe. This idea of the tropics exists today, alongside the reality that regions described as tropics are also the most disease- and poverty-stricken parts of the world.

Ideas of racial difference had an equally complex history. The concept of race is an old one. Although the word was used diversely in European languages in the Middle Ages, it was usually used to describe a group of humans, plants or animals descending from a common origin.[6] It was only from the seventeenth century that the word was used to describe ethnic groups almost exclusively. During the same period, race also became linked to geography and climate. In the seventeenth and eighteenth centuries, the predominant monogenist ideas of race influenced the question of the acclimatization of Europeans in the tropics. Monogenism was the hypothesis or doctrine of the common origin of the human species. The proponents of ideas of racial monogenesis, such as Johann Blumenbach (German physician) and Comte de Buffon (French naturalist), claimed that all humans shared a common ancestry with Adam and Eve, and differences in skin colour and physical appearance were due to climate, diet and lifestyle.[7] Blumenbach, for example, explained the flatter shape of the nose of Africans as caused by the habit of African women to carry their babies in a sack on their backs, and the pounding of a baby's head against its mother's back flattening its nose, rather than any inherent difference between Africans and Europeans.[8] Europeans thus believed that since all humans evolved from the same racial stock, Europeans could gradually adapt to different climates. The main distinctions were thus drawn between healthy and unhealthy environments, rather than between tropical and temperate regions.[9]

However, this environmental monogenism was not an entirely liberal or egalitarian idea. There was an inherent Eurocentrism in this. For example, in Blumenbach and Buffon's theories of monogenism, European racial features and climate were assumed to be the ideal or the norm from which all other races were assumed to have deviated due to different climatic and living conditions. This is evident in the explanation of the depressing of the African baby's nose.[10] They also believed that the primary or original colour of the human skin was white, which was darkened as humans settled in warmer climates.

The predominance of monogenist ideas also did not preclude racial distinctions or hierarchy in the seventeenth and eighteenth centuries. As Europeans expanded their colonies in the Americas and Asia there was a sense of awareness about their own racial and cultural identities and a perception of their difference from others. From the time of Columbus' voyages to the Americas, Europeans saw other races as distinct from themselves, sometimes even as 'savages'. According to Peter Hulme, savagery was the essential idiom of early European colonization of the Caribbean islands. This was because it was the original site of the encounter between Europe and the Americas. The Spanish were struck by the untamed and pristine nature of the islands, which to them evoked images of savagery. They thus identified the people who lived there as being similarly in a state of nature, or primitive. In addition, Europeans believed that the Caribbean islands were the sites of the practice of the essential act of savagery – cannibalism.[11] Cannibal (*Canibales* in Spanish) was the name given by the Spanish to the original Caribbean people, who they believed ate human flesh. The term later came to mean any person who eats human flesh.

I will not get into a discussion about the merits and demerits of this reference to original Carib races as cannibals here. It is important to note that from the seventeenth century the inhabitants of the Caribbean islands appeared as a primitive race to Europeans as they evoked ideas of the primal link between humans and nature. Even Africans who had been brought as slaves, who toiled in the plantations, who in their rebellion or to escape torture ran away and lived in the deep and impenetrable forests of the islands, who thus knew more of the plants and herbs of those places than European botanists, were seen by European settlers as savages. In their imagination the slaves and the vegetation of the Caribbean were both part of the same fecund natural entity, which Europeans possessed and exploited.

To the Jamaican planter Edward Long, the slaves of Jamaica reflected a state of nature, their women gave birth 'without a shriek, or a scream' like the 'female orang-outang', they ate raw flesh 'by choice' with their 'talons' (hands) and 'with all the voracity of wild beasts'. He similarly saw their knowledge of medicinal plants of the islands as instinctive, unrefined and undeserved: 'Brutes are botanists by instinct', and so in 'the operation of their *materia medica*, they have formed no theory'.[12] He explained that the slaves learnt the curative properties of nature either by chance or by observing other animals, such as the Amerindians who learnt the use of a particular herb as an antidote to the venom of the rattlesnake by observing animals.[13] Long recounted an incident narrated by Esquemeling. In Costa Rica, the Spanish settlers used to shoot monkeys to 'amuse themselves'. They noticed that the injured monkeys covered themselves with a particular moss from the trees or a specific styptic fungus, which stopped the bleeding. They also gathered particular herbs, chewed them in their mouths and then applied them as a poultice. This episode was used to explain that the monkeys and the slaves of the island performed medicine by the same instinctive understanding of and primitive and savage familiarity with nature.[14]

With the growth of the plantation economy in the eighteenth century, definitions of race and climate in the West Indies and South America were also linked to labour. Europeans often justified the use of African and Amerindian labour in the sugar, tobacco, coffee and cotton plantations that were growing in these regions by suggesting that these races were more suited to hard labour in the tropics. There was the belief that because these races were naturally from the hot tropical climates they were more suited to work in those conditions than Europeans who would need a lengthy process of acclimatization. The fact that these people were also portrayed as savages and thereby seen to be better suited to hard manual labour also reinforced this link between race and labour in the eighteenth century.

Race also featured in the patterns of early colonial settlements in Asia and the Americas. From the seventeenth century, colonial towns in the Americas were divided into White and Black Towns.[15] In India too, the British and the French colonial towns of Madras, Bombay, Calcutta and Pondicherry had similar Black and White towns. The Black towns were the parts where the indigenous population lived, as opposed to the White towns where the European population resided. In India, the White and Black towns had their own town centres and markets. In colonial Madras, the segregation between the White and Black towns was formalized in 1661 by a dividing wall.[16] In 1751 during the fortification of the White town in Madras, the Court of Directors ordered the Armenians in the town to leave the white residential areas, sell their houses to 'European Protestants' and move to the Black town outside.[17] These racial patterns of settlement also had an underlying commercial logic. By keeping such segments next to each other, the colonial government could draw together into one city the varied skills, capital and dominant social groups which would serve their economic purposes as well as provide political legitimacy.[18] Race was also a recognized category in commercial exchanges. European merchants' cards (business cards bearing images) in the eighteenth century reflected clear physical differences between Africans, Ameridians and Chinese.[19]

These ideas of race coexisted with a fear of the decomposition of European bodies in the tropical heat and in adopting local diet and habits. In the Americas, Europeans feared that eating Amerindian food or 'savage trash' might cause character mutation and undermine their social superiority.[20]

However, in the predominant monogenist ideas of race in this period, there was a lack of fixity about racial characteristics. Europeans often believed that, although inferior and different, these races were a product of their cultural, climatic and geographical factors, which could be 'corrected' by improving their lifestyle and diet, and by the introduction of European education and religion. On the other hand, this engendered the optimism that by adapting to local climate and adopting local food and habits, Europeans could themselves likewise acclimatize to the tropics. This optimism also provided the opportunity for greater interaction between different races as European settlers in this period considered it essential to learn about local cultures, customs and traditions.

These ideas of acclimatization were related to eighteenth-century European attitudes towards the tropics or regions of hot climates. Although Europeans were aware that the tropics were different, they often did not see these regions as fundamentally different from Europe, or as essentially unhealthy and unsuitable for European habitation, ideas which became much more dominant in the nineteenth century. English physicians such as Hans Sloane, who visited the Caribbean islands during the late seventeenth century, did not believe that the diseases there were fundamentally different from those found in Europe. He found very little that was exotic about the majority of illnesses that he encountered in Jamaica and he concluded that very little difference existed in the manifestation of illnesses in different climates.[21] This was the time when diseases such as plague and malaria were still widespread in Europe and mortality rates there remained high. Europeans even considered parts of India to be healthier than home. Only in the nineteenth century, when epidemic diseases declined in Europe through the adoption of preventive medicine, did the British come to view the Indian climate as essentially unhealthy and very different from that of Europe.[22]

Colonialism and the diseases of hot climates

As Europeans travelled to different parts of the world and encountered different climates, European physicians reworked their traditional medical theories to understand and explain the diseases – particularly the various 'fevers' – that they faced there. The stress was on adopting local remedies and cures, and an assimilation of European and non-European curative and therapeutic ideas. Thus European physicians and surgeons took a great deal of interest in indigenous remedies, read and translated local medical texts, and interacted with local doctors. Spanish physician Monardes, who wrote the earliest Spanish text on American diseases and medicines, endorsed the adoption of local remedies and foods, such as chilli pepper for the treatment of diseases in the Americas. He experimented with New World plants and published a medical treatise about them, as did Portuguese physician Da Costa in Asia.[23] Monardes also encouraged the importation of American plants to Europe to be used in European medicine, setting a precedent for Hans Sloane's importation of several Jamaican and American plants to Britain in the early eighteenth century.

From the sixteenth century, theories of fever in Europe underwent important changes. Increasingly, physicians during this period linked fever with heat – bodily and external environmental.[24] Fever was the means for the body to get rid of excess heat. This is most evident in the works of French physician Jean Fernel (1497–1558), who wrote at length about fevers. He located different types of fever in different organs of the body, but he also linked these to some 'occult' quality. By doing so, he shifted the traditional attention from poisonous miasmas to the idea that different types of fever were generated by different species of venom. He also linked bodily putrescence with external putrefaction that caused miasma. These influenced the seventeenth-century linking of fever with environmental miasma in the works of English physician and natural philosopher Thomas Willis (1621–75).[25] These physicians attributed the causes of fevers to putrefaction – the corruption of animal and vegetable matter which gave rise to morbid effluvia which could enter the body through respiration.[26] This notion of fermentation and putrefaction formed the basis of discussion about fever and later of identifying fevers with hot climates. Physicians in hot climates – and the tropics in particular – noted the rapidity of putrefaction, and it seemed reasonable to them that this process must play a central part in the production of disease.[27]

In English medical thinking the link between hot climates and fevers emerged from the seventeenth-century reinvention of the Hippocratic corpus of 'airs, waters and places'. Thomas Sydenham (1624–89) provided a clear shape to the European understanding of fevers. He was often referred to as the 'English Hippocrates' as he reinvigorated the old Hippocratic ideas to explain the links between fevers and environment.[28] The basic tenets of classical medicine were the interplay of humours and surrounding elements. The four bodily humours of Hippocratic medicine were blood, yellow bile, black bile and phlegm, which interacted with the four external elements: earth, water, fire and air. According to the Hippocratic medical corpus, airs represented winds and climatic effect; waters were the spring waters from ground, rain and snow; and places were the sites of human habitation. Followers of Hippocratic medicine believed that the secrets of health lay in balancing the humours and elements. The health of the human body was contingent upon its symbiotic relationships with the immediate environment. The theory of miasma was based on these principles. The miasma, which was believed to be noxious vapour rising from putrescent organic matter, which polluted the atmosphere, was caused by a disharmony between the human body and the environment, and led to diseases.

From the seventeenth century, with colonization and increasing encounters with different fevers in different regions, Sydenham reinstated this inherent significance of geography and environment of Hippocratic medicine in his treatise on fevers (*Methodus curandi febres, propriis observationibus superstructa* or 'The Method of Curing Fevers, Based on Original Observations'), which was published in 1666. He made climate an essential element of understanding putrid fevers. He studied the epidemics in London from 1661 to 1675 and kept detailed notes of the several cases that he observed. He also drew up the first scientific study of meteorological data, linking them with diseases. He argued that a particular fever and epidemic was caused by the exhalations of a particular atmosphere, and that different fevers were caused by different morbific matters as fever was an attempt by nature to rid the body of morbific particles. Sydenham created this link between fevers and specificities. Now Hippocratic airs, waters and places had a new meaning for Europeans as it opened up the possibility of explaining that the specific airs, new and specific waters, and new places that they experienced in the colonies caused specific fevers. This marked the origins of theories of diseases of the hot climates. A new form of colonial medicine thus emerged by creating the links between theories of race, disease and climate.

During the eighteenth century, as European colonialism expanded in the tropics, putrefaction and miasma, from a universal idea, gradually acquired specific characteristics as they came to be linked more with hot climates. Physicians now suggested that tropical climates and strong rays of sunlight were ideal for the putrefaction, both of the air and of the human body, plants and animals, leading to miasmas, specific to hot climates. James Lind played an important role in this linking of putrid fevers with hot climates. During this period, naval surgeons such as Lind, due to their vast experiences in different climatic zones in the empire and with different kinds of fever, became prominent proponents of theories of diseases of hot climates.[29] In his *Essay on Diseases Incidental to Europeans in Hot*

Climates, Lind firmly linked the phenomenon of putrefaction with hot climates. His book was the first attempt to synthesize the knowledge of diseases that were common in hot climates.[30] He believed that hot climates were more deadly to Europeans. Extreme heat and moisture of the tropics caused putrefaction and fevers.[31] Bodily putrefaction, according to him, was usually caused by exposure to putrid effluvia, which affected Europeans or newcomers to the tropics most. According to him, such putrid fevers were most common in the West Indies and in the Guinea coast of Africa.[32]

Despite identifying putrid diseases with hot climates, there was a sense of optimism in Lind's writings about the possibilities of European settlement in these areas. He was aware of the threats to European life in the tropics, but he believed that they could become acclimatized to such environments: 'By length of time, the constitution of Europeans becomes seasoned to the East and West Indian climates, if it is not injured by repeated attacks of sickness, upon their first arrival. Europeans, when thus habituated, are generally subject to as few diseases abroad, as those who reside at home.'[33] Therefore, although he identified the tropics as putrid zones, he did not identify any essential differences between the races that resided in different regions. The tropics were putrid but not pathological (peculiarly prone to or reservoirs of disease) and any race could adapt there.

According to him, Europeans who arrived in the tropics were initially susceptible to diseases but could be acclimatized through better planning, knowledge and changes to diet and lifestyle. This was in parallel with his instructions for the navy, helping it to acclimatize the new recruits to hot climates. Acclimatization was seen as part of a larger set of medical, cultural and social practices of colonial settlement. Physicians such as Lind believed that in order to acclimatize themselves to the tropics, Europeans had to also acquaint themselves with local cultures and practices, food, clothes and diet. These ideas gradually changed in the nineteenth century.

Nineteenth-century colonialism and theories of race

Patrick Brantlinger has suggested that a sense of melancholy and darkness pervaded British literature and public discourse on the empire throughout the nineteenth century.[34] This is striking since it was precisely during this period, as we shall in chapters 7–9, that European colonialism was its height. Yet as imperialism expanded in Asia and Africa, there was a growing sense of anxiety in both moral and physical terms about such massive territorial acquisition and the effects that this had on European bodies and souls, and on those they now ruled over. The nineteenth century was indeed an age of contrasts. On the one hand it was a period of relative peace and prosperity in Europe, which led to the growth of ideas of 'free trade' and laissez-faire, capitalism and technological revolution in industrial production and transport, aggressive colonial expansion and a growing confidence in European superiority. On the other, it was marked by anxieties about the fragility of empires, the weakness of the European race, loss of control and moral dilemmas. It was also a time when the diseases of the colonies infected the inhabitants of Europe. From the 1830s, waves of cholera epidemics appeared in Europe, which were believed to have been brought over by imperial commerce from Asia. These contrasting views of the empire were most evident in the racial anxiety and the growing fear of the tropics. By the nineteenth century the optimism about the adaptability of Europeans in the tropics became rare. Now there was a lack of the earlier colonial confidence, which marked the Age of Commerce, about large-scale transplantation of humans and species across the world.

These ideas were most evident in the writings of British surgeon James Johnson, who served in India. His *Influence of Tropical Climates on European Constitutions* (1813) displayed pessimism towards acclimatization and a critique of colonization, which was not evident in the writings of Lind and other eighteenth-century medical writers. Johnson showed much less faith in European habitation in the tropics and believed that tropical climate would lead to physical degeneration among Europeans; they would lose their racial characteristics and show apathy and laziness; they would droop and their offspring would be racially degenerate. Johnson's writing was a reflection of the general medical thinking of the age, encapsulating the growing pessimism among Europeans about their ability to colonize tropical regions and to adapt to the climate. Johnson's writings also reflected a move away from climatic theories of disease to climatic determinism of disease and race. It was typical of contemporary medical opinion to see a hot climate as pathological, not just different or 'torrid'.

Similar changes took place in French thinking about climate and race by the end of the eighteenth century. While earlier medical thinkers cited creolization as a mode of acclimatization in the tropics, nineteenth-century French scientists redefined 'Creolity' to highlight its negative aspects. The word 'Creole' holds a clue to understanding early ideas of acclimatization in the Spanish and French colonies. The Spanish originally used the word in the seventeenth century to identify a person of European or African descent born and naturalized in the Americas or in the West Indies. Over the next two centuries the word became commonly used for life in the colonies and came to mean various aspects of acclimatization and hybridization in colonial settlements: racial mixture and adaptation to local cultures, customs and technologies. Late eighteenth-century French physicians such as Pierre-Jean-Georges Cabanis were opposed to Creole identities or Creolization. They linked climate with moral and mental capabilities and suggested that acclimatization and Creolization would lead to mental and moral degeneration.[35]

French palaeontologist Georges Cuvier, on the other hand, linked acclimatization with evolution. His works represented some of the main features of nineteenth-century polygenism. He invoked Christian neo-Platonism to argue that each species was the original and unalterable creation of God. Cuvier put strict limits on environmental influence and stressed the fixity of species – the anatomical and cranial measurement differences in races and that human races were all distinct from each other.

These changes in medical theories were borne out of the climatic and racial determinism of the nineteenth century. In nineteenth-century medical theories, climate became a distinct and fixed category. Climates of certain regions of the world were seen to be fixed, homogeneous and the dominant characteristics of that place. Along with that the tropics were seen as pathological – that is, essentially unhealthy and ridden with diseases, and inherently different from Europe, which was seen as healthy. These ideas of geographical fixity and the pathologization of the tropics contributed to the rise of non-contagionist medical theories in the nineteenth century, which we will study in detail in Chapter 5. Non-contagionists believed that disease was a product of the climate and environment in which they were endemic. Outside such climatic zones the same disease could not spread or infect.

Within this theoretical framework, humans were seen as essentially products of their climate. Inhabitants of certain climates were inherently different from those of another. Indians and Africans were now seen as inseparable products of their native climates, as were Europeans. Climate determined human health and even their moral character. It made the tropical races slothful and heedless, while Europeans were careful, observant and industrious. This is the origin of the modern ideas of race and racism, which is based on the belief that certain races are inherently and essentially different from others. Scientists now believed and recommended that these races, their climates and their environments were to be avoided rather than embraced by Europeans. Acclimatization lost its optimism. Along with it was gone the early colonial confidence about open transplantation of humans and species across the world, which had been predominant in the Age of Commerce.

At the heart of this transformation and the emergence of modern ideas of race was the question of empire. Colonialism defined these transformations in the ideas of climate and race in the nineteenth century. As we shall see in Chapter 7, in the nineteenth century a major expansion of European imperialism took place, particularly in the interior of Africa. In the European medical and popular imagination of the nineteenth century, the whole of Sub-Saharan Africa appeared as a singular geographical and climatic entity. It was the land of dense dark forests and unhealthy swamps; its heat was unbearable for European constitution; its savannahs were infested with insects and vermin; and its inhabitants were devoid of civilization, enterprise and rationality. At the same time, in the process of colonization, Europeans themselves fell prey to diseases such as malaria, which they now believed to be endemic to these parts of Africa. In many ways the colonization of Africa fixed the link between environment and race in European thinking in the nineteenth century.

Related to this was the factor of colonial power. European colonization and rule over vast tracts of tropical lands in the nineteenth century established the fundamental divide between the ruler and the ruled. In the process, colonial physicians and administrators saw themselves as fundamentally different from and superior to the peoples over which they ruled. At the same time, they suffered from a deep sense of fear and racial anxiety, and gave up adopting native modes of life, asserting and even inventing their distinct European lifestyles in the colonies. As Europeans now believed themselves to be physically and culturally superior to Africans and Indians, they viewed acclimatization with caution, as something which could lead to 'degeneracy'.

It is important to note two qualifications to this history. First, it is not always possible to see this clear eighteenth- and nineteenth-century divide in theories of race and acclimatization. Secondly, ideas of race differed from one colonial setting to another. There were different ways in which theories of race and acclimatization and race developed in different colonial contexts. European racial attitudes were more pronounced, or at least different, towards Africans compared with Asians. There were also differences in opinion between the works of the physicians in Europe and those working in the colonies, particularly in the plantations of the West Indies.

Despite the increasing rigidity about racial differences between Europeans and others, 'race' remained a relatively ambiguous term and varied from one region to another. In India, nineteenth-century discussions of race referred much more to the constitution than to skin colour.[36] This idea of the constitution was thus informed by cultural factors, such as diet and rituals, rather than just medical ones. Meanwhile in Africa the darkness of the African skin was more clearly seen as a racial and derogatory feature and linked, at least metaphorically, to the imagined darkness of the continent. James Johnson viewed Indian skin pigment in a positive light, suggesting that it made Indians better suited to tropical heat and humidity. Thus although ideas of race seem to have been much more homogenized in the nineteenth century, there were important differences regarding how race was viewed in different regions.

In the eighteenth century there was a difference in the ideas of race and acclimatization between French metropolitan ideas of Buffon and those of the colonial physicians in the West Indies. French colonial physicians in the eighteenth century were generally less optimistic about European acclimatization to the Caribbean climate. Colonial physicians such as Antoine Bertin believed that the climate of the Antilles hastened the degeneration of the body. The solution was to protect the body from the environment through personal hygiene and bodily regimen. Bertin cautioned against exposure to the sun, encouraged moderate eating, drinking citrus liquids and wearing suitable clothing, which ultimately was similar to the regiments of acclimatization that British physicians had recommended. French physicians such as Pierre Barrére suggested that African bodies were inherently different from European ones and discouraged any encounter with slaves, sexual or otherwise. At the same time, physicians such as J.-B. Dazille also defined the illnesses among the slaves, such as yaws and dirt-eating, not through racial categories but as caused by the brutal physical labour and living conditions that they were subjected to on the plantations.[37]

However, by the middle of the nineteenth century, the climatic and racial divide between Europeans and other races had become predominant and it determined the divide between 'Settler Colonies' and 'Non-Settler Colonies'. Large-scale European settlement started in this period in North America, southern Africa and Australia. Meanwhile, in the tropics, Europeans confined themselves to remote hill stations, spas and sanitary enclaves. Efforts to acclimatize did not disappear but they changed character and were defined in more segregational terms.

Such racial anxieties continued in the twentieth century, particularly in places such as Australia. Warwick Anderson has studied how 'whiteness' as a racial and an essential European category was invented in Australia during the period when the colonial settler society came to refashion itself as a nation. The idea of whiteness was a complex mixture of cultural, racial and genealogical identities. This idea changed from the nineteenth century, when being 'white' in the Australian colonies usually meant being 'British'. In the twentieth century it assumed a more scientific and biomedical character. Now medical scientists and public health officials, aided by germ theory, tropical medicine and modern biomedicine, undertook more intricate methods of assessing the modes and consequences of the habitation of white races in hot climates. This coexisted with a paralysing fear of degeneration of white races in the northern, warmer parts of the continent. Thus the whiteness of the land was asserted to assuage such fears of degeneration of Europeans in Australia. From a predominantly 'British' identity the settlers professed 'whiteness' as the essential character of Australia. This concept of whiteness was distinct from 'Britishness' and had a clear biological meaning. These ideas were then cultivated through a range of scientific and medical discourses in Australia. In the twentieth century, Australian Aborigines became a focus of biological, medical and anthropological curiosity for researchers. The University of Adelaide's experiments and theories contributed to the new policy of absorption, by concluding that Aborigines actually formed an 'archaic' part of the white race that should be reabsorbed to 'breed out the colour' and inferior genes of Aborigines.[38]

Acclimatization was an enduring concern of white settlement in the colonies. Michael Osborne has described it as an 'essential' aspect of colonization.[39] The question of the future of European empires depended on this question. A number of European scientists, medical men, geographers, administrators, military personnel and civilians explored the possibilities and perils of European life in hot climates. The questions varied with regional and cultural differences. The answers similarly varied from the West Indies to the Americas and from Australia to India. The main emphases on aspects of acclimatization also changed over the centuries that spanned the debates. In the Americas, in the seventeenth century, the Spanish intermixed with the indigenes and survived in a medley of intersecting racial and cultural worlds. In Malaya or India, there was no sustained settlement of the land as in the settler colonies: civilians and military personnel, planters and doctors, left gradually after decolonization. In Queensland, Australia, in the nineteenth century, the whites settled and worked as sugarcane labourers and reinforced the idea and practice of 'white Australia'. The while settlers used medical and racial ideas to assert the idea and reality of Australia as a 'white' colony and thereby displaced the Chinese immigrant labourers. In the nineteenth century, ideas of acclimatization underwent profound changes. The changing notions of acclimatization from the eighteenth century to the nineteenth reflect the historical transformations in colonialism during that period. They also show that medical theories determined how Europeans came to see themselves with respect to others in the Age of Empire.

From the seventeenth century, as Europeans travelled to different parts of the world, they experienced different climates and encountered different races. This led to the reinvention of the Hippocratic medical ideas of airs, waters and places. Thomas Sydenham believed that fever was caused by environmental factors, which led to the linking of fevers in the tropics with tropical climate and environment. This shaped the emergence of studies of diseases of hot climates by European physicians. Yet, at the same time, the predominant belief among European surgeons was that it was possible for Europeans to be acclimatized to hot climates as long as they adapted gradually and adopted local food, dress and customs. These ideas changed in the nineteenth century as European colonial settlers, administrators and physicians saw themselves as fundamentally different from and superior to other races over which they ruled. There was pessimism in this period about European acclimatization to hot climates. The rise of this pessimism was due to a combination of medical and imperial factors. This was supported by the rise of polygenist ideas of race, racial segregation and settler and non-settler colonies. Therefore at the height of imperial expansion in the nineteenth century, particularly in Tropical Africa, Europeans were driven by racial anxiety and fear of degeneration. Dane Kennedy has studied the climatic debate and the resurgence of concern about the tropical climate in the nineteenth century as a commentary on the political choices and constraints of Western imperialism. A range of factors led to the rise of polygenist ideas of race, such as the idea that different races of people belonged to different places.[40]

Conclusion

How do we understand the history of racism from this history of race, climate and colonialism? As we have seen, historians have generally argued that the idea of race underwent significant changes from the eighteenth century to the nineteenth. In the eighteenth century, despite a strong awareness of racial difference, there was a greater tolerance towards different races, which was based on monogenist ideas about race. Modern anthropological and scientific research tends to confirm the monogenist theories of race – the idea of a single origin of the human race, perhaps in Africa. Some researchers have even identified a cave in South Africa from where *Homo sapiens* seems to have emerged and spread throughout the world, adapting to different climatic and geographical conditions.[41]

The search for a single origin of the human race could be driven by, and therefore be more acceptable to and compatible with, our contemporary ideas and experiences of multiculturalism and multiethnicity. However, the history of the human race is more complicated than the rise and fall of ideas of monogenism and polygenism. The idea of a single origin of the human race does not explain or eliminate the problem of human exploitation and marginalization by fellow humans, of which racism is one particular manifestation. Exploitation, domination and torture of one group of humans by another have often taken place due to economic, political and social factors. Race assumed a particular shape at the time of colonialism and after as Europeans became, and saw themselves as, dominant over other races. Exploitation and torture have taken place, and continue to do so, within the same race, religion, tribe and nation as and when one group or class assumes more power over the other. Thus the history of racism is about more than the history of race itself.

The history of race and European acclimatization in the tropics also shows the close proximity between power and anxiety. Fear and anxiety were at the heart of European colonialism. The history of colonialism and racism reflects the interplay of power and fear. It helps us to understand that fear and anxiety often lurk behind power and authoritarianism in general. These determined the policies of European colonial settlement and habitation patterns in the nineteenth century. Fear, anxiety and power led to segregation in the United States and to the Apartheid system in South Africa, and defined the settler policies and attitudes towards the aboriginal populations of Australia.

Notes

1 P.J. Marshall, *The Making and Unmaking of Empire: Britain, India, and America c. 1750–1783* (Oxford, 2005) and 'Britain and the World in the Eighteenth Century: I, Reshaping the Empire', *Transactions of the Royal Historical Society*, 8 (1998), 1–18, 'II, Britons and Americans', ibid, 9 (1999), 1–16, 'III, Britain and India', *Transactions of the Royal Historical Society* 10 (2000), 1–16, 'IV the Turning Outwards of Britain', *Transactions of the Royal Historical Society*, 11 (2001), 1–15.

2 Mark Harrison, ' "The Tender Frame of Man": Disease, Climate, and Racial Difference in India and the West Indies, 1760–1860', *Bulletin of the History of Medicine*, 70 (1996), 68–93.

3 Dane Kennedy, 'The Perils of the Midday Sun: Climatic Anxieties in the Colonial Tropics', in MacKenzie (ed.), *Imperialism and the Natural World*, pp. 118–40.

4 Valesca Huber, 'The Unification of the Globe by Disease? The International Sanitary Conferences on Cholera, 1851–1894', *The Historical Journal* 49 (2006), 453–76.

5 Julyan G. Peard, *Race, Place, and Medicine: The Idea of the Tropics in Nineteenth-Century Brazilian Medicine* (Durham, NC, 1999).

6 For a detailed analysis of the idea of race in Western thought from ancient times, see Ivan Hannaford, *Race: The History of an Idea in the West* (Washington, DC, 1996).

7 Schiebinger, 'The Anatomy of Difference', p. 394.

8 Ibid, p. 393.

9 Harrison, *Climates and Constitutions: Health, Race, Environment and British Imperialism in India 1600–1850* (Delhi, 1999) p. 123.

10 Scheibinger, 'The Anatomy of Difference', p. 394.

11 Peter Hulme, *Colonial Encounters; Europe and the Native Caribbean, 1492–1797* (London & New York, 1986), p. 3.

12 Edward Long, *The History of Jamaica; Or, General Survey of the Antient and Modern State of that Island: With Reflection on its Situations, Settlements, Inhabitants; In Three Volumes* vol. 2 (London, 1774), pp. 380–2.

13 Ibid, p. 380.

14 Ibid, p. 381

15 Jane Landers, 'Gracia Real de Santa Teresa de Mose: A Free Black Town in Spanish Colonial Florida', *The American Historical Review*, 95 (1990), 9–30; Farhat Hasan, 'Indigenous Cooperation and the Birth of a Colonial City: Calcutta, c. 1698–1750', *Modern Asian Studies*, 26 (1992), 65–82.

16 Henry Davison Love, *Vestiges of Old Madras 1640–1800, Traced From the East India Company's Records Preserved at Fort St. George and the India Office, and From Other Sources*, vol. 1 (London, 1913), p. 387.

17 Chakrabarti, *Materials and Medicine*, p. 93.

18 Susan M. Neild-Basu, 'Colonial Urbanism: The Development of Madras City in the Eighteenth and Nineteenth Centuries', *Modern Asian Studies*, 13 (1979), 217–46, p. 246.

19 See Elizabeth Kim, 'Race Sells: Racialized Trade Cards in 18th-Century Britain', *Journal of Material Culture*, 7 (2002), 137–65.

20 Trudy Eden, *The Early American Table: Food and Society in the New World* (Dekalb: IL, 2010/2008), pp. 20–1.

21 Wendy D. Churchill, 'Bodily Differences? Gender, Race, and Class in Hans Sloane's Jamaican Medical Practice, 1687–1688', *JHMAS*, 60 (2005), 391–444.

22 Mark Harrison, ' "Tender Frame of Man" '.

23 Karen Ordahl Kupperman, 'Fear of Hot Climates in the Anglo-American Colonial Experience', *The William and Mary Quarterly*, 41 (1984), 213–40.

24 For a detailed account of early European encounters with tropical fevers, see Harrison, *Medicine in an Age of Commerce*, pp. 28–63.

25 Ibid, pp. 37–8.

26 Iain M. Lonie, 'Fever Pathology in the Sixteenth Century: Tradition and Innovation', *Medical History*, Supplement (1981), 19–44, pp. 28–34.

27 Harrison, *Medicine in an Age of Commerce*, pp. 64–74.

28 Gordon Low, 'Thomas Sydenham: The English Hippocrates', *Australian and New Zealand Journal of Surgery* (1999), 258–262.

29 Geoffrey L. Hudson (ed.), *British Military and Naval Medicine, 1600–1830* (Amsterdam & New York, 2007), pp. 3–4.

30 James Lind, *An Essay on Diseases Incidental to Europeans in Hot Climates with the Method of Preventing Their Fatal Consequences* (6th edition, London, 1808), pp. 91–4.

31 Ibid, pp. 10, 12.

32 Lind, *An Essay on the Most Effectual Means of Preserving the Health of Seamen, in the Royal Navy* (2nd edition, London, 1762), p. 49.

33 Ibid, p. 114.

34 Patrick Brantlinger, *Rule of Darkness: British Literature and Imperialism, 1830–1914* (Ithaca & London, 1988).

35 Eric T. Jennings, *Curing the Colonizers; Hydrotherapy, Climatology and French Colonial Spas* (Durham, NC, 2006) p. 11.

36 Arnold, 'Race, Place and Bodily Difference in Early Nineteenth-Century India', *Historical Research*, 77 (2004), 254–73.

37 Sean Quinlan, 'Colonial Encounters; Colonial Bodies, Hygiene and Abolitionist Politics in Eighteenth-Century France', *History Workshop*, 42 (1996), 107–26.

38 Warwick Anderson, *The Cultivation of Whiteness: Science, Health and Racial Destiny in Australia* (Carlton South, Victoria, 2002).

39 M. Osborne, *Nature, the Exotic and the Science of French Colonialism* (Bloomington, 1994), p. xiv.

40 Kennedy, 'The Perils of the Midday Sun'.

41 Chris Stringer, *The Origin of our Species* (London & New York, 2011).

5

Imperialism and the globalization of disease

The Age of Commerce led to an era of global migration: of humans, botanical specimens, animals, ideas, cultures and diseases. The main human migrations started across the Atlantic when the Spanish and the Portuguese, followed by the French, Dutch and British, settled in different parts of North and South America. They took with them Africans as slaves to work in the newly established sugar, cocoa, tobacco and cotton plantations in the New World. The other major migration was that of the Europeans to tropical colonies in Asia and Africa, as well as to the settler colonies in Australia, Canada, South Africa and New Zealand in the eighteenth and nineteenth centuries. On the one hand, we can see in these global migrations the roots of modern multiracial societies and multicultural experiences. On the other, there was a grave human cost to this global migration as millions of lives were lost in the voyages or at the sites of the new colonial settlements. Disease was the unwelcome companion of global human migration. The most disastrous impact of this human movement was on the native population of the Americas. When in the sixteenth century Europeans and Africans crossed the Atlantic Ocean, they infected the Amerindians, who died of diseases such as smallpox, measles and typhoid. Following the Columbian voyages, it is estimated that up to 90 per cent of the indigenous population in certain parts of the Americas were wiped out. The mortality was most striking and drastic in the densely populated parts as disease could spread more quickly.

In this chapter I will examine the transmission of disease through global migrations in four different contexts of colonial history: first, the dissemination of Old World diseases to the New World in the sixteenth century; second, the rise in mortality among Europeans and other races in the eighteenth century; third, the political and medical debates over the transmission and control of cholera from Asian colonies to Europe in the nineteenth century; and finally, the history of yellow fever as it spread across the Americas from the early days of colonialism to the twentieth century. These four different episodes reflect four different instances of colonial migration of humans and diseases. They also highlight the fact that migration was only part of this story. The social, economic and cultural consequences in the regions where these migrations occurred are equally important. From these histories of migration and disease we learn more about the history of colonialism itself.

Inherent in these histories are narratives about when disease took epidemic forms. Epidemics by their very character affect large sections of populations and communities, and spread over vast regions and different climatic and geographic zones. Thus the history of epidemics enables us to understand the wider and deeper social, economic, political and cultural histories that lead to disease and death. It also helps us to understand the diverse ways in which the same disease can affect different communities,. In this chapter I will study two processes: how imperialism led to the global spread of epidemics and how the control of epidemics became an important rationale for imperial expansion.

Disease travelled with migrating humans across vast geographical regions even before the modern age of global empires. Emanuel Le Roy Ladurie introduced the concept of 'Unification of the Globe by Disease' to explain how, historically, disease accompanied two factors: first, human migrations, economic and social changes, and intensification of human contact; and second, the social and economic conditions in the countries in which disease took epidemic forms. He used the case of the Black Death of 1347–8, which originated in China but soon spread to Europe. Why did plague spread across Europe and Asia when it did? According to Ladurie, the answer was in the wider economic and social transformations taking place in this period. Between 1200 and 1260 the Mongols under Genghis Khan and his successors strengthened political and trade links between Asia and parts of Europe such as Russia, setting up new trade routes. Hordes of travellers moved through these routes, through Central Asia carrying Chinese silk to the bazaars of Constantinople, creating what Ladurie calls a 'common market of bacilli' and a 'community of disease' through contact between different populations: Chinese, Mongols, Europeans and Arabs. Contagion was also spread by the plague-stricken soldiers of the Tartar army in 1346, first to Eastern Europe and then through the Mediterranean to Western Europe by 1347–8. This, according to Ladurie, helped to spread plague in Europe between 1300 and 1600.

In Europe, on the other hand, ideal conditions existed for the spread of plague, since medieval urbanization had accelerated close human habitation. In crowded places, rats, the carriers of plague, could travel with humans in clothes, blankets and containers. Several parts of Europe, particularly France, experienced a demographic rise in the thirteenth century, leading to deforestation, a lack of firewood and greater exposure to harsh winters, food shortage and famines, and an increase in people living in shacks close to each other. Ladurie describes the spread of plague in Europe as the 'outcome of a culture of poverty, dirt and promiscuity'.[1] The epidemic affected the poor more than the wealthy and led to the stigmatization of the poor by the wealthy as carriers of disease, much like in the case of cholera.

The next major migration of plague and other diseases was from Europe across the Atlantic, following the Spanish Conquest of the Aztec and Incan empires, which, according to Ladurie, led to the microbial unification of the Atlantic world. His analysis helps us to understand that disease transmissions have taken place in times of major human migration, economic change, and growing social and economic inequalities. These are important frameworks which help us to understand how diseases travelled in the colonial period along with colonial migrations, economic transformation, the setting up of new trading routes, and devastated indigenous societies and economies.

Smallpox in a 'virgin' body: Disease and the depopulation of the Amerindians

The import of Old World diseases into the previously isolated communities of the New World caused catastrophic devastation.[2] A range of diseases spread from the Old World to the New World, such as measles, whooping cough, chickenpox, bubonic plague and, most serious of all, smallpox, to which the Amerindian populations were immunologically defenceless.[3] Although it is established that there was a collapse in the populations among Amerindian societies after 1492, it is difficult to determine the size of pre-Columbian populations and thus the exact extent of depopulation. It is likely that disease travelled faster than the conquering armies did so, as a result, estimates of Amerindian population sizes may unintentionally be inaccurate, as the community may have already contracted Old World diseases.[4] Estimates for the size of pre-Columbian populations vary widely from as low as 8 million to significantly higher estimates of over 110 million people.[5] The higher estimate suggests a greater devastation of population, cultural life and social practices of the indigenous population. Moreover, if the population was over 100 million it is unlikely that a mere few thousand conquistadors could have caused such an abrupt population decline, and therefore it seems only logical to assume that it was the epidemiological factor that had the most fundamental impact on the health and social life of the indigenous population.

Historians have debated the exact magnitude of the depopulation, but it is generally accepted that in the first 150 years after the Columbian voyages, on average, 80–90 per cent of the Amerindian societies were decimated by disease. There is also some evidence of how it affected their social life. Contemporary Spanish observers often commented that the family and community nursing care ceased to function among the Amerindians when the unfamiliar and devastating Old World diseases, such as smallpox, struck their communities. Experienced Amerindian healers were utterly unfamiliar with Old World diseases. Consequently, when the epidemics reached a critical point, many curers often fled, leading to the collapse of their health system.[6]

From a different perspective, Massimo Livi-Bacci has challenged what he refers to as the 'epidemiological paradigm' or 'monocausal' explanations – the sole focus on the devastation caused by smallpox and other diseases. He argues that it was highly improbable that Amerindian societies would either succumb to complete extinction or lose their social systems of care entirely.[7] He highlights the extreme difficulty in finding accurate estimates of population decline due to the lack of quantitative evidence. Livi-Bacci claims that historians' estimates of mortality rates caused by epidemics are extreme and unrealistic as they do not factor in the likelihood that a significant faction avoided contagion due to chance and by gaining immunity after the first wave of epidemics.

He argues that contemporary chronicles suggest a variety of factors for the depopulation apart from diseases, such as environmental destruction, deforestation and the impact of livestock brought from Europe, loss of cultivable land to the Spanish and the consequent decline in food supply, warfare, social dislocation, forced migration and lack of labour. In doing so, Livi-Bacci presents a more comprehensive picture of the impact of Spanish colonization on Amerindian societies. In parts of the Caribbean islands, population decline took place even before disease arrived as the Spanish, in their ruthless hunt for gold, killed and enslaved the people, leading to a decline of indigenous subsistence, famine and death. He suggests that different parts of the New World experienced different rates of population decline. The worst sufferers were the inhabitants of the Caribbean islands and the coastal parts of Peru and Mexico, who were almost extinct by 1550, while those in the low-lying internal regions of South America survived, only to be enslaved in the growing plantations. On the other hand, in Paraguay, the native populations who were protected by the Jesuit missionaries in the sixteenth and seventeenth centuries suffered depopulation in the eighteenth century, when the Jesuit efforts collapsed.

It is arguable that the ecological and economic impacts of Old World diseases fundamentally changed the lives of the indigenous populations in the New World. For instance, the temporary abandonment of land by Amerindians fleeing from disease, and the subsequent occupancy of the land by Spanish ranchers, changed their societies completely. Being restricted to marginal land not only reduced the ability of Amerindian societies to grow basic foodstuff and thus had a detrimental effect on the size of the native population, but it also increased Amerindian social dependency on Europeans. After they lost their lands, Amerindian communities often became destitute and submitted to the protection of Spanish missionaries or farmers, a factor which in turn affected their social life.

It is likely, to give one notable instance, that the great Aztec state and its capital, Tenochtitlan in Mexico, would not have been captured by the Spanish if it were not for military assistance from the city-state of Tlaxcala, or more importantly the outbreak of smallpox in the area.[8] For as Hernán Cortés, a Spanish conquistador, and many of his followers who arrived in Mesoamerica in 1519 commented, the capital of Tenochtitlan was not only magnificent but fully functioning, and in theory was well placed to resist the small Spanish expeditionary force that challenged it in 1521. Yet Aztec accounts testify that the ravages of disease made the defence of the city impossible, as they described how their 'spears lie broken in the streets' and that the 'illness was so dreadful that no one could move or walk'.[9] This adds credence to the assertion that disease, particularly smallpox, was the 'conqueror's best ally'.[10] The introduction of Old World disease in the Americas by the Columbian exchange from 1492 led to the decapitation of the ruling classes and the collapse of social order as seen in Tenochtitlan, and thus allowed European domination.[11]

This breakdown of the social order among the American indigenous population is evident in the Inca Empire, which fell victim to smallpox after the collapse of the Aztec state, as it allowed the disease to penetrate its territories through pre-existing trade networks.[12] It has been suggested that smallpox killed Huayna Capac, leader of the Inca Empire, and his legitimate heirs, thus precipitating a long and furious civil war, which arguably saw the collapse of social networks and the political structures of Incan society.[13] The civil war had a significant impact on the indigenous populations' health as well. In the provinces of Arequipa, for instance, the population rapidly declined due to the abandonment of cultivated lands during the war period.[14] Furthermore, by studying the ratio of men to women in Peru in 1573, it is clear that there was a shortage of men, possibly reflecting the losses during the civil war, which had ceased only 20 years earlier.[15] To conclude, although disease may not always have been directly responsible for the detrimental effects on the health and social life of Amerindians, it nevertheless caused economic and social destabilization and civil war, which was responsible for the rapid and gradual devastation of Amerindian societies.

Over the course of the next three centuries, around 15 million slaves from Africa were shipped to the Americas to fill the demographic gap created by the Amerindian depopulation, which was felt acutely in the seventeenth century as Spanish, then English and then French settled in these regions and sought to introduce plantations. African immigration in turn brought diseases such as malaria and yellow fever to the Americas. Especially in the tropical lowlands, many groups died out before they had an opportunity to build immunity against the joint assault of African and European diseases. The mortality was most striking and drastic in the densely settled population of the Greater Antilles, where the Spanish first landed, and to which they brought the first African slaves. Indians on the South and North American mainland fared somewhat better, and so did the Caribs of the Lesser Antilles, though they, too, had almost disappeared by the seventeenth century. In this case, isolation seems to have protected them from simultaneous attack by the full range of Old World diseases. With the highland peoples of Middle America, Colombia and the Andes, a cooler environment prevented the spread of malaria and yellow fever. Thus, though these peoples sustained steep declines in population over the sixteenth and part of the seventeenth centuries, they were able to maintain themselves and ultimately to recover once new immunity had been acquired. Epidemiological factors were thus responsible for depopulating some of the best agricultural land in the tropical Americas, which enabled the European exploitation of these resources. The most obvious reason for locating the productive centres in the American tropics was that land was there for the taking, and it was far better land for intensive agriculture than any that was available in Europe. The depopulation of the Americas following the Spanish Conquest was not just about disease and migration but also about a much greater transformation of society, economy culture and ecology.

The study of the Amerindian depopulation helps us to understand another less familiar but equally devastating example of depopulation in colonial history – that of the Polynesian population in the Pacific islands. By the most conservative estimates, these peoples declined from 250,000 in 1790 to 50,000 in 1890. Thus precisely in the period when the region was being colonized by European powers, the indigenous population declined by 80 per cent. This offered a decidedly major advantage to the colonizers in occupying and exploiting land and other resources in the region, as it did to the Spanish in the Americas. Similar to the point made by Livi-Bacci, Stephen Kunitz argues that disease was not the only, or even the most crucial, factor in the depopulation of the Polynesian islands. He has placed far greater importance on the impact of 'settler capitalism' – the appropriation of lands and other resources from the indigenous populations, first by the colonizing forces and then by the highly mobile and landless Europeans, leading to economic and social disruption among indigenous communities. Due to the differential nature of European contact with and settlement in these islands, different islands experienced different patterns of population decline.[16] At the core of these two instances of large-scale population decline was the colonial hunger for land and other resources, which started the cycle of social and economic disruption, starvation, fertility decline, disease, death and further loss of land.

Colonial migration and mortality in the eighteenth century

Europeans themselves were victims of disease during colonial journeys and migration. From the seventeenth century, as large numbers of European sailors and traders passed into 'new epidemiological zones', they contracted previously unknown or unfamiliar diseases in the newly colonized territories.[17] During the voyages and in the new colonies, the death rates of Europeans who travelled to the tropics were sometimes seven times those in Europe.[18] As we saw in Chapter 3, this made medical discipline within the structure and framework of empire urgent and necessary, for without this discipline the major European powers could not hope to settle effectively into their colonies throughout the world.

Another population that suffered terrible mortality and morbidity rates during the transatlantic voyages were the Africans sold in the slave trade. The so-called and dreaded 'Middle Passage' (the journey from Africa to the Americas across the Atlantic) has attracted substantial historical attention in recent times as historians search for the beginnings of the experiences and consequences of slavery.

On average, 300 slaves were shipped in each vessel from Africa to the West Indies. Owing to the obscure nature of the evidence it is difficult to get a clear idea of the mortality rates of the slaves and crew in the ships. Many slave ships never reached the Americas or the West Indies due to bad weather, slave rebellion and wreckage. These ships and the stories of suffering of their human cargo have largely been obliterated from history and leave a major gap in our understanding

Table 5.1 Crude death rates (CDRs) on ocean voyages, 1497–1917.

	Nature of voyage	Period	Number of voyages	Average voyage length (days)	CDR per month per 1,000
A	Slaves to Americas	1680–1807	728	67	50.9
B		1817–43	591	43	61.3
C	Dutch to Batavia	1620–1780	3914	218	15.3
D	Portuguese to India	1497–1700	1149	180	20.4
E	Portuguese from India	1497–1700	781	200	25.1
F	British convicts to North America	1719–36	38	60	56.5
G		1768–75	12	60	12.5
H	British convicts to Australia	1788–1814	68	174	11.3
I		1815–68	693	122	2.4
J	German emigrants to Philadelphia	1727–1805	14	68	15.0
K	European immigrants to				
L	1. New York	1836–53	1077	45	10.0
M	2. Australia	1838–53	258	109	7.4
N		1854–92	934	92	3.4
O	3. South Africa	1847–64	66	75	4.8
P	African indentured labour to West Indies	1848–50	54	29	48.7
Q		1851–65	54	29	12.3
	Indian indentured labour to Mauritius, Natal, West Indies and Fiji				
R	1. From Calcutta	1850–72	382	88	19.9
S		1873–1917	876	65	7.1
T	2. From Madras	1855–66	56	62	5.6
U	Chinese indentured labour to Americas	1847–74	343	116	25.5
V	Pacific islander indentured labour to				
W	1. Fiji	1882–1911	112	117	3.6
X	2. Queensland	1873–94	558	111	3.0

Source: Robin Haines and Ralph Shlomowitz, 'Explaining the Modern Mortality Decline: What can we Learn from Sea Voyages?', *Social History of Medicine*, 11 (1998), 15–48, p. 23.

of this part of the past. Nevertheless, Herbert S. Klein and Stanley L. Engerman have carried out a massive statistical analysis of the mortality rates of slaves in the slave ships. Their findings indicate that transatlantic slave ships had much higher mortality rates than other transoceanic ships in the same period. Between 1590 and 1700 the mortality rates were 20.3 per cent, then between 1701 and 1750 this fell slightly to 15.6 per cent. The real decline took place in the last quarter of the eighteenth century. Between 1750 and 1800, the mortality rate of slaves in all European ships was 11.8 per cent, and by 1820 it had fallen to 9.1 per cent. The decline was due to the restrictions imposed in this period on the number of slaves loaded in each vessel, better provisions for surgeons onboard and bonuses offered to captains who achieved lower mortality rates.[19]

Robin Haines and Ralph Shlomowitz have compiled an enormous set of data relating to global mortality patterns in sea voyages from the late fifteenth century to the early twentieth.[20] Their analysis presented a variety of statistics covering different groups and races of people involved in global migration (see Table 5.1).

We can draw two broad conclusions from the above table. First, there was an overall decline in mortality rates in the nineteenth century. Second, class and race are important with regard to global mortality rates. The decline from the eighteenth century to the nineteenth was most evident among Europeans (L, M, N and O). These indicate that improvements in medical science and sanitation did reduce mortality rates. Such advances and the general conditions of sea voyages differed, sometimes quite drastically, according to the class of the people and their race. The poorer class of Europeans, such as convicts from Britain who were sent to the Americas, often suffered as much as the slaves from Africa, at least in the early part of the eighteenth century (F). On the other hand, Chinese, Indian and African labourers suffered the worst mortality rates whether in the eighteenth or nineteenth century (R and U).

Africans were the worst sufferers on sea voyages, whether as slaves in the seventeenth to eighteenth centuries or as indentured labour in the nineteenth (A, B, P and Q). Migration to the New World exposed Africans to new diseases – notably tuberculosis, pneumonia, measles, influenza, smallpox and dysentery – which were not widely found in Sub-Saharan Africa. The low endemicity of these diseases meant that many slaves had not built up their immunity to them during childhood but encountered them for the first time during their forced march to the coast of Africa, while being held in coastal barracoons, during the Middle Passage or after their arrival in the Americas and the Caribbean islands. Resistance to disease may also have been reduced by an inferior nutritional status among African slaves and labourers prior to embarkation. We also need to bear in mind the extent to which international migration stimulated increased psychological stress, which itself may have exerted an impact on nutrition and mortality. This is evident in a particular disease known as 'dirt-eating', which was common in British Caribbean slave societies. Dirt-eating or *Cachexia Africana* was a practice in which slaves regularly ate baked clay cakes ('aboo'). European surgeons remained uncertain and provided different explanations for the practice. Most suggested that it was a disease of African origin which was brought over by the slaves. Some suggested that it was a product of Caribbean slavery, of the physical and psychological impact of life in the plantation, hard labour and inadequate diet.[21] West Indian surgeon R. Shannon suggested that this physical symptom had emotional roots as it was caused by despair and nostalgia: 'They [the slaves] are supposed to give into it [dirt-eating] at first from other motives; such as discontent with their present situation, and a desire of death, in order to return to their own country; for they are well aware it will infallibly destroy them'.[22] Dirt-eating continues to be seen as a 'deficiency disease' caused by either nutrition deficiency or depression, either way being acutely linked with the conditions of slavery.[23]

All of these factors combined to bring about the extraordinarily high mortality suffered by African slaves. Other migrating populations suffered from new diseases too, such as the Pacific Islanders who were recruited as indentured labourers in Vanuatu and the Solomon Islands. These were also 'virgin' populations who lacked prior exposure to the other infectious diseases of Eurasia.

There is a dichotomy here which can answer the differential nature of mortality decline among blacks and whites in the late eighteenth century. It is based on economic and mercantilist factors. Economically it was much cheaper for West Indian planters to buy fresh slaves than to settle and feed slave families to encourage reproduction.[24] At the same time, as we saw in Chapter 3, it was cheaper to look after the hygiene and health of European troops than to recruit and train new ones.

As is also evident from Table 5.1, Britain was not the only early colonial power that experienced a significant impact from foreign diseases and infections in the process of consolidating its empire; the Dutch, French, Spanish and Portuguese all suffered from disease and a chronic lack of medical supplies during distant voyages. Dutch figures from the seventeenth century put mortality rates in its slave trade network at roughly one slave in every six shipped. Of about 550,000 African slaves shipped across the Atlantic in Dutch ships between 1500 and 1850, about 75,000 died onboard.[25]

There was a drastic fall in European mortality rates from the eighteenth century to the nineteenth. Not only did death rates drop in Europe but they dropped even more dramatically among Europeans who settled overseas. In England from the late 1860s, the mortality decline first occurred among young adults and older children. From the turn of the twentieth century, infant death rates also commenced a long-run decline. This mortality transition was preceded by a marked decline in the death rate of the British armed forces, both at home and abroad, between the 1840s and the 1860s, when a 58 per cent drop in mortality occurred in non-combat troops, and by an even earlier mortality decline among some seaborne populations, most notably on convict voyages to Australia from 1815. Historian Philip D. Curtin attributes this decline to preventive measures adopted by the army: greater access to clean water, improved sewage disposal, less crowded and better ventilated barracks, the relocation of colonial troops to hill stations when epidemic disease threatened, and more widespread use of quinine as a treatment for malaria in Africa and Asia. From the 1870s to the 1890s, military medicines, tropical medicine and tropical hygiene experienced a revolutionary change. According to Curtin, this profoundly altered Europe's epidemiological relationship with the rest of the world. Following this, Europeans were free to move into the tropical world at far less risk than ever before, providing a greater impetus for colonial growth.[26]

Nevertheless, there is a danger in making too easy a connection between mortality decline and colonialism by assuming that low mortality rates, aided by better medical provisions, facilitated colonial expansion. This historiographical approach tends to identify medicine as a deciding factor not only in the spread of colonization but also in modern warfare, particularly colonial warfare. Modern medicine has been seen to play a vital role in this modernization and in securing military successes for the British and other European nations from the eighteenth to the twentieth century.[27] The factors highlighted are better management of resources, increased appointment of medical personnel, greater efficiency, better understanding of diseases, use of prophylactic treatment and an increase in professionalization and professionalism within medical personnel.[28]

Although these are significant points in understanding the relationship between the control of disease and the consolidation of Western empires, it is important here to remember that in the seventeenth and eighteenth centuries, medical facilities in the colonies remained rudimentary and did not lead to any significant decline in mortality rates. Despite that, a major part of colonial expansion took place in this period under the Spanish, French and British, particularly in India and the Americas.[29] As we have seen in the colonization of the Americas, and as we shall see in Chapter 7 in the case of Africa, often the connection was the other way round: colonial expansion provided resources, which improved European military and civilian provisions, leading to better mortality rates.

The global contagion of 'Asiatic' cholera

The history of nineteenth-century sanitarian changes leads us to our next problematic in examining disease and migration. This will be through the history of cholera and the institution of quarantine and preventive medicine in Europe. The politics of enforcing quarantine on ships from Asia, especially to prevent epidemic 'Asiatic' cholera from spreading to Europe in the nineteenth century, was imbued with several factors: medical constructions of the 'unhealthy tropics', medical debates about contagion, and economic and political rivalry between European nation states. The high mortality rates and the swift course of the epidemic 'Asiatic' Cholera, its causes, modes of transmission, methods of prevention and cure and, most crucially, the role of quarantine in halting its dissemination was the grand global theatre on which the political and economic rivalries among European nations were enacted.

Although the quarantine of leprosy patients was common in medieval Europe, international and maritime quarantine was first introduced in Italian port cities in the Mediterranean following the Black Death in the fourteenth century. The city-states of northern Italy developed the first systems of national and international quarantine system in their attempts to control recurrent episodes of the disease.[30] During the spread of plague to Italy between the fourteenth and seventeenth centuries, merchants urged for quarantine to protect their commercial practices. Yet even at this early date, political pressure to prevent epidemic disease tended to go against the needs of trade. Trade and quarantine were the two difficult sides on the nation states of the plague epidemic.

In the early seventeenth century, Italian cities such as Pistoia imposed a sanitary cordon around themselves, guards were placed at the gates and restrictions were imposed on the movement of goods and people. They also expelled all foreigners, including 'mountebanks, and Jews' from the city. Yet, from time to time, the city authorities also had to remove the quarantine to facilitate trade.[31] Often such measures were based on vague or mistaken ideas and were thus ineffective. The debates were between the three interest groups – the city authorities, the physicians and the merchants – each having their own interests and priorities. Italian physicians had little idea of how plague spread. They regarded silk worms for their connection with China as the culprit and banned the production of raw silk and the keeping of silk worms within the city. Such measures did not help to control the spread of plague but hampered commercial activities in cities, which were predominantly dependent on trade.

More than 200 years later, the quarantine system was revived on a large scale in the wake of cholera outbreaks in Europe. Cholera is an illness caused by the bacterium *Vibrio cholerae* invading the human bowel. The disease is usually spread via contaminated water supplies. The main symptom is watery diarrhoea, which leads to fluid depletion and death from dehydration. From the early nineteenth century, European medical authorities concluded that Asia or, specifically, lower Bengal in India was the 'home' of the disease. Medical discourse termed it 'Asiatic Cholera', believing that the air and water of the Gangetic belt of Bengal provided the ideal conditions for it to breed. This nomenclature was used to distinguish it from a form of severe diarrhoea called 'cholera' that existed in Europe for centuries. By the nineteenth century, when negative British medical perceptions of the tropics generally and India specifically had taken full form, the medical focus was on the climatic and environmental conditions of this region in which cholera was believed to thrive. When, in the nineteenth century, Europe experienced repeated episodes of pandemic 'Asiatic' cholera, political, diplomatic and medical discourses coalesced to prevent its transmission from India to Europe, but European nations (and medical authorities) were deeply divided over the means that would make that possible.

Between 1817 and 1870, several cholera pandemics appeared in Europe and the United States. By 1830, cholera had appeared in Russia, and then rapidly swept through Europe and reached Britain in 1831. In 1832 it appeared in Quebec, Montreal and New York and became a global contagion. It ravaged Africa at the same time: Tunisia (1835), Egypt (1823), and Ethiopia and Zanzibar (1821) (see Figure 5.1).[32] Cholera caused huge mortality rates wherever it visited. Between 1816 and 1832, millions of people died in Asia; during the 1832 outbreak, 55,000

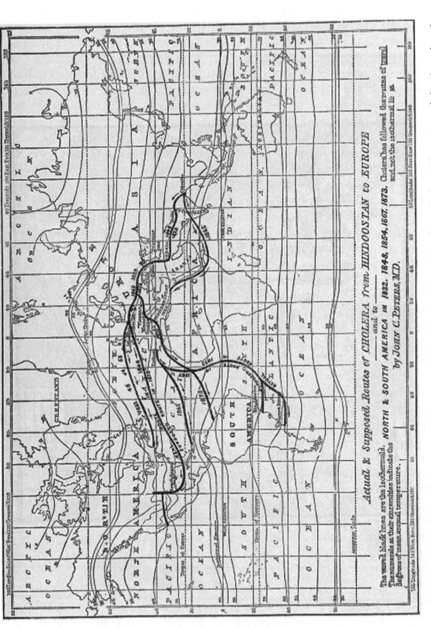

Figure 5.1 A map showing the likely routes of transmission of cholera from India to Europe and the Americas in the nineteenth century (by Edmund Charles Wendt, 1885). Courtesy of the Wellcome Library, London.

people died in Britain and 3,515 died in New York alone; during the second pandemic in 1849, over 33,000 people died in three months in Britain. In Europe, cholera appeared as a frightening disease and aroused public fear. It caused death in a few hours and patients died a pathetic death.[33]

The nineteenth-century pandemics of cholera occurred at a time when print had become an instrument of mass media. Therefore the dramatic news of the pandemics spread much faster and in greater detail, leading to public panic. Long before cholera reached the United States, for instance, newspapers and popular magazines carried long and graphic accounts, including images, about its advance through Russia and Europe from the East.

The nineteenth century also oversaw new technology and an emergent international trade across the globe, when far-flung parts of the European empires were integrated into the global economy. Improvements in sailing technology, especially the development of the clipper ship with its high speed, enabled the opium trade to expand, thus solidifying the economic links between Europe, the Americas and Asia. Steamships and railways carried waves of migrants within Europe and from there to the Americas. By the 1830s, steamships had started to operate on all major imperial routes to India, Africa and Australia from Europe. Till the 1840s, sailing ships going around Africa took more than five months to reach India from Britain. Steamships covered the distance in just two months. The Suez Canal reduced the journey time even more drastically. Similarly, from the 1830s, steamships reduced the journey time across the Atlantic from more than two months to 22 days. The railway networks (including the Trans-Caspian railway) in Russia, for example, brought subsequent waves of cholera to the country in the second half of the nineteenth century.[34]

Although there was a great deal of awareness and even panic regarding the spread of cholera in Europe, medical knowledge there about the disease was meagre and fragmented when the first pandemic appeared. The situation was strikingly similar to plague in the seventeenth-century Italian cities where physicians had little knowledge of the disease or means of combating it. While most medical men believed that cholera had travelled from Asia to Europe, they could not establish exactly how it travelled. It was more than half a century after the first pandemic before Robert Koch discovered the cholera bacterium in 1883. This intermediate period (and even after Koch's discovery) was one of fierce medical debates about the origins, causes and, most importantly, the precise mode of transmission of cholera. These were vital to the imposition of preventive measures, and therefore a manifestly medical puzzle took on a vital political and economic dimension.

In the early nineteenth century, essentially before germ theory was established, the miasmatic theories of disease offered opposing views regarding the transmission of cholera. One group was the contagionists who (mostly in France and Germany) believed that cholera spread by human-to-human contact. Contagionists in France and Germany believed that the increased trading and human links with Asia (particularly through British trade) had introduced cholera into Europe. The traditional European response to invading infections was thus quarantining any ship at the port that might have a patient with the disease.

The other group were the non-contagionist medical theorists (mainly British), who believed that cholera (like the Black Death in the fourteenth century) was spread by unhealthy vapours and environmental factors. They therefore staunchly supported sanitation in the overcrowded urban spaces as the solution to the cholera pandemic and opposed any quarantine measures, which they believed were ineffective in the face of diseases caused by local environmental factors. Some of the non-contagionists believed that cholera was not readily communicated by human contact but prevailed among people who lived in low and poorly ventilated places that were prone to miasmatic conditions. Others believed that cholera was produced by decaying organic matter in the 'putrefying' tropical atmosphere. The transmission of this disease to Europe, according to non-contagionists, occurred through trade winds and the unhygienic conditions of the trading vessels. They believed that cholera spread to Europe because unhygienic and insanitary conditions were being replicated there, particularly in Britain because of industrialization and urban poverty in the nineteenth century. Both non-contagionists and contagionists agreed on one point: that cholera originated in India.

While the medical debates continued, European nations resolved to tackle the pandemic at the diplomatic level. France proposed a meeting in 1834 at which the international standardization of quarantine could be discussed. The meeting did not eventuate and it was only in 1851 that the first International Sanitary Conference was convened in Paris. Several European nations met at ten formal sanitary conferences between 1851 and the end of the century, but it was not until 1893 that an agreement was reached concerning detection of the disease and the minimum and maximum periods of detention in quarantine.

At the first International Sanitary Conference in 1851, the majority of European countries favoured quarantine as the means of preventing seaborne diseases. However, Britain opposed it, particularly because of the apprehension that quarantine would adversely affect its trade with India. Long detention periods of its shipping vehicles and goods from India would become an immensely costly mode of preventing cholera. Britain was the leading maritime and trading nation in the world in the middle of the nineteenth century. Its rapidly growing industries needed colonial raw materials and markets. Thus, politically and economically, it was in favour of free trade. Although British sanitary medical traditions had their own trajectories, it is not entirely a coincidence that the majority of the publicly supported medical theories of cholera favoured non-contagionism. Non-contagionists suggested measures other than quarantine, such as sanitary 'cleansing' of urban ghettoes to eliminate cholera. Medical and local government authorities in Britain carried out sustained sanitary campaigns during this period. In Exeter in the 1830s, cholera was treated as an airborne disease and various fumigating methods were deployed, including the lighting of fires with tar and tar barrels, and the burning of vinegar in the most confined parts of the city to 'purify the air'. Often, as an observer noted, 'the smell of all these things [disinfectants] was worse than the Cholera smell itself'.[35]

Similar measures were undertaken in the British colonies. In 1850, when cholera struck Kingston, Jamaica, for the first time, 4,000 residents died in the city alone. The desperate Board of Health borrowed a canon from the ordnance stores and fired blank cartridges in the streets of Kingston 'to destroy the morbific power that lurked in the dark alleys of the pestilence-stricken city'.[36] In India, to prevent troops from being infected by cholera, which was believed to be in the air, the colonial military authorities ordered its troops, when attacked by cholera, to march at right angles to the wind.[37]

The airborne theory of cholera was challenged by John Snow in 1854, when he linked the disease with contaminated water supplies; – in the case of his research in particular, to a single water pump in Soho, London. His discovery moved the focus from air to water, but it actually strengthened the position of the non-contagionists who had an effective and relatively simple solution to the problem – a clean water supply – rather than the complicated method of quarantine.

Both the air and the water borne theories of the transmission of cholera vital-ized British public health movements, particularly in the overcrowded, unhy-gienic, industrial slums. The identification of cholera with local conditions also prompted the emergence of modern public health and epidemiology in Britain. In 1842, Edwin Chadwick, an English social reformer and civil servant, published his *Report on the Sanitary Condition of the Labouring Population*, which demon-strated that poor living conditions, overcrowding and foul air predisposed urban populations to epidemic disease.[38] The main achievement of the report was the massive statistics that it collected regarding the living and sanitary conditions of the poor in England. It provided a visibility to the conditions of living of the poor and it connected cholera, and disease in general, with the social and economic conditions of people.

This evidence also prompted state intervention into the lives of the unhealthy urban poor. Cholera outbreaks proved critical in public health movements in Europe and in providing clean water for the public from the mid-nineteenth century. In 1848 the Public Health Act was introduced in Britain. For the first time, ratepayers' money was used to clean up the streets and slums where the urban poor lived in industrial Britain. In 1875 a new Public Health Act forced new local authorities to provide adequate drainage, sewage and water. In effect, cholera epidemics stimulated public health measures in Europe. The triumph of nineteenth–century medicine was prevention, not cure; and the key was the provision of clean water for military cantonments as well as for civilian popu-lations in metropolitan cities, such as London and Paris, initiated by sanitar-ians such as Chadwick and Louis-Rene Villerme, respectively.[39] Better supply and drainage of water was arguably the single most important cause of the great mortality improvements from the mid-century in Europe. Along with that was the rise in class consciousness among the working-class populations in Britain

who demanded better wages, better conditions of service and a general improvement in the their lifestyle. Of course, these were extended, and very partially for
many, especially for the civilian population in the tropical colonies. Chapters 6
and 7 will show that public health facilities in the colonies remained rudimentary throughout the nineteenth and early twentieth centuries.

The enduring influence of non-contagionism and contingent contagionism
in British medical thinking, the successes of the public health movements in
the nineteenth century and, finally, the prospective losses to the British imperial
economy in the event of quarantining its vessels from India – all of these contributed to the obstinate stance against quarantine in Europe by the British government. The contemporary British press depicted quarantine as useless, obnoxious
and even immoral. However, most of the (continental) medical authorities
participating in the 1851 conference believed that cholera transcended national
boundaries and required international cooperation.

Another factor that contributed to the European insistence on quarantine was
the general fear of the Orient – now a dark and insanitary place in the European
imagination. This was particularly manifested in Europe's attitude towards the
Haj pilgrimage (the annual pilgrimage of Muslims to Mecca). As the French and
British became dominant forces in North Africa and parts of the Arab world from
the early nineteenth century, they increasingly assumed that Muslim pilgrim-
ages to Mecca from Asia brought cholera to the Middle East, which then spread,
through mercantile networks, to mainland Europe. Therefore both the French
and the British agreed that all Haj pilgrims should be subjected to quarantine,
and this was largely enforced.[40] Although from 1851 there were discussions about
Haj pilgrims as carriers of cholera, matters became urgent when the first major
cholera outbreak took place in Mecca in 1865. European nations assumed that
the epidemic was caused by the increased number of pilgrims from the East. This
made the issue pressing in the 1866 Sanitary Conference held in Constantinople
where all European nations demanded sanitary regulation of Haj pilgrims, especially from the 'Orient'. European sanitary commissions now clearly saw Islam
and the Orient as the blocks against the European progressive sanitary movement. As a result, brutal detentions and various kinds of obstacle were placed in
the way of the pilgrims. The most important quarantine station was the lazaret at
al Tur in the Sinai Peninsula. The pilgrims were kept there for 15 to 20 days before
they were allowed to enter Egypt. In addition, anyone suspected of cholera was
kept for three to four months in extreme heat and cold. Such rigorous quarantine
in the eastern borders of the Mediterranean continued well into the twentieth
century (see Figure 5.2). Only in 1933 was quarantine made more lenient when
pilgrims were subjected to vaccination.

The subsequent sanitary conferences held in the following decades were
concerned with developing international consensus on the control of not only
cholera but also plague and yellow fever from Asia and Africa to Europe and
the Americas. Agreement among the European nations was limited, and fraught
negotiations continued into the twentieth century. One of the key reasons why

it took so long to reach any agreement on a standardized quarantine was because for each of the states that attended the International Sanitary Conferences, quarantine fulfilled diverse political and economic needs, as well as providing protection against disease.[41]

The sanitary conference in Rome in 1885 explains how political and economic agendas prohibited open negotiation about quarantine and why a consensus remained elusive for such a long time. This particular conference was concerned with enforcing the quarantine system on the Suez Canal. Briefly, in 1854 and 1856, French diplomat and administrator Ferdinand de Lesseps (1805–94) obtained a concession from Muhammad Sa'id Pasha, the ruler of Egypt, to create a company to construct a canal connecting the Mediterranean and the Red Sea, open to ships of all nations. The Suez Canal Company (Compagnie Universelle du Canal Maritime de Suez) was to operate the canal for 99 years from its opening. De Lesseps had used his friendly relationship with Sa'id, which he had developed while he was a French diplomat in Egypt during the 1830s. The Suez Canal Company was established in December 1858 and work started on the site where the Port Said was built in the spring of 1859. The canal was opened in 1869. Although initially planned by the French, the canal proved vital for British international trade as it provided an immensely faster maritime route. The shipping distance between London and Bombay was cut by 41 per cent and that between London and Calcutta by 32 per cent. British tonnage through Suez in 1880 comprised nearly 80 per cent of the total tonnage through Suez. Britain, which had acquired a major interest in and control of the Suez Canal in 1875, had wanted to abolish quarantine altogether and make do with medical inspections on the ships. Moreover, in 1883 the British had undertaken plans to construct a second canal parallel to the first to ease the heavy traffic on Suez. The British realized that any imposition of quarantine would hamper free passage through the canal, and if free passage of ships from India and Australia were not secured, the value of a second canal, for which the British government had anticipated investing as much as £8 million, would be drastically reduced.

Meanwhile, disaster struck in Egypt, the very site of the canal. In 1883, cholera broke out there. Around 50,000 Egyptians died in three months between June and September of 1883 and Europeans feared that the epidemic might soon spread to Europe. At this time the British government appointed surgeon general William G. Hunter to study the disease in Egypt. He took the non-contagionist position, insisted that cholera was indeed endemic to Egypt and denied that the disease was imported from India through Suez. His report concluded that the cholera outbreak was due to inadequate local sanitation and a poor water supply in Egypt.[42]

Around the same time, the French and the Germans also appointed their own cholera commissions in Egypt. Germ theory, expounded by Robert Koch and others, held that outbreaks of cholera were not exclusively endemic to certain areas but were universally communicable. Koch visited Egypt in 1883 as part of the commission and discovered *Vibrio cholerae* (the comma bacillus). He even travelled to Calcutta from Egypt to find the same strain of bacillus in a tank

THE QUARANTINE HOUSE AT PORT SAID, OCCUPIED BY EGYPTIAN SOLDIERS.

Figure 5.2 An engraving of a quarantine house for Egyptian soldiers suffering from the plague, Port Said, 1882 (by Kemp). Courtesy of the Wellcome Library, London.

in the imperial city to stress the contagionist position that cholera was indeed caused by the specific comma bacillus and had travelled from Calcutta to Egypt. In his famous lecture to the Imperial German Board of Health in Berlin in 1884, Koch asserted that cholera travelled from Indian ports, such as Bombay and Calcutta, through the Red Sea and the Suez Canal to Europe. Via this route it could travel from India to Egypt in 11 days, to Italy in 16 days and to Southern France in 18 days. The main carriers, in his opinion, were the crowded British 'coolie ships' carrying Indian indentured labourers to the West Indies through the Suez.[43] Naturally, this provided a new impetus behind the demand for quarantine, particularly in the Suez Canal.

The 1885 conference, held in the backdrop to these events, therefore focused particularly on the canal and the diminished distance between Europe and Asia. A proposal was offered relating to the inspection and quarantine of vessels from India intending to traverse the canal. Britain objected and stressed that the free movement of its trading vessels was paramount and that such precautions would be extremely costly. France, at the same time, protested against Britain's unilateral assumption of power over Egypt and wanted to limit the growing British dominance of the canal. France insisted on an independent international inspection of vessels entering Suez, knowing that British ships constituted the greatest proportion of the canal's traffic. They suggested that a quarantine station at Port

Said would provide a buffer for Europe against disease from 'the East'. The British delegation objected vehemently to this proposal and withdrew from the debates, so the conference was adjourned.

The first International Sanitary Convention was not signed until the 1892 conference, and even then the agreements were about a limited quarantine system. The delegates decided that all ships passing through the canal were to be classified according to whether they had a case of cholera onboard. They agreed to subject all pilgrim ships, even those without cases of cholera onboard, to quarantine. They also agreed that shipping vessels without any case of cholera onboard could pass through the canal, while suspected ships would be inspected by a doctor and a disinfecting machine would be placed onboard.

In 1903, on the initiative of the Italian government, the eleventh International Sanitary Conference was convened in Paris. One of the main achievements of this conference was the unification of earlier conventions (1892, 1893, 1894 and 1897) in the light of contemporary scientific knowledge, and the result was the International Sanitary Convention of 1903, which was ratified by most of the participating states in 1907. This was the first convention to introduce some measure of international uniformity against the importation of cholera and plague. It was superseded by the conventions relating to maritime traffic of 1912 and 1926, and the latter was modified in 1938 and again in 1944. Therefore the International Sanitary Conference was the precursor to future international health organizations.

International cooperation was difficult to achieve because quarantine policies are a reflection of issues other than a state's desire to protect itself from infectious disease from outside. Quarantine is linked to questions of political sovereignty and rights. At one level it represents questions about the degree to which a state chooses to intervene in the activities of its citizens, in restricting their movements in and out of the territories. It also plays an important role in the types of regulation that govern the movement of foreign persons or goods across its borders. In doing so the state also defines and determines borders or territories under its control. Thus quarantine systems were determined by imperial power and influence. European nations debated about adopting quarantine in Egypt, a region well beyond their nation state, but within their colonial control. On the other hand, in Asia, the growing imperial power, Japan, imposed its maritime quarantine against cholera in Busan (Korea) in 1879.[44] Quarantine in the modern era has been closely connected with the development of restrictive immigration policy and the protection and control of trade. It has been used as an effective tool in international relations and as a means to define the sovereignty of a state.

The several International Sanitary Conferences showed how infections and dissemination of disease were related to imperial economy, diplomacy and politics. They also highlighted the cultural and political differences between not only Europe and Asia but even among European colonial nations, and made clear that the spread of infectious diseases could not be checked without international cooperation. The conferences were Eurocentric and viewed Asia as backward, filthy, corrupt and ridden with disease, and Asian passengers, whether labourers or pilgrims, as the carriers of disease to Europe. They also contributed widely to the general public view in Europe, the United States and Australia that unless checked stringently, their nations were liable to be 'invaded' by Asian diseases. This had future resonance globally, particularly in twentieth century Australia, which enforced a strict quarantine system to protect the continent from Asian diseases, flora, fauna and people. The conferences also established the fact that diseases needed to be controlled in the colonies through intrusive measures, strengthening in the process the hegemony of Western therapeutic and sanitary systems. This consensus overrode the differences regarding contagionist and non-contagionist positions.

Yellow fever: From the slave trade to the Panama Canal

Yellow fever is a unique disease in imperial history as it links the Age of Commerce, the Age of Empire, with the Era of New Imperialism and Decolonization. It also reflects the various trajectories of imperial history in the Caribbean islands and in North and South America over 300 years. At the same time, yellow fever continues to cause high rates of morbidity and mortality on both sides of the Atlantic, in Africa and in South America, and is classified as a 're-emerging' disease by the WHO. For these reasons it has attracted substantial attention from historians.

The history of slavery is entrenched in the history of the global spread of yellow fever. Scientists and historians believe that the disease was introduced into the Americas from western Africa, along with its vector, the *Aedes aegypti* mosquito, by sailing vessels carrying slaves. The earliest references to yellow fever in the Caribbean and on the Eastern Seaboard of North America were in the mid-seventeenth century.[45] The first clear reference to yellow fever in the New World was in 1647 in Barbados.[46] Scholars have devoted a great deal of attention to the question of the 'origin' of yellow fever, stressing that slaves imported it to the New World from Africa.[47] They have also studied at length how the yellow fever virus mutated from its 'jungle' forms (from apes) to its human forms in the forests of Africa before it became a major human epidemic.[48] As historians, however, the important questions that we need to engage with in the case of yellow fever, as well as for other global epidemic diseases such as malaria, plague, cholera, SARS and HIV/AIDS, are not just their geographical, racial and zoological origins but how and why they devastated the populations that they affected.

Frequent warfare and movement of troops, the expansion of the sugar plantations and the great increase in the importation of slaves, and the activities of buccaneers allowed the disease to spread rapidly throughout the Atlantic region. By the 1690s the disease had spread to the American mainland; in the south to Brazil and in the north to Charleston, Boston and Philadelphia. In all of these places it caused high mortalities.

The spread of yellow fever was facilitated not only by human migration. Much like in the case of the spread of malaria and sleeping sickness in Africa in the early twentieth century (see Chapter 8), historians have identified the ecological changes caused by colonialism that led to the spread of yellow fever in the Americas from the seventeenth century. The sugar plantations of the region provided the ideal breeding grounds for *A. aegypti*. The mosquitoes could breed in the pots, kegs and barrels used for the storage of water in the plantations and in the ports and ships. The sugarcane juice was also believed to be a good source of food for them.[49]

I will explore three major episodes of yellow fever outbreaks, each of which was marked by a distinct phase of imperial history in the Atlantic region. The first was the outbreak of yellow fever epidemics in 1792–3, which affected Saint-Dominique, Havana and Philadelphia. The second episode was in the period between 1850 and 1880s and the last one from the 1890s to the opening of the Panama Canal in 1914.

The outbreaks of yellow fever in the 1790s took place in the middle of polit-ical turmoil, slave rebellion and colonial warfare. Two regions in particular – Saint-Dominique and Philadelphia – suffered the worst epidemic outbreaks, sparked by the Haitian Revolution. Between 1791 and 1804, the slave leaders in that French colony led the revolution against the French rulers and the slave system established on the island. The French attempts to repress the rebellion led to influxes of fresh troops who had no immunity to the disease. In 1792 around half of the French troops sent out to crush the rebellion in Saint-Dominique perished within a year to yellow fever. It soon spread to other Caribbean islands, such as Jamaica, which also saw a great movement of troops. Some 50,000 British soldiers and seamen died of yellow fever on this and other West Indian islands, such as Martinique, between 1793 and 1798.[50] The epidemics then spread to the north, to the southern plantations and port cities of the United States. The 1790s were a period of great expansion of the plantation system in the southern United States. Following the hiatus during the American Revolution, planters began to import slaves, and trade with the Caribbean region expanded, leading to increased traffic of ships, humans and goods, and increasingly crowded ports. This facili-tated the spread of yellow fever in the region.[51] Philadelphia suffered the most: the 1793 outbreak killed about 10 per cent of the city's population and thousands more fled. The epidemic of 1792–3 marked a severe blow to French power in the Caribbean and to the newly formed American republic, and provoked major attempts at establishing extensive quarantine systems in the port cities, such as Philadelphia.[52] Although the disease continued to affect populations in the Caribbean islands and southern American ports, such as New Orleans, by 1815 it ceased to be a major threat in the northern cities, mostly due to the quarantine and isolation measures adopted there.

The second episode in the history of yellow fever was in the Age of Empire, from the 1850s to the 1880s. This was the period when there was a major resur-gence of yellow fever epidemics with increased frequency and severity. It had also become a more global phenomenon with its appearance in New Orleans, Rio de Janeiro, Buenos Aires and Europe, most severely in Portugal in 1857. These were marked by two historical events: first, the expansion of plantation economies and urbanization in South America and the growth of informal empires (the large-scale investment of European, particularly British, capital chiefly in South American plantations; see Chapter 8, pp. 150–1); and second the rise of the germ theory of disease. By the middle of the nineteenth century the growth of rubber, coffee, cocoa and tobacco plantations in South American countries such as Brazil and Columbia led to massive migration of people, from rural to more heavily populated areas, but also from Europe, who lived in unhygienic industrial and urban slums. The growth of plantation economies and the expansion of road and rail networks also led to the clearing of forests and ecological changes, which helped the breeding of mosquitoes and spread of yellow fever. To add to this was the major warfare in South America in the 1870s, which mobilized troops and helped in the spread of disease.

In 1850 there was a major outbreak of yellow fever in Rio de Janeiro, which killed more than 4,000 people. Between 1850 and 1908 there were regular outbreaks of the disease in the city, which caused large-scale mortalities.[53] The epidemic outbreak in Buenos Aires in Argentina in 1871 claimed more than 15,000 people.

On the other side of these events, during the same period, germ theory and vaccine research were heavily employed in studying most tropical diseases (see Chapter 9). Pasteurian science brought a new enthusiasm about vaccine research on yellow fever in the 1880s. In the late nineteenth century there was rapid growth and influence of bacteriology in South American countries. The Brazilian capital city, Rio de Janeiro, became the ideal site for the hunt for the yellow fever virus in the 1880s. Brazilian scientist Domingos Freire suggested that the yellow fever bacillus (*Cryptococcus xantogenicus*) was present in the blood and internal organs. Following Pasteurian methods, he attenuated the Cryptococcus to produce a vaccine with which he undertook large-scale vaccinations in 1884–5.[54] Mexican doctor Carmona y Valle developed his own yellow fever vaccine in 1885. Following his lead, Julio Uricoechea, a Columbian physician, conducted vaccinations in Cúcuta in 1886.[55] Thus at a time when South American economies were closely linked to global capitalism, their diseases and pathogens too were subjected to international medical investigations and intervention.

The third episode was marked by the growing imperialist designs and influence of the United States in South America and the Caribbean. In 1892 there was a major epidemic in the Brazilian port city of Santos. In 1897 another serious outbreak took place in Havana, Cuba (at that time part of the Spanish Empire). This epidemic spread to the southern parts of the United States. In the United States, although outbreaks of yellow fever had mostly disappeared, fears of it reappearing, particularly from regions such as Cuba, remained strong. The United States at the same time had vested economic interests in the region. These paved the way for increasing American intervention in Cuban affairs and the 1897 outbreaks provided the impetus. A consensus emerged between the medical authorities and the federal government that the United States needed to end Spanish rule in Cuba. In 1898 the United States government declared war on Spain and occupied Cuba. As American medical officers believed that yellow fever thrived in the unhygienic living conditions of the city of Havana – 'the sanitary condition of Havana is a perpetual menace to the health of the people in the United States' – they started a large-scale cleaning programme for roads, houses and the harbour as part of the occupation.[56]

The discovery of the mosquito vector (*Anopheles aegypti*) of yellow fever was a culmination of two historical developments: the Cuban invasion by the United States and the laboratory investigations into yellow fever. Scientists in the United States keenly followed the bacteriological research on yellow fever in South America. The occupation of Cuba provided an ideal 'field' for American scientists to study the disease, and in 1900 Walter Reed, an American bacteriologist and pathologist, headed the US Army Commission in Cuba. He studied the disease closely and suggested that *Anopheles aegypti* was the vector for yellow fever. However, historians have suggested that this was in fact based on the findings of Cuban physician Carlos J. Finlay, who first proposed the mosquito vector theory in 1881.[57] Thus the intrusive sanitary surveillance in Cuba under American dictates, and the cleansing of the streets, ports and homes of Havana, had the cumulative effects of germ theory, vector theory and American expansionism in the region.

The final phase of this episode was the opening of the Panama Canal in 1914. The construction of the canal was first proposed in the 1870s as a means of connecting the Pacific and Atlantic oceans. The French had started construction in the 1880s but the digging of the canal, which took almost two decades, was plagued by fears and the actual spread of tropical diseases from Central to North America. The Americans took the initiative of digging the canal in 1904, soon after the discovery of the mosquito vector of yellow fever, in 1900. The key concern was the high death rate of workers, caused by yellow fever. The large-scale human migration and labour settlements around the canal also gave rise to fears in the United States about the spread of epidemic diseases from South America further north. The construction of the canal was disrupted by outbreaks of yellow fever and malaria around the region. During and after the opening of the canal, the United States government adopted quarantine measures and undertook sanitary interventions and field surveys in Central America and Cuba to prevent the spread of yellow fever and malaria. The American authorities used their Cuban sanitary experiences against yellow fever in dealing with the threat of the epidemics in the canal region. They appointed William Gorgas, who had served as the chief sanitation officer in Havana under Reed, in charge of the sanitation programme of the canal construction project in 1904. He started cleansing the canal zone, with a view to destroying mosquitoes. By 1906, as a result of his sanitation activities, yellow fever disappeared from the area. However, fears of the disease spreading north, particularly to the United States, remained strong.

The opening of the Panama Canal in 1914 led to the beginning of the involvement and intervention of the Rockefeller Foundation in South American public
health. Wyckliffe Rose, the director of the Foundation's International Health
Board, launched a major campaign to prevent the spread of yellow fever to the
Asia-Pacific region through the canal. The newly formed Yellow Fever Commission
visited Ecuador, Peru, Colombia and Venezuela in 1916.[58] Research on yellow
fever remained the main concern of American tropical medicine in the twentieth
century. It also paved the way for continued American intervention in the sanitation and health matters of Cuba until the 1930s.[59]

Each of these episodes of yellow fever epidemics were marked by major historical events: the Haitian Revolution; the expansion of the plantation economy
and informal empire; the American invasion of Cuba; and the construction of the
Panama Canal. These help us to understand the history of yellow fever within a
broader historical context. In turn, they also help us to understand that disease
is often a mirror of history, in this case imperial history. In the same way, disease
and epidemics tell us about our contemporary social and economic realities. The
history of the spread of yellow fever is a reflection of the long imperial history
of the Atlantic world and its contemporary legacies. Yellow fever has now disappeared from the northern ports and cities but it continues to be a major concern
in South America and Africa.

Conclusion

Disease and death were the undesirable but inescapable consequences of imperialism. Global migration of human beings, flora and fauna led to economic prosperity, cultural hybridity and social transformations. However, these came with
a cost, as diseases began to spread much more rapidly and widely than before.
The global spread of epidemics also transformed modern medicine and led to
the birth of modern public health. The cholera outbreaks in nineteenth-century
Britain shifted the attention of the state towards the living conditions of the
poor and to a new sanitarian regime, leading to the birth of British public health.
Cholera and yellow fever outbreaks also initiated the modern quarantine system
in Europe, Asia and the Americas. Quarantine continues to be a major mode of
controlling the spread of epidemic diseases globally. Debates have long existed
about the nature and extent of impact of disease in colonial migration and on
depopulation: whether malaria and venereal diseases existed in the Americas
before Spanish invasion, and whether Europe suffered from cholera epidemics
before the 1830s.[60]

The main divide in the medical and historical understanding of the global
spread of disease has been on whether disease spread due to human migration and
contact or due to the local conditions of the places where such diseases became
endemic. These have been broadly known as contagionist and non-contagionist

positions. This divide has also shaped the way in which historians have looked at this history. The question is important because not only do the issues of international politics and economy depend on it but also the questions of political and personal freedom and local healthcare.

The debates about the geographical and zoological origins and the contagiousness of diseases assume relevance in contemporary debates about the origins and spread of SARS, HIV/AIDS and swine flu. However, a preoccupation with the origins and carriers of diseases can reflect the cultural prejudice that shaped the debates of the International Sanitary Conferences and the sanitary surveillances of the nineteenth century. It can also obfuscate other vital issues, two of which we have seen in this chapter. First, disease was part of a larger complex of changes introduced by colonialism. The crucial factor in the transmission of disease, if any, was the colonial appropriation of territories, which started the cycle of death and depopulation. The occupation of land also facilitated the introduction of strict sanitarian surveillance by foreign powers on indigenous populations, as done by the Americans in Cuba against yellow fever or by the French and British in Egypt against cholera. This connects us to the second point, that disease did not affect everyone in the same way. Amerindians died most from diseases, whether they were of New World or Old World origins. Cholera spread among the urban poor in Europe in the nineteenth century due to their unhygienic living conditions. Africans suffered the highest mortalities in the eighteenth century. This social and economic discrepancy in the occurrences of disease and death is true for non-contagious diseases as well.

Notes

1 Emmanuel Le Roy Ladurie, 'A Concept: The Unification of the Globe by Disease', in Ladurie, *The Mind and Method of the Historian* (Brighton, 1981) pp. 28–83, p. 50.

2 N. Nunn and N. Qian, 'The Columbian Exchange: A History of Disease, Food, and Ideas', *Journal of Economic Perspectives,* 24 (2010), 163–88, pp. 163–4

3 Ibid, p. 165.

4 W.G. Lovell, 'Heavy Shadows and Black Night: Disease and Depopulation in Colonial Spanish America, *Annals of the Association of American Geographers,* 82 (1992), 426–43, pp. 429–30.

5 Nunn and Qian, 'The Columbian Exchange', p. 166.

6 Sheldon J. Watts, *Epidemics and History; Disease, Power and Imperialism* (New Haven, 1997), pp. 102–3.

7 M. Livi-Bacci, 'The Depopulation of Hispanic America after the Conquest', *Population and Development Review,* 32 (2004), 199–232, pp. 206–11

8 Lovell, 'Heavy Shadows and Black Night', p. 429.

9 Ibid.

10 Crosby, *Germs, Seeds & Animals: Studies in Ecological History* (New York, 1994), p. 5.

11 Crosby, *The Columbian Voyages*, p. 24

12 Ibid.

13 N.D. Cook, *Born to Die: Disease and the New World Conquest, 1492–1650* (Cambridge, 1998) p. 76.

14 Livi-Bacci, 'The Depopulation of Hispanic America after the Conquest', p. 217.

15 Ibid.

16 Stephen J. Kunitz, *Disease and Social Diversity: The European Impact on the Health of Non-Europeans* (Oxford, 1994), pp. 44–81. For population decline, see Figure 3–2.

17 James C. Riley, 'Mortality on Long-Distance Voyages in the Eighteenth Century', *Journal of Economic History*, 41 (1981), 651–6, p. 652.

18 Curtin, *Death by Migration*, p. 1.

19 Herbert S. Klein and Stanley L. Engerman, 'Long-Term Trends in African Mortality in the Transatlantic Slave Trade', *A Journal of Slave and Post-Slave Studies*, 18 (1997), 36–48.

20 Robin Haines and Ralph Shlomowitz, 'Explaining the Modern Mortality Decline: What can we Learn from Sea Voyages?', *Social History of Medicine*, 11 (1998), 15–48.

21 'An Account of the Cachexia Africana', *The Medical and Physical Journal*, 2 (1799), 171.

22 R. Shannon, *Practical Observations on the Operation and Effects of Certain Medicines, in the Prevention and Cure of Diseases to Which Europeans are Subject in Hot Climates, and in These Kingdoms* (London, 1793), p. 375.

23 K. Kiple and Virginia H. Kiple, 'Deficiency Diseases in the Caribbean', *Journal of Interdisciplinary History*, 11 (1980), 197–215, pp. 207–9.

24 Sheridan, 'The Slave Trade to Jamaica, 1702–1808' in B. W. Higman (ed.), *Trade, Government and Society: Caribbean History 1700–1920* (Kingston, 1983), p. 3.

25 P.C. Emmer, *The Dutch Slave Trade 1500–1850* (Oxford, 2006), p. 4.

26 Curtin, *Death by Migration*, pp. 1–6.

27 M.S. Anderson, *War and Society in Europe*; John Brewer, *The Sinews of Power: War, Money and the English State 1688–1783* (London, 1994).

28 See Mark Harrison, 'Medicine and the Management of Modern Warfare: An Introduction', Harrison, Roger Cooter and Steve Sturdy (eds), *Medicine and Modern Warfare* (Amsterdam, 1999), pp. 1–22.

29 Chakrabarti, *Materials and Medicine*, pp. 52–110.

30 Carlo M. Cipolla, *Fighting the Plague in Seventeenth-Century Italy* (Madison, 1981).

31 Ibid, pp. 51–88.

32 Myron Echenberg, *Africa in the Time of Cholera: A History of Pandemics from 1817 to the Present* (Cambridge, 2011).

33 For a broad historical overview of cholera, see Christopher Hamlin, *Cholera: The Biography* (Oxford, 2009).

34 Charlotte E. Henze, *Disease, Health Care and Government in Late Imperial Russia; Life and Death on the Volga* (Abingdon & New York, 2011), pp. 21–66.

35 Thomas Shapter, *The History of the Cholera in Exeter in 1832* (London, 1849), p. 178.

36 Ernest Hart, 'The West Indies as a Health Resort: Medical Notes of a Short Cruise Among the Islands', *British Medical Journal*, 1920 (16 October 1897), 1097–9, p. 1098.

37 Leonard Rogers, 'The Conditions Influencing the Incidence and Spread of Cholera in India', *Proceedings of the Royal Society of Medicine*, 19 (1926), 59–93, p. 59.

38 Edwin Chadwick, *Report to Her Majesty's Principal Secretary of State for the Home Department, From the Poor Law Commissioners, on an Inquiry into the Sanitary Condition of the Labouring Population of Great Britain* (London, 1842).

39 Simon Szreter, 'Economic Growth, Disruption, Deprivation, Disease, and Death: On the Importance of the Politics of Public Health for Development', *Population and Development Review*, 23 (1997), pp. 693–728.

40 William R. Roff, 'Sanitation and Security: The Imperial Powers and the Nineteenth-Century Hajj', *Arabian Studies*, 6 (1982), 143–60. For a detailed study of the experiences of quarantine for the Haj pilgrims from South Asia, see Saurabh Mishra, *Pilgrimage, Politics, and Pestilence; The Haj from the Indian Subcontinent, 1860–1920* (Delhi, 2011).

41 Huber, 'The Unification of the Globe by Disease'.

42 Mariko Ogawa. 'Uneasy Bedfellows: Science and Politics in the Refutation of Koch's Bacterial Theory of Cholera', *Bulletin of the History of Medicine* 74 (2000), 671–707.

43 Robert Koch, 'An Address on Cholera and its Bacillus, Delivered before the Imperial German Board of Health, at Berlin', *BMJ*, 1236 (6 September 1884), 453–9, p. 458.

44 Jeong-Ran Kim, 'The Borderline of "Empire": Japanese Maritime Quarantine in Busan c. 1876–1910', *Medical History*, 57 (2013), 226–48.

45 David Geggus, 'Yellow Fever in the 1790s: The British Army in Occupied Saint Dominique', *Medical History*, 23 (1979), 38–58, p. 38.

46 Kenneth F. Kiple, *The Caribbean Slave: A Biological History* (Cambridge, 1984) p. 20.

47 Kiple, *The Caribbean Slave*, pp. 17–20; J.E. Bryant, E.C. Holmes, A.D.T. Barrett 'Out of Africa: A Molecular Perspective on the Introduction of Yellow Fever Virus into the Americas,' *PLoS Pathogens*, 3 (2007) doi:10.1371/journal.ppat.0030075.

48 T.P. Barrett, Monath 'Epidemiology and Ecology of Yellow Fever Virus', *Advances in Virus Research*, 61 (2003), 291–315; Kiple, 'Response to Sheldon Watts, "Yellow Fever Immunities in West Africa and the Americas in the Age of Slavery and Beyond: A Reappraisal" ', *Journal of Social History*, 34 (2001), 969–74. For a critical history of how the concept of 'jungle yellow fever' came to be used and accepted, see Emilio Quevedo et al., 'Knowledge and Power: The Asymmetry of Interests of Colombian and Rockefeller Doctors in the

Construction of the Concept of Jungle Yellow Fever, 1907–1938', *Canadian Bulletin of Medical History*, 25 (2008), 71–109

49 John R. McNeill, *Mosquito Empires: Ecology and War in the Greater Caribbean, 1620–1914* (Cambridge, 2010), pp. 47–9.

50 Benjamin Moseley, *A Treatise on Tropical Diseases* (London, 1789), pp. 119–53; Geggus, 'Yellow Fever in the 1790s.

51 Peter McCandless, *Slavery, Disease, and Suffering in the Southern Lowcountry* (Cambridge, 2011) pp. 78–80.

52 Harrison, *Contagion: How Commerce has Spread Disease* (New Haven, London, 2012), pp. 50–5.

53 Teresa Meade, '"Civilizing Rio de Janeiro": The Public Health Campaign and the Riot of 1904', *Journal of Social History*, 20 (1986), 301–22, pp. 305–6.

54 Ilana Löwy, 'Yellow Fever in Rio de Janeiro and the Pasteur Institute Mission (1901–1905), The Transfer of Science to the Periphery', *Medical History*, 34 (1990), 144–63, p. 147.

55 Mónica García, 'Producing Knowledge about Tropical Fevers in the Andes: Preventive Inoculations and Yellow Fever in Colombia, 1880–1890', *Social History of Medicine*, 25 (2012), 830–847, p. 837.

56 Quoted from Mariola Espinosa, 'The Threat from Havana: Southern Public Health, Yellow Fever, and the U.S. Intervention in the Cuban Struggle for Independence, 1878–1898', *The Journal of Southern History*, 77 (2006), 541–68, p. 551.

57 Löwy, 'Yellow Fever in Rio de Janeiro', pp. 144–5.

58 Marcos Cueto, *The Value of Health: A History of the Pan American Health Organization* (Washington, DC, 2007).

59 Espinosa, *Epidemic Invasions: Yellow Fever and the Limits of Cuban Independence, 1878–1930* (Chicago, 2009).

60 For the question of the origins of cholera, see Chakrabarti, *Bacteriology in British India; Laboratory Medicine and the Tropics* (Rochester, NY, 2012) pp. 180–94.

6

Western medicine in colonial India

European medicine was introduced to India through a gradual process from the late seventeenth century. As European traders started their commercial activities in the Indian Ocean region, they brought European drugs with them. This was also an interactive process, as Europeans, starting with the Portuguese, took an interest in the drugs they found in the local markets and the medicinal plants they found in the forests and gardens, and they studied the classical and vernacular medical texts of India in search of local names and uses for these plants and drugs.

The distinctive feature of the history of medicine in colonial India is that it followed the various stages of medicine and empire that we have referred to in this book. Since it is possible to study the history of imperialism and medicine in India throughout the colonial period (1600–1950), a study of the history of medicine in India helps us to understand the links between the various stages of colonial medicine. Contact with Iberian parts of Europe started in India very early, with Vasco da Gama's arrival in Calicut on the south-western coast of India in 1498. The Portuguese merchants and the Jesuit missionaries introduced European drugs into India and incorporated Indian drugs into their own medicine. The British, French and Danish following the Portuguese traders had arrived in India by the seventeenth century and a wider medical encounter took place between India and Europe. This was not an exchange borne so much out of intellectual interest and sharing; Europeans in the eighteenth century were seeking to establish their trading dominance and monopoly, and to sell their own products in Asian markets and stop the flow of bullion from Europe to Asia, which had started with the spice trade.[1] The British became the dominant colonial power in India in the eighteenth century and, along with establishing their territorial power and market monopolies, they introduced their own medical institutions, practices and drugs and marginalized indigenous ones. In the nineteenth century, with the consolidation of the empire, colonial medicine was firmly established through colonial medical services, hospitals, public health institutions, medical colleges and vaccination campaigns. In the early twentieth century, as a response to these colonial medical interventions, indigenous medical practitioners and doctors reorganized and revived Indian medical traditions, which I will address in Chapter 10. To avoid repetition, this chapter will focus on the period starting from the late eighteenth century. For the early history of medicine in India in the

Age of Commerce, see chapters 1 and 2, and for indigenous responses to Western medicine, /see Chapter 10.

Colonization of India

India was colonized through a gradual historical process. Following Vasco da Gama's arrival, the Portuguese set up colonies along the west coast of India, in Goa, Calicut and Bombay, in the sixteenth century. Vasco da Gama's voyage to India (1498–9) brought scurvy to Indian shores, along with Portuguese plunder. The English arrived in the early seventeenth century, following the formation of the EEIC in 1600. The French followed them. Gradually, through the late seventeenth and early eighteenth centuries, Portuguese power declined in India and the main colonial competition was between the French and the English. The English colonial bases were in Bengal in the east, Madras in the south and Bombay in the west. The French had scattered colonies on the Coromandel and Malabar coasts, with the main base at Pondicherry (south of Madras), Chandernagore in Bengal and Mahe in the Malabar Coast. Throughout the mid-eighteenth century, the two European nations fought wars in India and the British gained decisive advantage over the French in the battle of Wandiwash in 1760. Around the same time, the British also made the first major territorial gain in India by occupying Bengal in 1757, defeating the Nawab of Bengal, Siraj-ud-Daulah in the battle of Plassey. Bengal was the richest province in the late-Mughal period. Control over its revenue proved vital for the EEIC in enhancing its commercial profits, and financing the wars against the French and the rulers of Mysore in the south the eighteenth century.[2] Over the latter half of the eighteenth century and the early nineteenth century, the EEIC annexed major territorial regions in the south and west. The occupation of the vast and agriculturally rich regions of the Punjab in 1849 and of Awadh in 1856 in the north completed the formation of the Indian Empire and the EEIC became the sovereign rulers of the country.

The eighteenth century holds the key to the British colonization of India. This was the period of the gradual transformation of India from a maritime colony to a territorial one. Moreover, the pivotal location of India in the Indian Ocean between South East and West Asia provides a clue to the broader changes in the region during this period. Thus the colonization of India has attracted a great deal of historical attention from both territorial and maritime perspectives. In the 1960s, historians such as Irfan Habib and M. Athar Ali, who were influenced by Marxist writings, saw this transformation as a broad structural change – the decline of the feudal agrarian economy of the Mughals and the rise of maritime economies on the coasts. For them this was broadly a period of economic decline as the mainstay of the Indian economy, which was agricultural revenue, gave way to smaller coastal economies.[3] Later scholars who viewed imperialism much more as a global exchange stressed that the overall relationship between the European corporate enterprises and the Indian and other Asian authorities was often an amicable one based on a perception of mutual advantage, which often led to an increase in trade and regional prosperity.[4] They have argued that although the Indian Ocean area was integrated into European trade, it retained its regional and cultural autonomy.[5]

Scholars have predominantly agreed that from 1750, European trade became dominant in India and in the region. This was also the period of the incorporation of Indian trade into the European world economy. The most conspicuous feature in this complex history of the early eighteenth-century Indian Ocean was the rise of the European traders.[6] This was reflected in the decline of traditional ports, such as Masulipatnam and Surat, and the concomitant rise of European ones, such as Madras and Bombay.[7] Even beyond India, in South East Asia, similar trends occurred in the eighteenth century. In the Malay and Thai ports by the 1760s, British 'country traders' were much more organized in syndicates and were also much more effective than the indigenous merchants.[8] Later studies have connected the history of maritime trade with the economic history of the hinterland to show how these changes transformed local economies and politics in Asia. For example, European trading companies denied the indigenous rulers in Arcot (an important traditional and indigenous commercial town and the seat of the Nawabs of Carnatic), which was situated inland, west of the coastal English town of Madras, vital commercial and tax revenues, thereby weakening them and reducing their regional prosperity.[9]

This rule of the EEIC over India in the nineteenth century was cut short by the revolt of 1857. The revolt, which started as a military mutiny among the Indian soldiers (*sepoys*) of the EEIC, soon became an agrarian and peasant revolt throughout eastern, central and northern India and shook the company establishment to its core.[10] It led to the abolition of EEIC rule and the incorporation of the Indian Empire under the direct rule of the British Crown. Late nineteenth-century India under the rule of Queen Victoria underwent major phases of modernization and economic transformation, with the introduction of railways and university education, and the establishment of scientific bodies, such as the Geological Survey of India (1851), the Indian Meteorological Department (1875) and the Indian Agricultural Research Institute (1905). The Indian economy in this period was also linked closely to the global economy, and the British exploited Indian mineral resources (particularly coal) and introduced large-scale commercial and plantation-based agriculture (jute, indigo, cotton, tea, sugarcane and coffee), which were sold on the global market. It was also the period when significant regions, such as Burma (with its huge resources of timber and oil), were annexed to the British Empire in 1886.

From the late nineteenth century, anti-British and anti-imperial movements started in India. Initially the Western-educated Indian middle class led these but by the early twentieth century they had spread to rural areas and took the shape of peasant and labour insurgencies against both British rule and the local landlords and elites. M.K. Gandhi and the Indian National Congress (INC) sought to bring these diverse movements together into a nationwide nationalist movement against British rule, although much of the subaltern struggles remained outside the nationalist purview and were often opposed to the nationalist agendas. India gained independence from colonial rule in 1947.

In chapters 1 and 2 we saw how European medicine was introduced in India in the Age of Commerce by a range of people: traders, missionaries, surgeons and travellers. Medical encounters and exchanges took place in a variety of places, such as markets, botanical gardens, missions, cantonments and hospitals. As a result of these, European medicine was transformed as apothecaries in Britain regularly stocked drugs of Indian or Indian Ocean origin, physicians there prescribed drugs acquired from these regions and British medical texts referred to these new drugs.

At the same time, with the growing dominance of British power in India by the end of the eighteenth century, European medicines began to play a more significant role in the medical practices in India. European military and commercial establishments depended mainly on drugs imported from Europe in this period. While Oriental drugs were imported into Britain, there was also increasing importation of European drugs into Asia.[11] The hospitals that the English built in India were the institutions through which European medicine became dominant in India from the late eighteenth century. The history of British hospitals in India also reflects the transformation of India from a commercial colony to part of the British Empire. It thus helps us to understand the establishment of colonial medicine in India.

European hospitals and colonial medicine in India

The EEIC established several permanent and semipermanent hospitals in India from the seventeenth century. Unlike the hospitals in contemporary Europe, which were products of predominantly humanitarian and military concerns of the state, these were shaped by the commercial and territorial interests of the trading companies. The English hospitals in India were located within the company's forts and were close to the factories (a European trading establishment at a foreign port or mart), local towns, ports and markets. The surgeons at these hospitals often procured their medicines from the local markets and from indigenous doctors. The first English hospital in India was established by the EEIC in Madras in 1664 inside Fort St George. The port of Madras was founded by the EEIC in 1639 and was the main trading and military post of the British in eighteenth-century South India. In Bombay on the west coast, the first English hospital was established in 1677. Faced with high mortality rates among the soldiers, the Old Court of Judicature of Bombay was transformed into a hospital. This was replaced in 1733 with a new building near the marine yard.[12] In Bengal, the company built its first hospital within its fort (Fort William) in Calcutta in 1707, at a time when it was expanding its base following the death of the Mughal emperor Aurangzeb in Delhi the same year.[13]

The eighteenth century in India was marked by continuous warfare, mostly among European military powers seeking to expand their colonial posts and territories. They also fought with regional Indian rulers who were themselves seeking to expand and consolidate their dominions following the fall of the Mughal Empire. Most of the wars were in southern, eastern and western parts of India. In South India the main wars were the First and Second Carnatic Wars (1744–8 and 1749–54) between the French and the British, and the four Mysore Wars (1767–9, 1780–4, 1790–2 and 1799) between the British and the rulers of Mysore, Hyder Ali and his son Tipu Sultan. Both of these sets of wars ended in British victory and in the establishment of British political supremacy in southern India. In the east the British fought two successful battles against the Nawabs (rulers) of Bengal in Plassey (1757) and Buxar (1764) to establish their dominance. In western India the EEIC fought continuous wars against the Marathas (a growing regional power that sought to extend its control following the decline of Mughal power in the region) between 1775 and 1818 (First Anglo-Maratha War, 1775–83; Second Anglo-Maratha War, 1803–5; Third Anglo-Maratha War, 1817–18). These ended with the defeat of the Marathas and the British gaining control over major parts of western and central India.[14] These wars meant that the British military establishment expanded massively. There were large numbers of military casualties and injuries, and consequently the medical establishment, particularly hospitals, expanded in this period and became integral to the EEIC forces.

As battles raged in distant provinces, the bodies of wounded and dead soldiers piled up and the hospitals in the cities, such as Madras and Bombay, were unable to cope. The EEIC established smaller regimental hospitals and field hospitals which were makeshift institutions that served in the battlefields. In western India during the Maratha campaigns, the British built field hospitals in Ajanta. Although it suffered from a lack of doctors or medical facilities, it cared for the thousands of EEIC soldiers (both British and Indian sepoys) wounded during the decisive battle of 1803.[15] In south India, during the Carnatic Wars, the EEIC established similar field hospitals.

From the late eighteenth century, with the growth of the empire, these hospitals became part of the British colonizing process and grand representations of imperial power. In most cases the expansion of English military hospitals followed important military victories and territorial expansion rather than preceding them in preparation for the campaigns. The hospital in Madras developed with the expansion of the British territories in India.[16] From the 1740s the EEIC was involved in military conflicts with the French in southern India in order to gain territorial and trading supremacy. The English wars with the French brought unprecedented numbers of European troops, military provisions and artillery to India.[17] In 1746 the French army occupied Madras for three years, and the hospital at fort St George was moved to Fort St David near Cuddalore, along with the entire British administration.[18] For the next few years during the subsequent Carnatic Wars, the hospital moved from site to site. The Treaty of Aix-la-Chapelle declared peace in Europe in 1748, but hostilities between the French and the

English continued in the colonies and Madras was restored to the latter only in 1749.[19] Following the war with the French, the EEIC decided to expand the Madras hospital. A naval hospital was also established in Madras at this time.

In 1752 the Court of Directors (which governed the EEIC from London) passed orders to expand the hospital in Madras to ensure proper treatment of the sick and wounded, regular attendance of surgeons and procedures for charging the sick and wounded soldiers. However, before these plans could be carried out, war broke out again with the French. In December 1758 the French laid siege on Fort St George and plans for the new hospital were postponed. In January 1760 the English achieved their most decisive victory in their territorial struggle against the French, in Wandiwash in South India. Following victory there, the military authorities decided that the medical logistics of the army needed more attention. Centralization of the scattered medical arrangements seemed necessary and the military authority decided that a general hospital was to be maintained in Madras, and all of the officers in provincial towns such as Wandiwash, Arcot and Chenglepet would send all of their wounded and sick from those garrisons to Madras.

However, the first Mysore War (1766–9) between the EEIC and Hyder Ali, the ruler of Mysore, led to the postponement of plans to expand the facilities of the Madras hospital. Expansion could only be undertaken once the wars had ended in British victory (temporary in the case of the Anglo-Mysore wars) and Madras had been secured. Finally, in 1772, Madras General Hospital was established, initially only to care for the company's servants.

The expansion of the medical and military establishment also brought about administrative changes within hospital management in India. From the 1760s, as the EEIC gained ascendancy over the French in southern India, the military establishment sought to centralize the medical administration of the province within the city of Madras, which had become the centre of English power. Madras had also become the main trading port of the Coromandel Coast, leading to the decline of traditional ports, such as Masulipatnam.[20] Militarily too, Madras became the centre of English power in south India. Thus the English hospital in Madras became the main medical institution in south India by the 1770s. In the 1790s, as the company army had become the supreme military power in southern India, it took over control of the hospitals from the hands of the doctors and surgeons.

In 1775, soon after the establishment of the Madras General Hospital, hospital boards were formed to administer the running of EEIC hospitals in India. These were run initially by physicians, who had autonomy in medical matters such as recruitment, promotion and supplies. However, this autonomy was lost with the establishment of the Medical Department in 1786, which was run by army officers. The surgeons lost the autonomy they had in using diverse forms of medicines and medical aides in running the hospital and in their medical practice generally. From 1802, following victory over the rulers of Mysore and the annexation of large parts of southern India, the military authorities curbed the role and autonomy of the surgeons in selecting local medicines, appointing local aides and treating different types of patient.[21] The hospitals now depended increasingly on medicines sent from Europe. They also severely curbed the roles of the indigenous medical practitioners ('black doctors') who served in the European hospitals, converting them into assistants and 'dressers' (those who dressed wounds and put bandages). Thus, with the establishment of colonial rule in India in the Age of Empire, European hospitals lost the earlier eclectic character that they had had in the early eighteenth century and became firmly imperial institutions.

At the same time, as the mercantilist monopoly of the EEIC grew in the region, it promoted the sale of English drugs in the colonial market towns. For example, in 1800, Thomas Evans, a surgeon of the EEIC establishment in Madras, proposed the establishment of an English dispensatory in the black town of Madras. Evans believed that an English dispensatory selling English drugs would help to promote European medicine among Indians, thereby earning profit for the company. It would also help to suppress all 'irregular practitioners and vendors of medicine in Black Town'. The Madras government adopted Evans' proposals enthusiastically as they would 'ultimately increase the demand for European Medicines to an object of commercial importance'. They saw the dispensatory as a solution to stopping the 'spurious' sale of local drugs as well as a way to 'introduce into India the knowledge of European Pharmacy, to remove by degrees the errors of ignorance and to extend the practice of scientific adepts'.[22] The dispensatory became part of the commercial and political monopoly of the EEIC in Madras.

In the nineteenth century the colonial hospitals also became the sites of medical training for Indians. The Madras Medical School, the first of its kind to be established in India, was established as an adjunct to Madras General Hospital in 1835 to develop a new class of local medical assistant. The school was established following the instructions in 1802 by the Medical Department to produce more dressers and medical assistants from among the locals. The school was established with the intention 'of having a well-instructed set of subordinates in the medical department of the public service'.[23] British medical and colonial officials saw the medical school as an institution, which would help to spread the benefits of British medicine and science across India. The medical authorities suggested that trained Indians would become 'warm advocates ... for the adoption of our superior modes, and desirous of rescuing their countrymen from the ignorance and empiricism of their own practitioners'.[24] Instructions in the school were to be given only in English, so that Indians could be 'weaned from the study of their own authors, from whom little but error and superstition is to be gained'.[25] This indicated a break from the traditions that marked the earlier period of acquiring medical knowledge by studying local vernacular texts and learning local languages. In 1850 the Madras Medical School became the Madras Medical College. It also trained medical students from other colonies.[26] The EEIC established similar medical colleges attached to its hospitals in other cities, such as the Grant Medical College in Bombay (1845) and the Medical College of Bengal in Calcutta (1835).

By the middle of the nineteenth century, as British colonial rule was firmly established in different parts of India, European hospitals and dispensatories became
the symbols of British rule and cultural superiority in India (see Figure 6.1).

Colonialism and public health in India

In the second half of the nineteenth century, as the British became the rulers
of India and peace was restored in most parts of the country, colonial medicine
moved from the forts, battlefields, field hospitals and barracks to the towns, provinces, streets and everyday lives of the common people. Colonial medical services
developed along with the growth of the colonial power of the EEIC. The mainstay of the British medical administration in nineteenth- and twentieth-century
India was the Indian Medical Service (IMS).[27] This was a unique colonial body of
medical men that emerged from the eighteenth-century military traditions of the
EEIC and controlled both civilian and military medical services. By the middle
of the nineteenth century, British medical men and members of the IMS became
leaders in the study of diseases of the tropics. As the company's territorial control
over the Indian subcontinent expanded by the middle of the nineteenth century,
British doctors belonging to the military establishment enjoyed large civilian
practices as well, in the hospitals, dispensaries and research establishments in

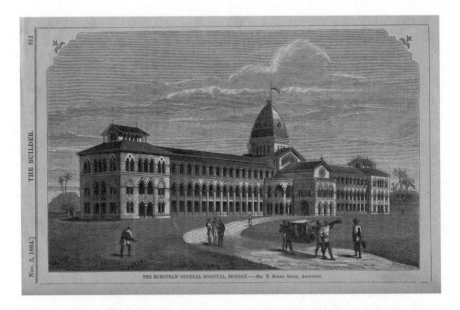

Figure 6.1 An engraving of the European General Hospital, Bombay, India (by
J.H. Metcalfe, 1864). Courtesy of the Wellcome Library, London.

various parts of the country. This dual role of the IMS was unique and crucial to its power and hegemony within the colony.

With the assumption of Crown rule from 1858, public health in India became an important concern of the government, thereby increasing the sphere of activity of the IMS officers beyond hospitals and military barracks. The establishment of the colonial medical service took colonial medicine into the provinces and smaller towns, and into more direct contact with Indian society. The IMS officers were those who had the onus of this major undertaking, which further enhanced their influence and status in India. Along with that the colonial government sought to retain the strong British character of the service. The entrance exams for recruitment into the IMS were held only in England and candidates were almost entirely trained in British universities. In Indian universities, on the other hand, training in medical science and research remained rudimentary. Very few Indians joined the profession until the turn of the century.

A transformation in colonial medicine took place in India in the nineteenth century, and this was also evident from the late nineteenth century in Africa. European medicine was now no longer just 'medicine of the hot climates', following the predominant eighteenth-century concern with saving and preserving the health of European troops and settlers in the tropics. Colonial medicine was now extended to large sections of the indigenous population. This meant that now there were two new phenomena: the imposition of medical ideas and practices of Europe among a wider and diverse population; and the negotiations and resistances from below in response to that.

The process of introduction of public health among the indigenous population in India started in the mid-nineteenth century. The Revolt of 1857 led to the abolition of the EEIC's rule in India and the British Crown then adopted direct responsibility for the governance of India. To appease the radical and revolutionary sentiments expressed by Indians during the revolt, the British Crown declared that colonial administration was responsible for the 'moral and material' wellbeing of its Indian subjects. Public health became one of the key features of this new colonial investment in Indian wellbeing. A Royal Commission was formed in 1859 to enquire into the sanitary conditions of the British army. Sanitary commissioners were appointed from the 1860s in each province, such as Bengal, Madras, the Punjab and Bombay, to oversee the health of the 'general population'.[28]

All major Indian cities, such as Madras, Calcutta and Bombay, underwent transition from commercial ports to imperial centres of administration in the nineteenth century. These three cities had been established by the EEIC as their ports and military bases in the seventeenth century. With the growing power of the company, these cities in the nineteenth century became the main urban centres of India, with huge migration and settlement of Indian labourers, clerks and traders. These eclipsed the prestige and economic significance of the precolonial cities, such as Delhi, Lucknow, Aurangabad and Vijaynagara. The colonial cities also became the administrative headquarters of the respective presidencies and the centres of imperial authority and power.[29]

One characteristic of the eighteenth-century colonial port cities was retained in the nineteenth century. In the eighteenth century, these were divided into 'white' and 'black' towns, with the European population living in the former areas, which were sparsely populated and had better civic amenities. Indians lived in the densely populated black towns.[30] In the nineteenth century, as the British introduced public health facilities in the Indian cities, they reflected this skewed pattern of settlement – the British parts enjoyed much better facilities of sanitation and drainage than the crowded black towns and slums.[31]

The responsibility for the sanitary administration included urban refuse and waste collection, along with long-term epidemic control measures – for cholera and plague in the nineteenth century and malaria in the twentieth.[32] One of the major initiatives undertaken by the British in Calcutta was the purification and supply of Hughly River water for domestic use in the city. Water supply in Calcutta had become an important concern by the mid-nineteenth century. In 1847, F.W. Simms, a British engineer, for the first time proposed a plan to supply treated water to the residents of the city. He identified the river at Pultaghat, 18 miles north of Fort William, as an ideal site for collecting water from the Hughly River and then transporting it to Calcutta via open canals.[33] Work started under Lord Dalhousie and in 1868 the waterworks at Pulta were built. By 1870 the major streets of the city were piped, providing water to the houses. Around the same time the sewage system of the city was established.[34] Pulta is still the major water supply system for the city of Calcutta (now Kolkata).

The plague outbreak in Bombay in 1896–7 and the measures to control are one of the most noted episodes of colonial public health policies in India. In September 1896, plague broke out in the densely populated Mandvi region of Bombay. To control it the British government passed the Epidemic Diseases Act in 1897. The governor general of India conferred special powers upon local authorities to implement the necessary measures for the control of epidemics. The Bombay government adopted policies of forceful segregation of infected persons, and often used ruthless policies of disinfection, evacuation and demolition of infected places and inspection of the private dwellings of Indians. This raised alarm among local populations, leading to riots in some areas. Plague proved to be a catalyst for the colonial administration to undertake wider sanitary improvements. It exposed to the authorities, in ways that cholera had in Britain in the 1840s, the living conditions of the urban poor in Bombay. The Bombay Improvement Trust was set up in 1898 to tackle the problem. Some investments were also made in improving working-class housing. The outbreak also led to the setting up of the Indian Plague Commission, which toured all over India and submitted its report in 1900. The commission recommended the reorganization of preventive health in India, the establishment of medical research laboratories and the modernization of Indian hospitals.[35]

The Bombay plague appears as a watershed in the history of colonial Indian public health. On the one hand, it ushered in an institutional change in public health in India through the recommendations of the Indian Plague Commission. The colonial government realized the need for laboratory research for the prevention of epidemic diseases and established plague research laboratories in Bombay.[36] On the other, the outbreak led to the implementation of the first intensive local sanitary regulations and measures in India.[37] It was also the first epidemic outbreak that triggered political and social strife between the colonial sanitary regime and Indians.[38] David Arnold has described plague and the subsequent interventionist medical and sanitary measures of the colonial government in Bombay as an 'assault on the body'.[39] According to him the Bombay plague was the critical moment in the colonization of the body: 'If there was a single moment when Western medicine in India appeared to have turned a corner, to become something more than just colonial medicine, that moment surely came in the aftermath of the first phase of the plague epidemic.'[40]

The other major episode of public health in India was the smallpox vaccinations. Smallpox was a major health problem in India with fatality rates ranging from 20 to 50 per cent. British troops suffered large numbers of fatalities and vaccination was started by the EEIC in the early nineteenth century. These were initially conducted among the EEIC troops and the European population. In the 1820s, Lord Elphinstone, the governor of Bombay, for the first time proposed the introduction of smallpox vaccination in rural parts of the Bombay presidency. Vaccination campaigns remained sporadic until the 1860s. The epidemic outbreaks of smallpox in 1865 in Calcutta, in 1876 in Bombay and in 1884 in Madras led to the respective provincial governments to introduce compulsory vaccination acts in these cities.[41] Around this time, smallpox vaccination also became a matter of public debate and resistance in India. Many high-caste Hindus objected to arm-to-arm vaccination using low-caste or untouchable vaccinators. Others protested against the painful consequences of the arm-to-arm vaccinations.[42]

In the early phases of smallpox vaccination, resistance to the vaccine forced the colonial authorities to adopt more conciliatory modes of popularizing the measures. They searched for indigenous roots of the technique in ancient India in Sanskrit texts. They also employed more Indian vaccinators and agents as the mainstay of the campaigns.[43] In the late nineteenth century, as vaccination was made compulsory, the colonial authorities showed much less appreciation of indigenous variolation methods and were far more keen to establish a 'medical monopoly' rather than cultural pluralism. This evoked a range of reactions from the subaltern sections of society; fear, anxiety and active opposition. At the same time the British were able to win the support of the Indian urban elites, who themselves urged for active vaccination among the lower classes.[44] While opposition to smallpox vaccination or vaccination in general in colonial India can be seen as instances of popular and subaltern agency and resistance, this was not just an Indian phenomenon. Smallpox vaccination in Britain was unpopular and had encountered a great deal of popular protest in the nineteenth century.[45]

By the end of the nineteenth century, Pasteurian science and germ theory introduced a new phase in colonial medicine. It led to the establishment of bacteriological laboratories and the introduction of prophylactic vaccination for diseases such as cholera, plague, typhoid and rabies. The British established several bacteriological laboratories, such as the Imperial Bacteriological Laboratory in Poona (1890), the Bacteriological Laboratory in Agra (1892), the Plague Research Laboratory in Bombay (1896), the Pasteur Institutes of India in Kasauli (1900), Coonoor (1907), Rangoon (1916), Shillong (1917) and Calcutta (1924), and the Central Research Institute (CRI) at Kasauli in 1905. The same year the Bacteriological Department, a special cadre of the IMS officers, was formed to staff the newly established laboratories. In 1911 the Indian government set up the Indian Research Fund Association to sponsor and coordinate medical research in the country. Originally catering predominantly to the European population, over time the Pasteur Institutes treated mostly Indian patients from various parts of the country.[46]

In the late nineteenth century, India became a popular site for bacteriological experimentation and vaccination campaigns. Several outbreaks of cholera and plague provided opportunities to scientists from Europe and Japan to conduct vaccination experiments there. Russian-born scientist Waldemar Haffkine came to India in 1893 to conduct experiments with his cholera vaccine. He also developed a vaccine against plague in 1897 when it broke out in Bombay. Several other prominent European bacteriologists, such as Robert Koch, Lauder-Brunton, Leonard Rogers, Paul-Louis Simond, A. Lustig and Alexandre Yersin, visited India during this time to conduct bacteriological research. Vaccines produced in Indian laboratories were used for civilian vaccination, as well as the vaccination of the British army serving in different parts of the world, particularly during the First World War. Mass vaccination campaigns against plague, cholera and rabies were conducted in different parts of India – in the cities, villages and pilgrim sites. Vaccinations were also conducted in the Pasteur Institutes in the remote hill stations where Indians travelled to be inoculated, particularly against rabies. Millions of people were vaccinated in India in the twentieth century, often without the kind of popular resistance as has been depicted in the case of smallpox vaccination.[47] Large-scale vaccination campaigns and malaria surveys were also carried out among the labourers on the plantations and industrial sites.[48]

The predominant focus of the historiography of colonial medicine in the nineteenth century has been on urban public health. There has been little work on areas such as industrial and agricultural labour health, which have received much greater attention in the historiography of colonial Africa, as we shall see in Chapter 8.[49] The introduction of modern sanitary methods, and the institutionalization of urban municipalities and public health, and of the practices of preventive vaccination in the nineteenth century, have been seen as the critical phase of colonial medicine in India. Some historians have seen this as a period when medicine became integral to colonial rule and governance. Others have suggested that the introduction of public health in India reflected the heterogeneity of nineteenth-century Indian social and political history.

Arnold argues that colonial rule in India used the human body as the site of its authority, legitimacy and control. He has described the phenomenon as the 'colonization of the body'.[50] This included two elements. First was the discursive process by which colonial medical writers from the eighteenth century accumulated a vast amount of information about the human body in the colonies, about the physical constitutions of Europeans as well as Indians in the tropics. The second was a more interventionist phase when the colonial state, armed with this knowledge and that of modern biomedicine, sought to define and govern the lives and behaviours of Indians. Through the study of three major epidemic diseases in India – smallpox, cholera and plague – Arnold showed that colonial medicine intervened in the lives of Indians in physical as well as psychological ways. He identified three stages of interaction between Western and non-Western medicine in British India: appropriation, subordination and denigration, through which Western medicine intervened in everyday lives in India and became dominant.[51]

In Arnold's work the focus is predominantly on the colonial state and its policies. Although it traces some Indian agencies, particularly in smallpox vaccination, the main argument is about the intrusive and hegemonic nature of colonial public health policies, which interfered with and determined the lives of the common people. However, the interaction between the colonial state and the people, as far as public health was concerned, was more nuanced – by instances of inner dilemma within the colonial regime as well as ambiguities of Indian responses to Western medicine. For example, during the Bombay plague, which Arnold studies at length, residents of Bombay, including the lower classes, often voluntarily and enthusiastically, participated in vaccination campaigns against plague. This was partly due to their growing faith in Western medicine and colonial health officials, and partly their seeking to escape from the harsh sanitarian measures.[52] Moreover, the colonial state did not show the same authoritarian and deterministic attitude towards all epidemic diseases. Colonial policies towards cholera in India did not reflect the same dominant and decisive colonization of the body. The colonial state and its medical authorities, as Arnold himself shows, often remained inactive, either faced with the enormity of the task of eradicating or treating cholera in India, or paralysed by their own fatalism, which was a product of their understanding of cholera as essentially an 'Asiatic' disease, uniquely and permanently 'at home' in India.

Mark Harrison shows the more diffuse nature of public health in colonial India. His book closely analyses the local politics and interests in the city of Calcutta. This was the capital of the British Raj throughout the nineteenth century. It was also seen as the 'home' of the cholera epidemics, which raged in India and globally throughout the nineteenth century. The city received significant medical attention from both colonial medical authorities and the international scientific community. Thus public health in Calcutta was an important colonial and international concern at the end of the nineteenth century. Harrison studied in depth how local politics in the city played an important role in defining public health policies there. On several occasions he shows that Indian elites thwarted the interests and intentions of the colonial state to introduce public health measures. Such measures remained inadequate, not just due to colonial apathy but because the Indian elites and ratepayers opposed the sanitary reforms because they meant an increase in their taxation. His main argument is that the Indian rentier class (broadly the Indian propertied class, but including a wide section of urban elites such as lawyers, absentee landlords and urban property owners) were often opposed to public health measures. The result was that municipality budgets remained unused while people continued to live in inhospitable conditions: 'Sanitary reform was largely opposed on economic grounds by the city's rentier class, and Indian ratepayers resisted moves to increase local taxation for sanitary purposes.'[53] The rentier class, driven by its own class and economic interests, derailed many of the proposed public health measures.

Harrison's book highlights the complex political and economic circumstances under which colonial public health measures functioned. It shows that neither was the colonial state a monolithic structure nor were public health matters within the sole authority of the state. It also points out that the British, despite their imperial priorities, did introduce public health measures and sanitary administration in India, including a clean water supply, drainage and sewage collection, and fumigation of streets and houses during plague. Harrison's argument is based on the rentier class and its interests. However, a close study of this class provides us with a deeper understanding of colonialism itself and of the links between colonialism and medicine. The rentier, a new class, which created its wealth from colonial revenue and commerce, was a product of colonialism itself.[54] The economic interests of this class, together with those of the Indian bourgeoisie, were closely linked to those of the colonial state. Colonialism was a regime propped up by powerful class and political alliances among Indians and the British, ultimately serving the imperial interest. In terms of their disinvestments in urban public health, the rentier class, who were the new privileged residents of nineteenth-century Indian cities such as Calcutta, inherited and adopted the deeply divided nature of colonial urban settlements and similarly distanced themselves from the slums of the cities. As we shall see in Chapter 9, a very similar phenomenon took place in late nineteenth-century Rio de Janeiro when the city's indigenous elites disassociated themselves from the problems of those living in the slums. The rentier class in Calcutta was much keener to associate itself with prestigious imperial institutions than invest in the cleaning of drains. It enthusiastically raised funds for the imperial medical institutions, such as the Calcutta School of Tropical Medicine (1921) and the Indian Pasteur Institutes, which were established by the British colonial officers and the colonial state. The lack of investment in the public health facilities in the slums of Calcutta was due to the inequitable pattern of economic and social growth in colonial India and the broader class alliances formed under colonialism. There is scope for more research on the history of class and colonial public health.

The books by Harrison and Arnold focus on the period until the First World War. There has been relatively less research on the subsequent period, between 1920 and 1950. Two significant changes took place in colonial medicine in India during this period. The first was organizational changes within health administration and the second was the rise of Indian nationalism. From the early decades of the twentieth century, the struggle for political concessions and rights launched by Indian nationalists led to several constitutional and legislative changes. The Government of India Act of 1909 – also known as the Morley-Minto Reforms – gave Indians limited roles in the central and provincial legislatures, known as legislative councils. In 1919 the GOI introduced the Montague-Chelmsford reforms, which further prompted the provincialization. The act introduced the system of 'Dyarchy': transferring functions such as education, health and agriculture (referred to as 'transferred' subjects) to provincial legislative bodies while retaining others such as finance, revenue and home affairs as 'reserved' or 'imperial' subjects. The post of the sanitary commissioner for the Government of India was replaced by the public health commissioner because of these changes. The main change in this dual model was that municipality and public health administration was now placed largely in the hands of Indians. However, the issues of funding and medical research were retained by the imperial government. This created a gap where many local projects and institutions suffered from a lack of funding.

Following the First World War, international health initiatives and funding intervened in colonial health policies in India, Africa and South America. In 1921 the League of Nations Health Organization (LNHO) was formed, which organized and administered Indian malaria-eradication projects. By the 1920s, American schools of public health, such as those of Johns Hopkins, Yale and Harvard universities, had become the centres of international public health instruction, attracting large numbers of foreign students, including Indians. Thus debates in Indian public health in the interwar period were increasingly informed by American and international public health philosophy. Internationally renowned scholars of public health, such as Henry Sigerist, visited India in the 1940s to discuss the future of Indian health planning. Organizations such as the Rockefeller Foundation also increasingly determined tropical health policies and research funding in Asia and Africa in the interwar period.[55] Victor Heiser of the Rockefeller Foundation visited India in the 1920s to survey the state of medical research in the country and determine the nature of funding and support to be offered. Therefore the devolution of public health administration into local control was accompanied by a greater internationalization of Indian public health and disease-eradication programmes.

Nationalism and medicine

The rise of Indian nationalism from the late nineteenth century led to two main changes in colonial medicine. The first was the growing competition and rivalry among British and Indian physicians for medical posts and to assume control of colonial health administration. The second was the rise of traditional medicine in response to Western medicine. Here I will focus only on the first development as the second theme will be elaborated in Chapter 10.

Medical education in India developed within the main hospitals that the British established in the major Indian cities. These colleges produced Indian doctors who joined the lower ranks of the colonial medical administration. University education became a key feature of gaining economic and social status for the emergent Indian middle class. Indian universities were historically sites of political mobilization among Indian intellectuals and nationalists. The first generation of Indian scientists who earned international recognition, such as P.C. Ray, J.C. Bose and C.V. Raman, were all products of Indian universities. The colonial government supported the growth of the Indian medical profession because it promised cheaper doctors, who were increasingly in demand in the empire. Indian doctors were also posted in other parts of the empire, particularly in East Africa.[56]

The Indian doctors asserted political pressure on the colonial government for greater rights, which coincided with the rise of Indian nationalism. Prior to the First World War, cadres of the IMS had served most of the medical research institutes and university professorships, apart from the military posts, while Indian graduates occupied the subordinate and provincial posts. There were obstacles in their joining the IMS, the most prestigious medical service. The IMS admission exams were held only in Britain, which put severe restrictions on Indians sitting them. In addition the General Medical Council of London often refused to recognize Indian medical degrees as being equivalent to British ones.[57] Until 1913, Indians comprised 5 per cent of the members of the IMS, and in 1921 this rose to only 6.25 per cent.[58]

Indian doctors established their own associations, such as the Bombay Medical Union (BMU) and the Calcutta Medical Club (CMC), in response to this lack of opportunities at the highest level. These groups had strong nationalist sentiments and their activities were mainly directed towards 'enhancing the status and dignity of the Indian medical profession'.[59] From the late nineteenth century the BMU, in partnership with the INC put forward demands against the monopoly of the IMS.[60] In 1913 the BMU sent its representations to the Royal Commission on the Public Services in India demanding equal status, privileges and emoluments for non-IMS doctors, especially those at higher grades. It argued that the so-called 'open' IMS examinations, held biennially in London, were practically 'shut' for the majority of Indians.[61] It suggested that the problem was with the colonial character of the IMS, which from a primarily military body had increasingly 'poached' all important civilian posts: 'It is high time now that the noxious overgrowth of this service were cut off and the numerous civil posts thus liberated from it were thrown open to the indigenous talent of proved merit and ability.'[62] Dr Jivraj N. Mehta (1887–1978) was an influential member of the BMU and one of the leading figures in this contest. He passed his MD examinations in London in 1914 and entered private practice in India. Mehta was the founder of Seth Gordhandas Sunderdas [GS] Medical College and King Edward VII Memorial Hospital in Bombay. He also took active interest in Gandhi's nationalist movement. Mehta accompanied Gandhi to London for the

Second Round Table Conference in September 1931, where he argued in favour of the Indianization of the IMS.[63]

The INC established the National Planning Commission (NPC) in 1938 to draw up a blueprint for the social and economic reconstruction of post-independence India.[64] The NPC had a subcommittee with S.S. Sokhey, an Indian physician (and a member of the IMS), in charge. The subcommittee was responsible for drawing up the plans for postcolonial public health and medicine.[65] The committee recognized poverty as the main cause of disease in India. It suggested that the community health service should be a cornerstone of the postcolonial Indian health service, with one healthworker for every 1,000 in the rural population.[66] In 1943 the colonial government set up the Health Survey and Development Committee under the leadership of Joseph Bhore, a member of IMS. The Bhore Committee report too urged for more widespread public health infrastructure in India along with centralized research institutes. This led to the establishment of the All India Institute of Medical Sciences (AIIMS, 1952) in Delhi following independence.[67]

Independence marked a period of nation-building in key sectors, such as economy, health and education. Health was and continues to be one of the key problems in modern India. In independent India there were concerted efforts to build nationwide medical institutions to provide medical training, and to improve the general health of the people through improved nutrition and diet, and vaccination campaigns. However, medical infrastructure, despite substantial investment in rural sectors through a planned economy, has tended to remain urban oriented and millions of people do not have adequate or affordable medical care. Epidemics of cholera, malaria and rabies continue to cause huge levels of mortality along with problems of child mortality and lack of proper nutrition.

Conclusion

The history of colonial medicine in India reflects the broad patterns of the history of the colonization of India. It also mirrors the broad trajectories of colonial medicine. As the colonization of India progressed, medicine changed from eclectic exchanges in coastal regions to being part of colonial administration and hegemony. In the Age of Commerce, medicine in the colonies developed through diverse modes of exchange between European missionaries, surgeons, traders and Indian physicians. In the course of these commercial exchanges, European trading companies also became dominant mercantile and territorial powers in South Asia and the Indian Ocean. Colonial hospitals reflected this growing colonial power and authority in India by the end of the eighteenth century. From the mid-nineteenth century, as the British established their empire in India, Western medicine became part of the everyday lives of ordinary Indians. In the twentieth century, colonial medicine was marked by two orientations. One was the nationalist resistances and negotiations. The other was the internationalization of colonial health concerns. Despite this rich history of Western medicine in India over several centuries, a large section of Indians remain without the benefits of modern medicine.

Notes

1 Om Prakash, 'Bullion for Goods: International Trade and the Economy of Early Eighteenth Century Bengal', *Indian Economic and Social History Review*, 13 (1976), 159–186.
2 For an analysis of the significance of the Bengal revenue to the EEIC, see Javier Cuenca Esteban, 'The British Balance of Payments, 1772–1820: India Transfers and War Finance', *Economic History Review*, 54 (2001), 58–86.
3 Irfan Habib, *The Agrarian System of Mughal India, 1556–1707* (New Delhi, 1963); M. Athar Ali, *The Mughal Nobility Under Aurangzeb* (London, 1966).
4 Lakshmi Subramanian, *Indigenous Credit and Imperial Expansion: Bombay, Surat and the West Coast* (Delhi, 1996).
5 C.A. Bayly, '"Archaic" and "Modern" Globalization in the Eurasian and African Arena', in Anthony Hopkins, ed., *Globalization in World History* (New York, 2002), pp. 47–73.
6 Sanjay Subrahmanyam, 'Asian Trade and European Affluence? Coromandel, 1650–1740', *Modern Asian Studies*, 22 (1988), 179–188, p. 185.
7 Sinnappah Arasaratnam, *Merchants, Companies and Commerce on the Coromandel Coast, 1650–1740* (Delhi, 1986). Ashin Das Gupta, *Indian Merchants and the Decline of Surat: 1700–1750* (Wiesbaden, 1979).
8 D.K. Bassett, 'British "Country" Trade and Local Trade Networks in the Thai and Malay States, c. 1680–1770', *Modern Asian Studies*, 23 (1989), 625–43.
9 Sanjay Subrahmanyam, *Penumbral Visions: Making Polities in Early Modern South India* (Ann Arbor, 2001).
10 For a detailed study of the agrarian nature of the revolt of 1857, see E. Stokes (edited by Bayly), *The Peasant Armed: Indian Revolt of 1857* (New Delhi, 1986).
11 See Chakrabarti, '"Neither of meate nor drinke, but what the Doctor alloweth": Medicine Amidst War and Commerce in Eighteenth Century Madras', *Bulletin of the History of Medicine*, 80 (2006), 1–38
12 *The Gazetteer of Bombay City and Island*, vol. 3 (1909), p. 181.
13 See, Chakrabarti, *Materials and Medicine*, pp. 91–2.
14 For a detailed account of the British military campaign during the Anglo-Maratha Wars, see Randolf G.S. Cooper, *The Anglo-Maratha Campaigns and the Contest for India: The Struggle for Control of the South Asian Military Economy* (Cambridge, 2003).
15 Ibid, p. 118.

16 See Chakrabarti, 'Neither of meate nor drinke, but what the Doctor alloweth'.

17 Holden Furber, *Rival Empires of Trade in the Orient, 1600–1800* (Minneapolis, 1976), pp. 149–50.

18 W.J. Wilson, *History of the Madras Army*, vol. 1 (Madras, 1882), p. 6.

19 Furber, *Rival Empires of Trade*, pp. 149–50.

20 Arasaratnam, *Merchants, Companies and Commerce on the Coromandel*.

21 See, Chakrabarti, *Materials and Medicine*, pp. 100–2.

22 See ibid, p. 212.

23 'The Madras Medical School', *Madras Journal of Literature and Science*, 7 (1838), 265.

24 Ibid.

25 Ibid.

26 Stella R. Quah, 'The Social Position and Internal Organization of the Medical Profession in the Third World: The Case of Singapore', *Journal of Health and Social Behavior*, 30 (1989), 450–66, p. 456.

27 For a study of the emergence of the IMS, see M. Harrison, *Public Health in British India*, pp. 6–35.

28 Mark Harrison, *Public Health in British India*, p. 61.

29 For the history of eighteenth-century Calcutta, see P.J. Marshall, 'Eighteenth-Century Calcutta' in Raymond F. Betts, Robert J. Ross and Gerard J. Telkamp (eds) *Colonial Cities: Essays on Urbanism in a Colonial Context* (Lancaster, 1984), pp. 87–104.

30 Susan M. Neilds-Basu, 'Colonial Urbanism: The Development of Madras City in the Eighteenth and Nineteenth Centuries', *Modern Asian Studies*, 13 (1979), pp. 217–46.

31 David B. Smith, *Report on the Drainage and Conservancy of Calcutta* (Calcutta, 1869), pp. 1–10.

32 For details of cholera control measures in nineteenth-century India, see Harrison, *Public Health in British India*, pp. 99–115.

33 F.W. Simms, *Report on the Establishment of Water-Works to Supply the City of Calcutta, With Other Papers on Watering and Draining the City* [1847–52] (Calcutta, 1853).

34 Partho Datta, *Planning the City, Urbanization and Reform in Calcutta, c. 1800–1940* (New Delhi, 2012), pp. 152–3.

35 *Report of the Plague Commission of India*, vol. 5 (London, 1901), pp. 409.

36 M. Harrison, *Public Health*, pp. 81, 155.

37 Prashant Kidambi, 'An Infection of Locality: Plague, Pythogenesis and the Poor in Bombay, c. 1896–1905', *Urban History*, 31 (2004), 249–67.

38 Ira Klein, 'Plague, Policy and Popular Unrest in British India', *Modern Asian Studies*, 22 (1988), 723–55.

39 Arnold, *Colonizing the Body*, 200–39.

40 Ibid, 238.

41 Jayant Banthia and Tim Dyson, 'Smallpox in Nineteenth-Century India', *Population and Development Review*, 25 (1999), 649–80.

42 Arnold has quoted the harrowing experiences of 'arm to arm' smallpox vaccination in India. See Arnold, *Colonizing the Body*, p. 142.

43 Niels Brimnes, 'Variolation, Vaccination and Popular Resistance in Early Colonial South India', *Medical History*, 48 (2004), 199–228. See also Dominik Wujastyk, '"A Pious Fraud": The Indian Claims for Pre-Jennerian Smallpox Vaccination', in Jan Meulenbeld and Dominik Wujastyk (eds), *Studies on Indian Medical History* (New Delhi, 2001), pp. 131–67.

44 David Arnold, *Colonizing the Body*, p.144–58.

45 Nadja Durbach, '"They Might as Well Brand us": Working-Class Resistance to Compulsory Vaccination in Victorian England', *Social History of Medicine*, 13 (2000), 45–63.

46 See Chakrabarti, *Bacteriology in British India*, pp. 65–9.

47 Ibid, pp. 55–60.

48 Nandini Bhattacharya, 'The Logic of Location: Malaria Research in Colonial India, Darjeeling and Duars, 1900–30', *Medical History*, 55 (2011), 183–202.

49 One of the few works on plantation labour health in India is by N. Bhattacharya, *Contagion and Enclaves; Tropical Medicine in Colonial India* (Liverpool, 2012).

50 Arnold, *Colonizing the Body*, pp. 7–10.

51 Ibid, pp. 55–9.

52 Chakrabarti, *Bacteriology in British India*, pp. 49–60.

53 Harrison, *Public Health in British India*, p. 226.

54 For details of the economic foundations of the Indian rentier class, see Barbara Harris, 'Agricultural Merchants, Capital and Class Formation in India', *Sociologia Ruralis*, 29 (1989), 166–79.

55 John Farley, *To Cast out Disease: A History of the International Health Division of the Rockefeller Foundation (1913–1951)* (New York, 2004); Shirish N. Kavadi, '"Parasites Lost and Parasites Regained" Rockefeller Foundation's Anti-Hookworm Campaign in Madras Presidency', *Economic and Political Weekly*, 42 (2007), 130–7 and *The Rockefeller Foundation and Public Health in Colonial India, 1916–1945; A Narrative History* (Pune/Mumbai, 1999).

56 John Iliffe, *East African Doctors: A History of the Modern Profession* (Cambridge, 1998) p. 28.

57 E.W.C. Bradfield, *An Indian Medical Review* (Delhi, 1938).

58 Roger Jeffery, 'Recognizing India's Doctors: The Institutionalization of Medical Dependency, 1918–1939', *Modern Asian Studies*, 13 (1979), 301–26, p. 311.

59 As quoted in Mridula Ramanna, *Western Medicine and Public Health in Colonial Bombay, 1845–1895* (Hyderabad, 2002), p. 3.

60 Ibid, pp. 217–21.

61 *Representation of the Bombay Medical Union to the Royal Commission on the Public Services in India* (Bombay, 1 May 1913), p. 1.

62 Ibid, pp. 3–4.

63 Jeffery, 'Doctors and Congress: The Role of Medical Men and Medical Politics in Indian Nationalism', in Mike Shepperdson and Colin Simmons (eds), *The Indian National Congress and the Political Economy of India, 1885–1985* (Avebury, 1988), pp. 166–7.

64 Sunil S. Amrith, *Decolonizing International Health: India and Southeast Asia, 1930–65* (Basingstoke, 2006), p. 61.

65 *National Planning Committee, Subcommittee on National Health Report* (Bombay, 1948).

66 Debabar Banerji, 'The Politics of Underdevelopment of Health: The People and Health Service Development in India; A Brief Overview', *International Journal of Health Services* 34 (2004), 123–142, p. 127.

67 *Report of the Health Survey and Development Committee* (Delhi, 1946).

7

Medicine and the colonization of Africa

Historians have described the nineteenth century as the Age of Empire, a period when from predominantly maritime enterprises and settlements, European nations established large-scale territorial empires in Asia and Africa. The description is not entirely accurate because major parts of the Americas, the West Indies and Asia were colonized in the eighteenth century. The above description of the Age of Empire fits best with the history of the colonization of Africa. European traders from the seventeenth century had maintained commercial links with the coastal parts of Africa, particularly the west and east coasts and the southern tip, in search of gold, ivory and, most important, slaves. However, almost suddenly, by the early nineteenth century, there was a major impulse among European nations to enter into the interiors of Africa, leading to the period of major colonial expansion there. This chapter will first explore the motives behind the colonization of Africa. It will then examine the role that medicine played in this territorial expansion and in the establishment of European colonial rule in Africa, and how far medicine can be seen, as historians have argued, as a 'tool of empire'. Finally, it will explore how Western medicine became part of colonial rule and the European explorations and understanding of African culture and society. Other aspects of the history of medicine and the colonization of Africa will be discussed in the next three chapters.

Why did European nations seek to enter the interiors of Africa in the early nineteenth century? There were two main reasons. First were the commercial and economic factors. The abolition of the slave trade and slavery (in the British Empire in 1807 and 1832, respectively, whereas France and Netherlands abolished slave trading in 1818, and both of these nations formed alliances with Britain in 1835 to abolish the international slave trade) meant that slave traders looked for items of trade and commercial profit from Africa, other than slaves. Second, with abolition, the moral fulcrum of European history moved from the Caribbean islands and the Americas to Africa. Abolition led to a humanitarian impulse, and abolitionists and missionaries now sought to eradicate slavery from its 'roots', which they believed was in the heart of Africa. They believed that there were two causes for the continuance of slavery in Africa: the existing Arab presence and trading practices, and the innate backwardness of African culture and society. Both of these could be eradicated by spreading the Gospel and 'Christian commerce'. The colonization of

Africa took place due to the shared interest of expanding European commercial, humanitarian and missionary activities and influence in the continent.

From their experiences of influence and prosperity in the eighteenth century in Asia and the Americas, which was chiefly brought about by commerce, European traders, politicians and missionaries in the nineteenth century viewed their commercial activities as an essentially liberating and civilizing force. On the other hand, missionaries involved in the abolition movement sought to preach the Gospel in the interior of Africa, to expunge the Arab influence and free Africans from their perceived mental 'thraldom'. Similar to the Spanish conquistadors who drew inspiration from the experiences of the Crusades to spread Christianity and European civilization in the New World in the sixteenth century, Victorian missionaries in the nineteenth century were stirred by the abolition movement to spread Christianity and civilization in Africa. Missionaries such as David Livingstone believed that the spreading of the Gospel would truly liberate Africans from mental and physical slavery. This led to the establishment of the Society for the Extinction of the Slave Trade and for the Civilization of Africa in 1839 by Thomas Fowell Buxton. In that sense the traders and missionaries were part of the same mission, which was to propagate 'lawful Christian commerce'.[1]

Arabs have had long commercial and cultural relationships with Sub-Saharan Africa. The Arab slave trade was a significant part of this commerce from the ninth to the nineteenth century. While the transatlantic slave trade from the west coast of Africa, which was carried out by Europeans from the seventeenth century, was much larger and more significant, the Arab slavers carried on the slave trade from the eastern parts of Africa in the Indian Ocean region until the late 1800s. Arab traders had also traded with the interior parts of Africa through old trade routes in goods such as ivory, cloth, slaves and firearms in alliances with African rulers. Due to this long historical connection, Arabs also made important cultural contributions, particularly in East Africa. For example, the predominant language of these parts – Swahili – was deeply influenced by Arabic.[2] There were also Arabic influences in Sub-Saharan African diet, dress and rituals.[3] In the nineteenth century, in the wake of abolition movements, Europeans saw this long Arab influence as essentially negative and one of the roots of African slavery and backwardness.

Historians have studied the role of medicine in the colonization of Africa from two different perspectives. The first was to describe the colonization as an instance when medicine became a 'tool of empire'. The other was to explore the role that medicine played in civilizing Africa, in spreading European culture and influence. I will start with the first approach. Daniel Headrick has shown that the use of quinine prophylaxis against malaria (along with steamships and guns) gave Europeans a decisive advantage in the colonization of the continent in the nineteenth century.[4] While it is true that quinine was vital in protecting European lives against malaria in Africa, this chapter will also show that this was not always the case and that the links between colonialism and medicine were more complex. Medicine helped in the African colonization by playing a role in the broader social, cultural and economic transformations of continent.

Entering Africa

The urge to promote new commercial activities led to the establishment of several new trading companies in Britain and the adoption of new roles by some of the old slave-trade companies. McGregor Laird and several Liverpool merchants who were involved in the slave trade started the African Inland Company in 1832. The African Lakes Company was established in 1889, the early aim of which was to replace the central African slave trade with 'legitimate' trade in ivory and other goods.[5]

There were three main points of European entry into Africa at the beginning of the nineteenth century. The first was in West Africa, on the Guinea coast; the second was through the southern tip, the Cape colony; and the third was on the east coast of Mozambique and was used by the Portuguese traders.[6] These coasts have long been familiar to Europeans, as they had established trading connections and colonial settlements here from the sixteenth century. The main slave-trade coast was the Guinea or Gold Coast, the west coast of Africa extending from Sierra Leone to Benin. This was the area from where slave-trading companies purchased slaves and shipped them to be sold to the plantations in the West Indian islands and North and South America. In this history of the Gold Coast, we go back to the history of mercantilism and bullion. West Africa, particularly the Guinea coast, was initially known to European traders for its gold from the sixteenth century. Some of the gold brought from West Africa was minted into coins, which were popularly known as 'guineas'. The traders soon became interested in another profitable 'commodity' available on this coast: slaves. The Guinea coast was thus most familiar to Europeans and they preferred to travel to the interior via the major rivers, such as the Niger. Even elsewhere in Africa, the preferred route of entry and exploration for Europeans was rivers, such as the Congo and Zambezi, to avoid the dense forests and the open savannahs, wildlife and swamps.

However, entry into Africa was not straightforward. The first difficulty to overcome was the fear of the unknown. Europeans were still largely unfamiliar with the interior regions of Africa. Throughout the seventeenth and eighteenth century, European habitation in Africa was mostly on the coasts, while the huge areas beyond the coasts remained unknown. This fear of the unknown was infused with imaginations and myths of Africa being the inhospitable and uncivilized 'Dark Continent'. While the spirit of adventure and the urge to spread commerce, Christianity and civilization helped them to overcome some of these perceived fears, disease and mortality posed a real threat.

From the late eighteenth century, European voyages into Africa via the rivers suffered from heavy casualties. In 1777–9, during William Bolt's expedition to Delagoa Bay up the Zambezi River, 132 out of 152 Europeans died. In 1805, during Mungo Park's expedition to the Upper Niger, only five survived, with the rest dying of fevers or dysentery. In 1816–17, during Captain James Tuckey's expedition to the Upper Congo River, 19 out of 54 Europeans died, and in 1833, during

McGregor Laird's expedition to the Niger valley, 37 out of 48 Europeans died. These experiences led to the belief that Europeans could not survive in coastal West Africa, which now came to be known as the 'white man's grave'.

Why was it called the white man's grave? Europeans noticed that Africans survived much better in the region from these fevers. They thus believed that Africa was climatically and pathologically unsuitable for European habitation. Although this supposed immunity of Africans contributed to ideas of racial immunity to diseases, African immunity (relative) to malaria (which was the main killer) was not a racial one. European explorers and surgeons at that time had little knowledge of African society and the diseases they suffered from. Africans acquired resistance to malaria in their childhood. Malaria infected and killed African children as much as it did European adults. Only those among them who survived acquired resistance and reached adulthood. Even today, mortality from malaria in Africa is highest among children and infants.

One reason for this amorphous nature of the fear of Africa from these deaths was the ambiguity in European medicine about 'fevers'. European physicians were still heavily influenced by Sydenham and Lind, and they referred to most remittent or intermittent affliction in the tropics (e.g., malaria, typhoid, yellow fever and kala-azar) with the generic term 'fevers', which they believed was caused by noxious miasmas of the hot climates. They often did not distinguish one from the other.[7] Based on this principle, European physicians explained European mortality in Africa through general miasmatic theories – that it was the burning heat, the fever-laden swamps, the swarming insects and miles of trackless forests that caused the fevers and high mortality rates. Throughout the eighteenth and nineteenth centuries, 'fevers' represented a large variety of malaise that Europeans dreaded while venturing into the tropics.

The early experiences of high mortality on the west coast of Africa led to the exploratory Niger Expedition of 1841 under captains H.D. Trotter and William Allen. In response to the pressing demands of the Society for the Abolition of the Slave Trade, the British government commissioned three ships – *Albert*, *Wilberforce* and *Sudan* – to explore and chart the rivers Niger and Benue. This expedition too met with huge casualties, as 35 per cent of the British members died. Of the 145 Europeans on the expedition, 130 fell ill with 'fever' and 40 died; but among the 158 Africans, there were only 11 cases of fever and 1 death.[8]

Despite these high mortalities, the Niger Expedition of 1841 was unique as it had a few surgeons in it who conducted the first significant experiments with quinine prophylaxis in Africa. Two of them, Dr T.R.H. Thomson and Dr James O. McWilliam, carried out systematic experiments with quinine and achieved significant results. They made two important discoveries: first that a specific fever ('the fever') caused most of the mortalities; and second that quinine given in correct doses could act as an effective prophylactic against that fever.[9] Thomson carried out the most extensive experiments both during and after the expedition and found that in earlier voyages, dosages of quinine that were given were either too small or too irregularly administered. He demonstrated that optimum protection could only be had with 6–10 grains taken daily. He experimented on himself and was not affected by the fever during his stay in Africa, even though he was ashore a great deal. When he returned to England, however, he stopped taking quinine and came down with malaria.[10] Thomson was the first to publish his results in a prominent journal on the prophylactic use of quinine against malaria, although similar experiences were reported by other naval surgeons as well.[11]

Back in Britain in 1847, naval surgeon Alexander Bryson studied the accumulated evidence of the Niger Expedition and showed that there was a close correlation between the incidence of regular bark or quinine prophylaxis and both mortality and morbidity. As a result of this work the British navy adopted quinine as the main prophylactic and new orders were issued extending its use by shore parties.[12] At the end of 1848 the director general of the Medical Department of the army sent a similar circular to all governors in West Africa, advising them to use quinine prophylaxis. It became common practice for the British forces on the Gold Coast to keep a bottle of quinine and take it as a prophylaxis against fevers. However, at this point, physicians had very little scientific knowledge of the aetiology of the fever (most commonly malaria) or its mode of propagation (which was through insect vectors), which only became known about half a century later in the discoveries of Patrick Manson and Ronald Ross. Until the 1890s, in colonial public health reports, malaria cases were not reported separately but were listed under the general category of 'fevers'. The medical advice in favour of using quinine was based purely on the statistical evidence available from the experiments.

Quinine and the 'tool of empire'

The use of quinine prophylaxis marks the link between the Age of Commerce and the Age of Empire. As we saw in Chapter 2, French scientists Caventou and Pelletier produced quinine in 1820 from cinchona bark, which was found in the Peruvian and Bolivian forests. Despite the discovery, physicians used quinine for a range of fevers, both remittent and intermittent. They believed that it was an effective remedy because it had powerful tonic properties, capable of restoring vital power to bodies debilitated by long residence in hot climates or exposure to miasma. It was also used as a tincture and as an antiseptic.[13] Following the Niger Expedition and the subsequent medical experiments, quinine was regarded as the specific treatment against intermittent fever.

At the same time, procuring cinchona bark for quinine, which was increasingly in demand in the tropical colonies, remained a problem. This led to a new era of colonial bioprospection of cinchona in the 1850s. A major rush started among European nations to acquire cinchona from the Peruvian forests and to grow it in plantations in their respective colonies. As Ray Desmond describes, 'Cinchona became a coveted plant for nations like Britain and the Netherlands with colonies in the tropics where malaria was endemic.'[14] In 1861, British geographer Clements Markham (1830–1916) procured saplings of cinchona from Bolivia and Peru, and introduced these into cinchona plantations in India, in the Nilgiri hills.[15] The Dutch similarly started large-scale cinchona plantations from the 1860s in South East Asia. By the early twentieth century, 80 per cent of the world's cinchona production was in Asia.

Due to the widespread use of quinine prophylaxis and the adoption of other preventive measures, such as better hygiene and lodging facilities, the mortality rates of Europeans in Africa dropped from the 1850s. As shown by Philip D. Curtin, between 1819 and 1836 the annual average mortality rate of European troops on the West African coast was 483 for enlisted men and 209 for officers per 1,000. Between 1881 and 1897 the annual average mortality rate in West Africa had fallen substantially to 76 on the Gold Coast and 53 in Lagos. He ascribes this to the use of quinine prophylaxis and the abolition of harmful treatments, such as bleeding.[16]

With this discovery of its prophylactic effect against malaria, quinine appeared to be the magic drug in the mid-nineteenth century. It introduced optimism against the pessimism that pervaded among the British public in the nineteenth century about European health and survival in the tropics. It provided the impetus behind the European colonization of Africa. The general impression among the British colonialists and public was that quinine prophylaxis had made Africa accessible to them. This optimism did not entirely abolish the image of the 'white man's grave' but it helped to introduce a new hope and impetus in the colonization of the continent.

The River Niger became the major route for the British into the interior of Africa through the West. British colonization of western Africa started rapidly. British forces captured Lagos in 1851 and a new colony was set up there in 1861. Traders achieved success too as the MacGregor Laird trading company sent another expedition through the Niger River in 1854 under Dr W.B. Baikie. The expedition sailed up the Niger and the Benue further than any Europeans had done before, and returned to the coast without a single fatality. These successes encouraged the missionaries to push along the Niger into the interior of Africa. This marked the beginning of the important role that Christian missionaries played in colonial Africa, which I shall return to later in this chapter. In general, the changes from the 1850s marked the character of the colonization of Africa. As traders and sailors were the agents of European colonialism in the Age of Commerce, doctors, travellers and missionaries were at the forefront of the European colonization of Africa in the nineteenth century. Headrick and Curtin have argued that the rapid colonization led to the 'Scramble for Africa' in the late nineteenth century. Although medical factors, such as the use of quinine prophylactics, did not directly contribute to the scramble, the decline in European mortality in Africa from the mid-nineteenth century certainly facilitated the rapid colonization.

However, we need to be careful in establishing a straightforward cause and effect connection between medicine and colonization. As William Cohen points out, it is possible to overestimate the impact of quinine prophylaxis in the African colonization. The decline in mortality rates did not play such a critical role in the expansion of French imperialism in Africa in the nineteenth century. Rather, it was the other way round: successful empire-building by the French eventually reduced the loss of French lives overseas and the adoption of better medical care.[17] The French did not use quinine as a prophylactic in a regular manner until the end of the nineteenth century, and French mortality rates in Africa remained high throughout that century. In the Upper Sudan in the 1880s, the French suffered mortality rates as high as 800 per 1,000. In the expedition in Madagascar in 1895, the French lost 5,592 lives out of a total force of 21,600 men. Similar figures are observable elsewhere as well. Most of these deaths were due to malaria.[18] There were various reasons for this slow uptake in the use of quinine. As we know, malaria was confused with several other fevers and the use of quinine among French physicians remained sporadic. The French also often used quinine as a cure for malaria, rather than a prophylactic, which reduced its effectiveness. Many French soldiers disobeyed orders in taking quinine, as it was a bitter drug and had side-effects.

The French Empire expanded in Africa despite high death rates. Cohen here sees continuity between the patterns of the nineteenth-century colonization of Africa and the earlier eighteenth-century colonial expansion in Asia and the Americas, which took place despite high mortality rates among European troops. What proved critical in the French colonial expansion was not medical advancement but superior military organization, strategies and arms. This allowed small columns of European-led armies to accomplish successful conquests while keeping French battlefield deaths low.[19] Cohen argues that the relationship between mortality rates and imperialism was the opposite to that suggested by Curtin; it was colonial conquest that reduced European mortality. After conquering territories and gaining control over certain sections of the indigenous population, the French were able to employ large numbers of indigenous soldiers and porters in their military expeditions. In addition, living and fighting in the colonies over a period helped French troops to acclimatize to Africa, eventually leading to better mortality rates. This too was in continuity with the eighteenth-century patterns of colonialism.[20]

Colonial expansion also brought more resources and better facilities for Europeans. For example, when the French first entered Sudan they marched on foot, leading to a large number of infections and high death rates. Later the French army was able to recruit mules, and Europeans travelled on these. As road networks were secured, communication became better, ensuring regularity of medical and food supplies, and enabling the ill and wounded to be moved to the colonial hospital or even to France. Improved housing meant that European troops lived in more hygienic conditions in the barracks rather than in the tents, and away from the swamps that bred mosquitoes.

The debate about whether medicine was a 'tool of empire' has sought to determine the role that medicine played in colonization. Imperialism was a complex historical process and modern medicine was often a product of imperialism rather than just being its catalyst. By linking this history of African colonization with that of the Age of Commerce, as is evident in the history of cinchona and quinine, we see that medicines and drugs were often resources and products that became available to Europeans through colonialism. These were then used in further colonial conquest. The relationship between mortality rates and imperialism too was far more complex than one of simple cause and effect. In the French colonization of Africa, colonial expansion and consolidation secured more resources and eventually led to better mortality rates in the European armies. Often this improvement was due to successful military campaigns, which in turn allowed access to better resources, such as diet, housing and indigenization of European armies, and this in turn reduced European military mortality in the colonies.

The appreciation of this complexity leads us to understand the role that medicine played in the next phase of African colonialism, the history of the medical missionaries. Christian missionaries went to Africa to spread the Gospel and the benefits of European civilization, such as modern medicine. Medicine became an important point of contact and interaction between African and European cultures and, in the process, Western medicine was established within African social, cultural and economic life.

Africa and the medical missionaries

Who were the medical missionaries? As we saw in chapters 1 and 2, Christian missionaries were part of the European colonization of the Americas and Asia from the sixteenth century. From the time of the Spanish Conquest, missionaries set up missions, established gardens, offered medicine and food to the sick and the starving, and conversed with local people about God, nature and spirituality. Missionary institutions were often the main sites where people took refuge and received solace in times of war, famine and epidemics. Missionaries played diverse roles in the colonies. While they administered European medicine to the locals, they also learnt about local practices. While they spread Christianity (often aggressively) and suppressed local practices, they also provided shelter and hope to indigenous people. There were both continuities and discontinuities to this history in nineteenth-century Africa.

Africa was colonized at a time of the expression of European racial and cultural superiority. With the expansion of colonialism in the nineteenth century, European physicians and missionaries felt the need to pass their medicine onto the colonized people. European medicine in the nineteenth century therefore did not remain only for Europeans to save European lives in hot climates; it also became the medicine for the colonized populations. Missionaries performed this task of taking European medicine to Africans, and their activities reflected this idea of the superiority of European civilization and medicine. Missionaries also realized that the use of medicine was often the best mode of propagating their religion. Medical missionaries played roles which combined spiritual and medical healing. In doing so they became the bridge between the colonizing forces and the indigenous societies in Africa. In the process of integrating European medicine with the everyday lives of African communities, the medical missionaries also competed with African healers, leading to hybrid medical traditions, which I will discuss in Chapter 10. Thus medicine was part of this broad project that Christian missionaries undertook in nineteenth-century Africa. As argued by Megan Vaughan, due to the wider role that missionaries played, Africans saw them differently from either the colonizing forces or government medical officers.[21]

Terence Ranger has identified four main features of the medical missionary activities in Africa. First, missionaries portrayed their activities as a living example of Christ's own work, thereby creating a close link between medicine and Christianity in Africa. Second, they were able to work within and penetrate indigenous communities; third, this link between religion and medicine also helped to establish the efficacy of Western biomedicine at the expense of traditional beliefs and practices; and fourth, this embedded ideas of 'time sense, work discipline, sobriety' though modern institutions such as hospitals and dispensaries in the African healthcare system.[22]

In this section I will start with a brief study of the life and work of the most famous and iconic nineteenth-century medical missionary, David Livingstone (1813–73). He was a Scottish Congregationalist who studied theology and medicine in Glasgow. He then joined the London Missionary Society (LMS) and became convinced of the need to reach the interiors of Africa and introduce Africans to Christianity, as well as free them from slavery. He also wanted to use medicine to heal them from their sufferings. He summarized these intentions as he set off on his Zambezi Expedition.[23]

Livingstone became the pioneer medical missionary and explorer in Africa. He reached Cape Town in South Africa in 1841. Much of the Sub-Saharan interior of Africa at this time was unknown to Europeans. Livingstone spent the first ten years of his time in Africa based at the mission station north of Cape Colony. From there he undertook several expeditions to the interior parts of Africa. After every expedition he returned to the Cape to dispatch his letters and collections to Britain. In 1842 he began a four-year expedition tracing the route from the upper Zambezi to the coast. This filled huge gaps in European knowledge of central and southern Africa. Between 1849 and 1851 he travelled across the Kalahari Desert, and on the next trip he reached the upper Zambezi River. In 1855, Livingstone discovered a spectacular waterfall, which he named Victoria Falls. He reached the mouth of the Zambezi on the Indian Ocean in May 1856, becoming the first European to cross the width of southern Africa from the Indian Ocean in the east to the Atlantic on the west. After a brief spell in Britain, he returned to Africa in 1858 as an official explorer, and for the next five years he carried out explorations of eastern and central Africa for the British government. In January 1866 he returned to Africa, this time to Zanzibar, from where he set out to seek the source of the River Nile.

Livingstone contributed greatly to the European 'discovery' of Africa. In his urge to find the sources of the Zambezi he became the first European to travel through the interior parts, map large regions of central and southern Africa and collect information about its natural world and inhabitants. His expeditions marked the European 'Age of Discovery' in Africa as he sent African botanical specimens to Kew Gardens and zoological specimens to the British Museum. These were then studied and displayed in these institutions in London. He sent home extensive geographical reports of his travels along the Zambezi and across the Kalahari.[24] Livingstone's accounts helped in the preparation of European maps of these parts of Africa for the subsequent waves of colonial armies and settlers. His collections, stories of adventures and the maps encouraged other travellers to undertake similar expeditions, thereby paving the course of the colonization of Africa. Felix Driver has shown that the expeditions of Livingstone and Henry M. Stanley helped the Royal Geographical Society of London to promote the heroic and imperial character of geographical knowledge.[25]

Apart from the making of imperial geography, Livingstone's main contribution in Africa was in linking commerce, Christianity and medicine. As he travelled through the Zambezi basin he looked to keenly exploit the commercial potential of the region. He wrote in his *Missionary Travels and Researches* about the valuable natural resources of the Zambezi valley, about coal and dyestuffs, as well as potentially exportable agricultural products such as tobacco, coffee, cotton and sugar.[26] He believed that the Zambezi basin could become a new British territory for growing cotton and thus replace the dependence of the English textile industries on American cotton. At the same time, he described the whole continent of Africa as 'sick' – suffering from the moral and physical effects of slavery – and thus in need of Christian religious and commercial regeneration. He believed that European commerce, religion and medicine would together serve to 'heal' the continent. In subsequent years this became the motto for European medical missionaries in Africa (see Figure 7.1).

Following Livingstone's expeditions and missionary activities in the Zambezi basin, two Scottish missions, Livingstonia and Blantyre, were established along Lake Nyasa and the Shire highlands in 1875–6. Both became crucial in the subsequent British occupation of what became Nyasaland and then Malawi, in setting up local contacts, collecting local information and spreading European influence. The town of Blantyre also became the commercial hub, particularly for British trade in ivory. By 1878 the LMS and the Church Missionary Society had set up their missionary and medical activities along Lake Tanganyika and in Buganda, respectively. Another prominent British medical missionary establishment in central Africa was the Universities' Mission to Central Africa (UMCA) along the shores of Lake Malawi, which was founded in 1875. Dr Robert Howard, the first full-time medical officer for the UMCA, expanded the role of the mission by mixing Christian piety with medical care. Despite the missionaries' plans to introduce Western medicine and Christianity, their activities

Figure 7.1 *A Medical Missionary Attending to a Sick African* (by Harold Copping). Courtesy of the Wellcome Library, London.

resulted in 'medical pluralism' in the negotiations between African and Western therapeutic systems.[27]

Historians have analysed the hegemonic roles played by missionaries in colonial Africa.[28] Terence Ranger, in his study of the history of the UMCA in Tanzania, shows that although it was a supposedly benevolent and paternalistic institution of healthcare, which cared deeply for the welfare of Africans and was opposed to the forces of industrial capitalism that ravaged the African economy and ecology, the UMCA nevertheless played an important part in extending British colonial influence across the continent. By portraying their activities as Christian healing, missionaries were able to take Western medicine and Christianity deep into indigenous communities. They also helped to establish the efficacy of Western biomedicine at the expense of traditional beliefs and practices, and established modern institutions such as hospitals and dispensaries within the African health-care system.[29] The missionary hospitals in Africa were also sites where Christianity could be preached. Missionary hospitals were the main institutions of healthcare in colonial Africa. Many were converted from old churches. Often due to a shortage of doctors, missionaries themselves treated the patients and prescribed medicines. Along with ordinary religious practices, such as holding prayers every morning and evening and services on Sundays, the missionaries used their hospitals in creative ways to serve their proselytizing objectives. Non-Christian patients were often placed next to converts to encourage religious conversations and conversion. All of the walls of the hospital were decorated with biblical images.[30]

Thus in Africa, missionary activities and imperialism went hand in hand. Missionaries helped in the exploration and exploitation of African natural resources, and also in establishing European cultural and economic hegemony. Medicine was an integral part of European missionary contact with and influence in Africa. The medical practices of the missionaries also highlight the complexity of the role that they played in African history. On the one hand, those like Livingstone by their geographical explorations and by developing local contacts paved the way for European colonization. They also helped to establish European cultural influence through their medicine and their religious preaching. This is why there was the paradoxical situation where the mission to liberate Africans from slavery contributed to their colonization and subjugation.

Yet, at the same time, missionaries were often opposed to the colonial exploitation of indigenous communities. Their missions were the vital sites of human compassion and of the expression of African agency within wider economic and social transformations of Africa. Their medicines and hospitals provided a healing touch to a continent increasingly ravaged by colonial warfare and exploitation. The medical missionaries, through their work in Africa, by their fundraising activities in Europe, by creating links among local communities and by taking a holistic view of healing and wellbeing, also set a precedent for future international charities and health agencies in Africa and other colonies in the twentieth century.

Western medicine and African 'otherness'

I now return to the question of difference in colonial medicine. European medicine in Africa had a strong moral tone as it was supposed to both heal African bodies and enlighten African souls. At the same time, the predominant nineteenth-century medical and polygenic views suggested that Africans as a race were inherently different from and inferior to Europeans. This raised new questions. Was it possible or even desirable to undertake such a fundamental transformation of the African people? Although modernizing the African economy and civilizing the African mind were two of the premises of the colonization of Africa, by the early twentieth century, faced with some of the consequences of the rapid economic transformation and the introduction of modern education and medicine in Africa, European missionaries, physicians and administrators were faced with a moral dilemma. What was modernization leading to in Africa? In seeking the answer to this question, exploring and understanding this 'otherness' became a preoccupation of colonial medicine.

Megan Vaughan has studied this dilemma in her analysis of the history of madness and psychiatry in Africa. To understand the broader intellectual context of her work, we will briefly see how madness was analysed by Michel Foucault in European history. In *Madness and Civilization: A History of Insanity in the Age of Reason*, Foucault argued that from the mid-seventeenth century, in the Age of Reason (which French scholars describe as the 'Classical Age'), madness was seen in Europe as 'unreason' or the 'other' (or opposite) of the rational, enlightened and 'normal' mind. In the earlier period, Foucault argued, there was a greater acceptance and tolerance of the mad or the insane. The insane was often seen as a wise person who had certain proximity to divine reason and was thus accepted in society.[31] However, in the eighteenth century, madness was seen more clinically and as being the 'other' of the rational mind or reason. Lunatics were ostracized and imprisoned in newly created mental asylums all over Europe, away from the public gaze. The divide between the sane and the insane became a fundamental feature of European enlightenment. Foucault explained the emergence of the science of psychiatry as an attempt to salvage the insane into the world of sanity.

According to Vaughan, this divide between madness and rationality in European psychiatry posed a problem in colonial Africa. Europeans had viewed Africans as culturally and racially savage, or the 'other' of the modern, scientific, industrialized Europeans. So to them a normal African was also childlike, immature and irrational, and thus insane within the modern rational paradigm. Since Europeans had stressed that Africans and their diseases were inherently different, they came to believe that African madness too was different from European insanity. The issue was in what ways it was different and what the cure could be.

The Zomba lunatic asylum was established by colonial authorities with the help of missionaries in Malawi in 1910. Initially designed along European lines to separate the insane from others, the project soon ran into a predicament. In 1935, two medical officers of the asylum – H.M. Shelley and W.M. Watson – wrote a report in which they noted an increase in madness among Africans. This posed the 'problem'. Why was there such a rise in Africa? The psychiatrists determined that there were two forms of madness there. One was the 'native' form of madness – the inherent irrationality of the African mind. The other was the more modern one, brought about by Western civilization. Africans were becoming insane because they could not cope with modernity. These conclusions coincided with contemporary anthropological studies on African culture, customs and traditions, which corroborated the ideas of inherent irrationality of African practices such as witchcraft. European contact then made Africans further disoriented and insane, as they were out of touch with their own ('simple') realities and modes of life. Colonial administrators blamed the missionaries for destroying the 'primitive innocence' of the natives by educating them.

Understanding both forms of African madness became important for colonial administrators. They needed to determine the 'normal' standards and customs in supervising the inherently irrational and increasingly insane Africans. Yet defining such a custom was a challenge. How could Europeans explain and administer cultural behaviour that appeared so different? Added to that was the problem that the supposed cure for such inherent irrationality, such as education and economic change, was, in turn, leading to a modern form of madness among Africans.

Thus colonial medicine, particularly psychiatry, was placed at the heart of colonial governance. Colonial doctors contributed to colonial administrative policies. Their investigations into African psychiatry and neurology contributed to racial studies within modern medicine and science. Dr H.L. Gordon (superintendent of Mathari Mental Hospital, Nairobi) and J.C. Carothers (British colonial psychiatrist in Kenya) contributed to the rise of the East African School of Psychiatry, which employed a combination of physiological and psychological analysis of African culture and mentality. They linked neurological and psychological studies of African brains and minds to identify the racial inferiority of Africans. At the same time, these psychiatrists explained that modern education made Africans alienated and rebellious. Unable to comprehend the political nature of African protests and insurgencies against British rule, Carothers explained the Mau Mau Rebellion in Kenya (an insurgency by Kenyan rebels against the British colonial rule, which ran from 1952 to 1960) as an example of such a new form of African madness.[32] He was subsequently employed by the colonial government in Kenya to 're-educate' Mau Mau rebel prisoners to turn them into compliant subjects. Psychiatrists suggested that the cure for this African madness was in keeping Africans pristine, simple and away from civilization. Thus, more than 100 years later, colonial officials arrived at a solution, which was contradictory to the premises of the colonization of the continent of the 1830s.

The real difference in the understanding of madness in Africa and Europe, as Vaughan suggested, was that madness in Africa was not seen as an individual problem but as a collective one. While in Europe, in the Age of Reason, the insane person was seen as an individual in divergence with the norm, in Africa insanity was seen as the collective norm. Individual Africans who were perceived to be lunatics were not as closely monitored or institutionalized as were the insane in Europe. In Africa there was no 'great confinement' such as in eighteenth-century Europe.[33] The focus was more on general governance. The government was often happy for African families to look after the 'insane'. The dilemma was in deciding whether modernity was the solution or the catalyst to the problem.

European psychiatrists faced similar challenges of madness in northern Africa although in different historical circumstances. Within European Orientalism, Arabs too were considered (as we have seen in Edward Said's analysis in the Introduction) to be distinct from the rational and enlightened Europeans. In the early twentieth century, with French colonization, the French established asylums in Morocco, Tunisia and Algeria. These French institutions became important sites of research and 'schools' of colonial psychiatry in the study of the 'Arab mind'. French psychiatrists such as Jean-Michel Bégué and Antoine Porot concluded that Arabs were inherently violent, criminalistic and unpredictable. Porot developed the Algiers School of Psychiatry, which put forward the theory that Algerians were inherently childish and devoid of curiosity. He concluded, as did British psychiatrists in Malawi for Africans, that Arabs were 'normally abnormal'.[34] French mental asylums too became sites of colonial power and the articulation of the French civilizing mission in North Africa. They provided the premises of viewing the Arab/Muslim/Oriental mind as incapable of rational thinking.[35]

Madness holds an important place in colonial African history because colonialism there was based on the idea of African 'otherness'. It also explains why insanity and witchcraft are such significant themes in colonial and postcolonial African studies. Curing African madness was both the motive and the predicament of European colonization. Insanity represented that 'otherness' of Africans in the European perception.

Yet, this idea of otherness and difference was fraught with deep inner contradictions. Johannes Fabian shows in *Out of Our Minds: Reason and Madness in the Exploration of Central Africa* that German and Belgian travellers in Africa in the nineteenth century, while they treated Africans and savages as irrational, were themselves often in various states of madness – at times induced by drugs, at others caused by diseases and exhaustion. This often led to their using brutal and irrational acts of violence against Africans.[36]

The question of African agency is integral to this history of madness and otherness in Africa. As ideas of African otherness, along with those of African madness, became the idiom of African colonialism, Africans themselves adopted and asserted these identities for various purposes. Africans often played active roles in using these colonial categories of madness to label and institutionalize those against whom they held grudges.[37] In the process, these categories were redefined and reformulated. In her study of psychiatry and madness in Natal and Zululand from 1868 to 1918, Julie Parle highlights the non-European agencies, both African and Indian, that defined both the diagnostic categories and the therapeutic modes of insanity in South Africa. On the one hand, the history of insanity in South Africa and the establishment of the Natal Government Asylum (NGA) is located within the wider economic and social changes in the region: the economic depression in the 1870s, the establishment of diamond and gold mines, and the migration of Indian and African labourers led to the increase in mental health problems. On the other, there were confusions within colonial psychiatry regarding different forms of mental disorders that doctors and missionaries faced, and their diagnosis of these shifted constantly, designating these at times as insanity and at others as criminality. These ambiguities allowed spaces for indigenous ideas of insanity and modes of treatment to be incorporated within the NGA. Parle thus argues that the NGA should not be seen as a 'Western' institution but one that was adopted by the local population to control and cure the insane living among them.[38] In the process of studying and curing African madness, the sciences of psychology and psychiatry were transformed as they incorporated ideas from anthropology, racial pathology and alternative notions of insanity. Historians have shown that African medical and cultural practices, such as witchcraft, which European doctors regarded as evidence of African irrationality, were similarly categorized and reinvented in medical, cultural and criminal terms by both Europeans and Africans.[39] It is possible to see parallels between the history of madness in colonial Africa and that of voodoo and obeah in the West Indies in the eighteenth century.

Conclusion

Europeans arrived in Africa hoping to spread European commerce, religion and civilization across the continent. The task was not easy. Death, disease and the fear of the unknown hampered their progress from the outset. European armies, merchants and missionaries overcame these issues with the help of their medicine, technology, and colonizing zeal and prowess. Then, in the process of introducing their modernity, civilization and commerce, they were faced with another dilemma – about the consequences of their civilizing mission, some of which were unfolding before them. The dilemma was partly because their ideas of modernity and progress were themselves contested in Europe. It was also because Europeans in the nineteenth century saw Africans as a race that was fundamentally different from their own. Thus they remained unsure whether modernity was achievable or even desirable. Furthermore, Europeans had lost some of the agency in introducing these changes. Africans had taken the task of reforming their land and culture upon themselves. Seeking to release Africa from its confinements, Europeans were faced with their own limits and failures in the continent. Alongside these dilemmas and predicaments, colonization marched ahead forcefully – almost the whole of Sub-Saharan Africa was colonized by the First World War, which transformed the continent; Christianity spread rapidly and became popular; modern medicine was widely employed; and plantations, cities, mines and industries transformed the African landscape.

The history of medicine in Africa narrates the broad contours of these physical and psychological transformations of the continent. It also reveals the crevices and paradoxes of imperialism. Colonial armies, politicians, settlers and missionaries sought to, and did to a large extent, transform African culture, agriculture and economy (which I will discuss in detail in the next chapter), even though the trajectories of those transformations sometimes did not follow the intended goals. In the process, Western medicine, alongside Christianity, became part of African social, cultural and economic life. For this preoccupation there has been relatively less work in the African history of medicine on topics such as urban public health, sanitation and municipal administration, which have dominated the history of colonial medicine in India.

Notes

1 Henry Rowley, *The Story of the Universities' Mission to Central Africa*, 2nd edition (London, 1867), p. 3.

2 David Robinson, *Muslim Societies in African History* (Cambridge, 2004), pp. 27–88. Ali A. Mazrul, 'Black Africa and the Arabs', *Foreign Affairs*, 53 (1975), 725–42.

3 Lyndon Harries, 'The Arabs and Swahili Culture', *Africa: Journal of the International African Institute*, 34 (1964), 224–9.

4 Daniel Headrick, *Tools of Empire; Technology and European Imperialism in the Nineteenth Century* (Oxford, 1981) Chapter 3: 'Malaria, Quinine and Penetration of Africa', pp. 58–82.

5 For a detailed study of British trading companies in the nineteenth century, see Geoffrey Jones, *Merchants to Multinationals: British Trading Companies in the Nineteenth and Twentieth Centuries* (Oxford, 2002).

6 The Portuguese were present on the east coast of Africa from the sixteenth century, but they had ceased to be a critical factor in the colonization of Africa in the nineteenth century.

7 It was only in the 1880s, with Alphonse Laveran's discovery of the malaria parasite, that malaria was identified as a specific disease.

8 William Allen and T.R.H. Thompson, *A Narrative of the Expedition Sent by Her Majesty's Government to the River Niger, in 1841*, vol. 1 (London, 1848), Appendix (Part 4).

9 James Ormiston MCWilliam, *Medical History of the Expedition to the Niger, During the Years 1841, 2, Comprising an Account of the Fever Which Led to its Abrupt Termination* (London, 1843).

10 Curtin, *The Image of Africa: British Ideas and Action, 1780–1850*, vol. 2 (Madison & London, 1973), pp. 355–6.

11 T.R.H. Thomson, 'On the Value of Quinine in African Remittent Fever', *Lancet* (28 February 1846), 244–5.

12 Bryson, 'Prophylactic Influence of Quinine', *Medical Times and Gazette*, 7 (1854), 6–7. See Curtin, *The Image of Africa*, p. 356.

13 Andreas-Holger Maehle, *Drugs on Trial: Experimental Pharmacology and Therapeutic Innovation in the Eighteenth Century* (Amsterdam, 1999), pp. 264–75.

14 Ray Desmond, *The European Discovery of the Indian Flora* (Oxford, 1992), p. 222.

15 Markham, *Peruvian*, pp. 34–5.

16 Curtin, *Image of Africa*, pp. 361–2.

17 Cohen, 'Malaria and French Imperialism', 23–36.

18 Ibid, pp. 23–4.

19 Ibid, pp. 31.

20 Ibid, p. 32.

21 Vaughan, 'Healing and Curing: Issues in the Social History and Anthropology of Medicine in Africa', *Social History of Medicine*, 7 (1994), 283–95, pp. 294–5.

22 Ranger, 'Godly Medicine', p. 259

23 Lawrence Dritsas, 'Civilising Missions, Natural History and British Industry', *Endeavour*, 30 (2006), 50–4, p. 50.

24 Timothy Holmes (ed.), *David Livingstone: Letters and Documents 1841–1872* (London, 1990), p. 5.

25 Felix Driver, *Geography Militant: Cultures of Exploration and Empire* (Oxford, 2001).

26 David Livingstone, *Missionary Travels and Researches in South Africa* (London, 1899), pp. 180–4, 241, 262–83, 359–93, 410, 430. Holmes (ed.), *David Livingstone: Letters and Documents*, p. 47.

27 Charles Good, *The Steamer Parish: The Rise and Fall of Missionary Medicine on an African Frontier* (Chicago & London, 2004).

28 Ranger, 'Godly Medicine'. See also Vaughan 'The Great Dispensary in the Sky'.

29 Ranger, 'Godly Medicine', p. 259.

30 Vaughan, *Curing their Ills*, pp. 62–5.

31 Foucault, *Madness and Civilization: A History of Insanity in the Age of Reason* (London, 1967).

32 J.C. Carothers, *The Psychology of Mau Mau* (Nairobi, 1954).

33 Vaughan, *Curing their Ills*, p. 101.

34 Richard Keller, 'Madness and Colonization: Psychiatry in the British and French Empires, 1800–1962', *Journal of Social History* 35 (2001), 295–326, p. 315

35 This discussion is largely drawn from Keller, *Colonial Madness: Psychiatry in French North Africa* (Chicago, 2007).

36 Johannes Fabian, *Out of Our Minds: Reason and Madness in the Exploration of Central Africa* (Berkeley, 2000).

37 Vaughan, *Curing their Ills*, pp. 106–7
38 Julie Parle, *States of Mind: Searching for Mental Health in Natal and Zululand, 1868–1918* (Scottsville, 2007).
39 See the essays in Henrietta L. Moore and Todd Sanders (eds), *Magical Interpretations, Material Realities: Modernity, Witchcraft and the Occult in Postcolonial Africa* (London & New York, 2002) and Luise White, *Speaking with Vampires; Rumor and History in Colonial Africa* (Berkeley & London, 2000).

8

Imperialism and tropical medicine

By the First World War, European colonial powers had acquired huge parts of the tropical world as their colonies. By 1878, European nations controlled 67 per cent of the world and by 1914 over 84 per cent.[1] The most significant expansion was by the British. Between 1874 and 1902, Britain added 4.75 million square miles to its empire, mostly in Africa, but in Asia and the Pacific islands as well.[2] This huge imperial expansion generated the need to invest in the healthcare of primarily European troops and civilian populations in the tropics. These new medical investments, research and surveys in the tropical colonies that developed in parallel with the expansion of the empire from the late nineteenth century, primarily in the British Empire, is often known as tropical medicine. Gradually this emerged as a medical specialization in which other European imperial powers, such as the Dutch, the Belgians, the Germans and the French, participated. By the twentieth century, what came to be known as tropical medicine was a composite and, to some extent, amorphous medical tradition. On the one hand, it incorporated the several medical, environmental and cultural experiences and acumen that Europeans had gathered in warm climates over the last 200 years of colonialism. One the other hand, it incorporated newly emergent germ theory and parasitology (the medical specialization concerned with parasites and parasitic diseases), which shifted medical attention from diseased environment to parasites and bacteria. Tropical medicine can therefore be defined as a medical specialization that developed by the end of the nineteenth century, which was based on the idea that certain diseases were caused by pathogens which were endemic or peculiar to the tropics.

How was such a link between tropical geographies and pathogens created? Why were the tropics seen as particularly pathological regions? Historians have debated at length on these complicated questions. I will investigate these by considering tropical medicine – both its historical development and its ontology.

What is tropical medicine?

Tropical medicine has dominated writing about the histories of nineteenth-century imperial medicine. Michael Worboys wrote in 1976 that 'during the past 75 years tropical medicine has been the main scientific expression of Western medical and health policy for the Third World'.[3] He showed that there were three key factors leading to the birth of tropical medicine: the emergence of a new science, the socioinstitutional aspect of that science and the influence of late nineteenth-century imperialism. Other scholars have worked on the various facets of British tropical medicine and imperialism as well. Three characteristics of tropical medicine are evident in these histories. First, they focus mainly on the late nineteenth century and the Age of New Imperialism. Second, they have often been based on the works of prominent scientists and doctors, such as Patrick Manson, Ronald Ross and Leonard Rogers. Third, they show that tropical research was conducted over large geographical areas encompassing the entire tropical world under European control, from Assam to Accra.[4]

The very phrase 'tropical medicine' leads to the problem of assuming that the whole of the tropics – that is, Asia, Africa and parts of the Americas – were all similar and uniform geographical regions. One reason why a more region-specific history of tropical medicine has remained difficult to write is the ubiquitous nature of the works of its proponents. Manson's career spanned China, Scotland and London. Ross worked his way through London, Netley, Madras, Vizianagram, Moulmein, Burma and the Andaman Islands in search of the malaria vector. After his discovery of the *Anopheles* mosquito as the insect vector of the malaria parasite, he continued to travel on field surveys in West Africa, the Suez Canal zone, Greece, Mauritius and Cyprus. The idea of a distinct medical specialization is also based on metropolitan scientific paradigms, which focus on the works of these scientists who sought to define a medical specialization to train doctors for colonial services in specialized institutes in London, Berlin and Liverpool. In the tropical colonies such as India and Africa, as we shall see, tropical medicine was not so distinct a medical speciality as it was in Europe.

The other problem is that historians who have written about British tropical medicine have followed the careers of these men and focused on what they worked on – that is, parasitology. This focus on parasitology in British tropical medicine was borne out of a certain professional expediency, as William Bynum has suggested:

> the extent to which bacteriology was widely seen as a French or German science gave the British some comfort: tropical medicine was still there to be colonized. Not only were other groups of organisms besides bacteria implicated, the chains of infection were often more complicated than the water, food, milk or airborne diseases of temperate climates.[5]

As a result it appeared that tropical medicine was essentially about parasitology. This is not necessarily true as a lot of research conducted in the tropics in this period was not on parasitology. Several British imperial scientists worked on bacteriology and parasitology simultaneously. Therefore there are several gaps in our understanding of the real nature of tropical medicine as practised in the tropical colonies.

This raises the question of what is tropical medicine and what is it not. What was the nature of medicine practised in places such as British India or Sub-Saharan Africa? Can these all be defined under the umbrella of tropical medicine? It is difficult to conceptualize tropical medicine as a distinct scientific speciality in the true sense of the term as it did not have a well-defined research methodology of its own. Tropical medicine, as we know it, was an invented tradition aided by institutional growth, a process particularly noticeable in two articles that Ronald Ross wrote in 1905 and 1914. In the first, his definition of tropical medicine was relatively vague. He admitted that 'the term tropical medicine does not imply merely the treatment of tropical diseases.'[6] What it implied was a 'science of medicine' and, more importantly, a medicine for the empire, where diseases were the 'great enemies of civilization'.[7] In 1914, when both the London and Liverpool schools of tropical medicine had begun their research in Britain and in the colonies, Ross provided a more therapeutically defined characteristic to tropical medicine. He now foregrounded the role of the parasitologists: 'The movement really commenced with the work of the old parasitologists rather than with that of the bacteriologists.'[8] However, the specific lineage that he then sketched was essentially a history of the establishment of the two schools in Liverpool and London, not a history of the science of parasitology or of any distinct research methodology. There is a reason for this. Scientists in the late nineteenth century used words such as 'germs' and 'parasites' interchangeably.[9] In contemporary scientific thinking, bacteria and parasites were not distinguished; both were considered to be surviving on living bodies and could be cultivated in the laboratory.[10] Scientists debated the nature of their origins and life cycle, not whether they were germs or parasites.[11] Koch widened his field of bacteriology to include other diseases in the tropics, particularly malaria and sleeping sickness. He even went to Batavia to conduct malarial investigations.[12] In his lecture at the Prussian Academy of Sciences in 1909, he provided a wider definition: 'bacteriology did not remain the exact definition, but different fields of knowledge are summarized under this name, because they use the same or comparable methods and have a common aim, investigation of infectious diseases and the fight against them'.[13] Even so, contemporary scientists used phrases such as 'malaria germs' and identified malaria as a 'germ disease'.[14]

In order to understand tropical medicine it is therefore more useful to see it not as a single and specific scientific tradition but a combination of several new and existing traditions in medical research, which was intrinsically linked to the establishment of various research institutes in or about the tropics from the late nineteenth century. The volume *Warm Climates and Western Medicine* highlighted the need to go beyond the individual and understand tropical medicine before Manson, the 'Father of Tropical Medicine'. The articles in the book depict the history of tropical medicine before Manson and Ross.[15] It is important to focus beyond Manson and Ross and to revisit the history of tropical medicine as practised in the tropical colonies. This will help us to understand the reasons behind the coexistence of diverse research methodologies and motivations in the colony and to develop a better understanding of tropical medicine in the tropics. Two factors are important in understanding the history of tropical medicine: the long and complex historical tradition of European medicine in hot climates, and its links with nineteenth-century imperialism.

The origins of tropical medicine

What came to be known as tropical medicine within British medicine in the twentieth century had an early and varied origin in British ideas of health, hygiene and climate in the tropics from the mid-eighteenth century.[16] As we have seen in the earlier chapters, European medicine in the hot climates, through the works of Lind, put emphasis on environmental and climatic factors, and established links between hot climates, miasmas and putrefaction. These played a crucial role in the debates about acclimatization – in understanding what impact hot climates had on the European constitution and whether white races could survive in the tropics. By the nineteenth century the general optimism about European acclimatization in the tropics was replaced by a sense of pessimism, and tropical regions such as Africa appeared something of a 'white man's grave'. The emergence of germ theory from the late nineteenth century and the discovery of the germs and vaccines involved in cholera, plague and typhoid brought new hope to medical research in the tropics, but it also complicated the understanding of diseases there. The tropical environment, which in the early nineteenth century appeared to be full of noxious foul air to Europeans, also appeared to be infested with invisible germs by the late nineteenth century. Meanwhile, despite the emphasis on germs as the causal factor for putrefaction and disease, miasmatic ideas persisted in the tropics. Thus a new medical thinking emerged there which combined germ theory with miasmatic ideas.

German hygienist Max Von Pettenkofer (1818–1901) insisted on incorporating environmental and climatic factors within germ theory in his theory of the ecology of germ theory. He proposed that in order for a cholera epidemic to occur, three factors, which he called X, Y and Z, had to play their respective roles. X was the specific pathogen, typically found in the soil. Y was the local and seasonal preconditions that allowed the pathogen to transform into a contagious miasma. Z was an individual's susceptibility to the disease. Pettenkofer's main belief was that germs had to transform or 'ferment' under favourable conditions before they could become contagious and cause an epidemic. Isolated from these circumstances, germs could not cause disease. When Robert Koch discovered the cholera bacterium in 1883, Pettenkofer asserted that the germ (X) required the ideal composition of the soil and its interaction with groundwater (Y), and depended on individual susceptibility (Z) to the pathogen. Pettenkofer famously demonstrated the strength of his belief in 1892 by drinking a quantity of water infected by pure cultures of the cholera bacillus. This was in order to demonstrate that without favourable climatic conditions and individual susceptibility, cholera could not spread. Though he felt a bit ill, he did not develop a full-blown case of cholera. He thus concluded that since cholera was linked to the peculiar environment of India, it could not be contagious or endemic in Europe. This was in accordance with the dominant view among British medical men in nineteenth-century India who believed that cholera was a disease of 'locality'.[17]

Due to his emphasis on these various local factors for the spread of disease or contagion, Pettenkofer's theory is also known as 'contingent contagionism' or 'localism'. This had particular significance in understanding diseases in the tropics as it provided a link between earlier Hippocratic ideas of airs, waters and places, and the more recent germ theory, and stressed that tropical climate and waters were ideal conditions for the spread of pathogenic diseases. Colonial medical officers received his work with great enthusiasm.[18] Scientists and doctors working in the tropical colonies now believed that the hot climates of the tropics were ideal for the growth of germs and parasites. Despite Koch and Pasteur proposing the universality of germs in their germ theory, suggesting that germs could survive and be active anywhere, tropical colonies were seen as dirty, unhygienic, disease-prone and peculiarly disposed towards germs. The tropics were seen as reservoirs of germs and parasites.

The breakthrough in malaria research came from French laboratory investigations. In 1880, Charles Louis Alphonse Laveran, a French military physician, discovered the protozoan (single-celled parasite) cause of malaria while working in Algeria. He convinced French bacteriologists such as Louis Pasteur and Émile Roux that a protozoan and not a bacterium caused malaria. Laveran received the Nobel Prize in 1907 for his work.

Despite the discovery of the parasite, scientists needed to explain how it spread from one human to another. The clue had been provided earlier, from beyond malaria research, by British physician Patrick Manson. He received his medical degree from Aberdeen and travelled to different parts of the British Empire in Asia and Africa. In 1877 he demonstrated in his research on lymphatic filariasis that mosquitoes transmitted filarial worms.[19] His discoveries were significant for two reasons. First, they opened up the possibility of linking malaria with the vector: the mosquito. Second, they brought vectors, and thereby the environmental condition in which they survived, into the focus of medical research in the tropics. Following Laveran and Manson's findings, parasites and vectors became the focus of tropical medicine.

Ronald Ross created the formal link between the malaria parasite and the mosquito vector. Ross was an officer of the IMS. In 1897 he identified the *Anopheles* mosquito as the vector that transmitted the malarial parasites to the host. He also identified the life cycle of the malaria parasite, *Plasmodium*. He discovered some black pigmented cells in the wall of the mosquito's stomach where the parasite lived.[20] He further traced the parasite from there to the mosquito's salivary glands, which provided valuable support for the hypothesis that transmission occurs through the bite of the mosquito. Ross received the Nobel Prize in 1902. He was the first Briton to receive the award for medical research.

These discoveries of parasites and vectors of malaria marked the emergence of the modern specialism of tropical medicine. It combined laboratory research on parasites with field surveys on vectors in various parts of the tropics. Doctors and scientists now believed that they could prevent or even eradicate diseases endemic to the tropics through these modern medical interventions. The discovery of the microbe or the parasite of a particular disease, or its vectors, could lead to their eradication. This gave rise to a new optimism to European imperial ambitions in the tropics and rescued them, to an extent, from the climatic pessimism that had overshadowed them in the nineteenth century.

The distinctive characteristic of tropical medicine is that it developed as a result of the convergence of two different fields of medicine. On the one hand, it incorporated the several medical, environmental and cultural experiences and acumen that Europeans had gathered in the colonies over the last 200 years. On the other, it borrowed from germ theory, which shifted medical attention from diseased environment to parasites and bacteria. Merging these two traditions marked a return to the geographical determinism of diseases along with a new confidence about germs and vectors. It also resulted in an apparently contradictory theory of germs and geography.[21]

An important feature of this new specialization was the institutional developments that accompanied it. Manson established the London School of Tropical Medicine in 1899. Ross, seeking to make his own mark in the emerging field, founded the Liverpool School of Tropical Medicine in 1900. Other colonial nations with major interests in Africa established similar institutes of tropical medicine. In 1906, King Leopold II of Belgium founded a 'school for tropical illnesses' in Antwerp. In 1900 the Institute for Maritime and Tropical Diseases was started in Hamburg, Germany, with Bernhard Nocht as the director. The School of Tropical Medicine was founded in 1902 in Lisbon and the Royal Tropical Institute was established in Amsterdam in 1910. In the United States the American Society of Tropical Medicine was established in 1903 in Philadelphia. This focused predominantly on yellow fever, the major threat to the United States, and William Gorgas, who was active in anti-yellow fever operations in Cuba and the Panama Canal, played a leading role in its activities in the early days

There was, however, no similar institutional and cognitive development in the tropical colonies. To begin with, India, Africa or any other colonies did not have any institute devoted to tropical medicine until 1920, when the Calcutta School of Tropical Medicine (CSTM) was set up, long after the discipline had established itself in Europe. The new school had as much to do with the interplay of various interests of the Indian government, the IMS officers and the Western-educated Indian doctors as with tropical diseases.[22] Moreover, it remained an isolated medical research institution, and its research emphasis was not exclusively on tropical diseases. A major part of the research carried out at the CSTM was on indigenous drugs under its director, R.N. Chopra. Thus tropical medicine as an institutional and cognitive discipline developed outside the tropics, fuelled by imperial ambitions.

Despite the establishment of institutions of tropical medicine and the emergence of the medical specialization, medical research in the colonies remained a hybrid tradition which combined germ theory with research on vectors, laboratory medicine with field surveys, and Pasteurian scientists working closely with parasitologists.

The hybrid traditions of tropical medicine

As we will see in Chapter 9, following the establishment of Pasteur Institutes in Asia, Africa and South America, bacteriology became increasingly important in medical research in the tropics. Pasteurian science and the global network of Pasteur Institutes provided a critical moral imperative as well as an institutional motivation in colonial medicine, even in the British colonies, such as India.[23] This suggests the need to reject the rather neat national divisions of late nineteenth-century tropical medicine, which tended to assign parasitology to the British and bacteriology to the French and Germans. It is important to understand the hybrid character of tropical medicine.

In India, several Pasteur Institutes were established in Kasauli, Coonoor and Shillong from 1900. These became popular and performed the majority of medical research and vaccination in the colony. The British scientists who worked there did not necessarily identify with or have training in tropical medicine. In Britain around the 1890s, prominent medical scientists such as Leonard Rogers, David Semple, George Lamb, W.F. Harvey and Stevenson Lyle Cummins, who went to serve in the tropical colonies, were trained under Almroth Wright at the Army Medical School at Netley, not by Ross or Mansion. Wright was a physician and a pathologist, and he did not specialize in tropical medicine. The British scientists who worked under him were often known as 'Wright's men' and they conducted significant research on typhoid, tuberculosis, rabies and snakebites in India and Africa.[24] Other medical scientists who conducted research in the Indian laboratories, such as A. Barclay, W.A. Crawford Roe, Kenneth McLeod and Major R.W.S. Lyons, were also trained in the bacteriological laboratories in Berlin, Paris and Lille.

The medical research conducted in late nineteenth- and early twentieth-century India was not in conventional areas of tropical medicine. The Pasteur Institutes of India focused mainly on researching and producing vaccines for cholera, plague, rabies and typhoid based on Pasteurian bacteriological research. The treatment of leprosy in India meanwhile remained a combination of Western and traditional methods. The use of gurjon oil, a treatment developed by surgeon Dougall of the Madras Medical Service in the early 1870s, remained popular throughout the early part of the twentieth century. Another indigenous remedy, chaulmoogra oil, remained the main treatment for leprosy until the introduction of sulphone drugs in the 1940s.[25] As far as smallpox was concerned, the efforts were towards vaccine production and eradication programmes, developing modes of storing, preserving and transporting the vaccines, rather than on smallpox as a subject of tropical medicine.[26]

Medical research in India in the twentieth century was not all about vectors. Research on the main tropical disease in India, cholera, did not follow the methodology of Manson and Ross of identifying the vector as the key aspect of a disease. Cholera, as Harrison described, remained a disease shrouded in mystery: 'No disease was more important, and no disease so little understood, as the "epidemic cholera".'[27] Various theories about the causality of cholera were thrown up. The non-contagionists, stressing the atmospheric causes, attributed it to lunar influences, to pandemic waves and to endemic constitutions of the air.[28] As we saw in Chapter 5, British physicians suggested a range of environmental factors to identify India as the home of cholera.

The breakthrough in cholera research came from bacteriology. After Koch found the cholera bacterium in Egypt, he presided over the German Cholera Commission. Its members arrived in Calcutta in 1883, collected evidence and found vast quantities of a particular bacterium in the intestines of the people suffering from the disease. However, the dominant view among the Indian medical men was that cholera was essentially a disease of the 'locality'.[29] Around this time, Haffkine, who was then at the Pasteur Institute in Paris, began to study the bacteriology of cholera. In 1892 he published a paper in French in which he demonstrated that immunity could be induced in animals by inoculating them with attenuated cholera bacilli.[30] News of his work spread quickly and soon came to the notice of Lord Dufferin, a former viceroy of India. Dufferin wrote to the secretary of state for India in London, requesting that Haffkine be permitted to pursue his studies in what was regarded by many as the 'home' of the disease. Haffkine arrived in Calcutta in 1893. After injecting himself and four Indian doctors, he was able to induce some villagers in the cholera belt of Bengal to come forward for inoculation. Indeed, bacteriologists considered the eminently tropical disease, cholera, to be their subject of research.[31]

Malaria research in India, definitely the focus of specialists in tropical medicine, was very much a component of metropolitan and international research programmes, driven as they were from London or Liverpool. As mentioned previously, despite their colonial connections, both Ross and Manson were particularly international figures and did not conduct any sustained field or laboratory research in India or Africa. The only substantial research on malaria in India was Samuel Rickard Christophers' work on the *Anopheles* mosquito at the CRI. Christophers was sent to India by the Royal Society in 1900 to conduct experiments on malaria eradication. At any event, after initial research at Mian Mir (Lahore Cantonment) in India, he concluded that large-scale distribution of quinine as a prophylactic measure would be more effective than local knowledge-based anti-malarial sanitation in tropical regions.[32] The principal malaria eradication programmes in the colonies as well as in parts of Africa in the twentieth century were first undertaken by the LNHO and later by the WHO.

Research in kala-azar (of a parasitic disease caused by the protozoan *Leishmania donovani* and transmitted by sand flies) in India perhaps most closely followed the conventional trajectories of tropical medicine in the sense that the research traditions were British and were based on parasites. However, here the breakthrough came in Netley rather than in Calcutta or Liverpool. William Leishman entered the Army Medical Service in 1887, and in 1890 he was posted to India, where he spent the next seven years. In 1900 he recorded the discovery of the parasite of kala-azar, or dum-dum fever, near Calcutta. In 1899 he was posted back to Netley, where he discovered the parasite. However, this work was not published until 1903, when Lieutenant-Colonel Charles Donovan of the IMS confirmed it in Madras. Leonard Rogers was successful in culturing the recently discovered parasite of kala-azar to the flagellate stage. Drawing an analogy with sleeping sickness, he suggested that antimony administered intravenously as tartar emetic could be an effective treatment.

Research on trypanosomiasis (sleeping sickness) in Africa too was a hybrid tradition in which bacteriologists and parasitologists worked together. In 1895, Scottish bacteriologist David Bruce discovered the parasite *T. brucei* as the cause of cattle trypanosomiasis. In 1902 he provided evidence that sleeping sickness is transmitted by tsetse flies. The subsequent medical research and surveys into sleeping sickness were heavily influenced by German and Italian medical researchers who came from diverse medical backgrounds, such as microbiology, pharmacology and veterinary medicine. In 1909, German military surgeon Friedrich Karl Kleine (1869–1951) demonstrated the cyclical transmission of *T. brucei*. In 1904 and 1905, Belgian physician Alphonse Broden (1875–1929) and German naval doctor Hans Ziemann (1865–1905) discovered two other pathogenic trypanosome species: *T. congolense* and *T. Vivax*, respectively. Italian bacteriologist Aldo Castellani went to Uganda in 1902 and discovered the cause and means of transmission of sleeping sickness. He also discovered the Spirochaete (the bacteria) of yaws in 1905. Koch too conducted research on trypanosomiasis in the north-west of Lake Victoria in Tanzania.

Therefore medical research based in the colonies did not follow the strict paths of tropical medicine as institutionalized in early twentieth-century Europe. Nor was it a simple derivative of the research agendas of any particular school or tradition. Tropical medicine developed out of a strong metropolitan involvement, albeit a diverse one. It was conducted with the close involvement of institutions such as the Army Medical School at Netley, the Pasteur Institutes of Paris and Lille, Koch's Institute for Infectious Diseases of Berlin, the British Institute of Preventive Medicine, the School of Tropical Medicine, scientific journals such as *The Lancet* and the *British Medical Journal*, and men of diverse scientific lineage and specialization. Because of this medical ambiguity, we have to look at the political and imperial context to identify the essential character of tropical medicine. It was a scientific discipline that emerged specifically to serve imperial purposes and interests.

Tropical medicine and 'New Imperialism'

The essential character of tropical medicine can be understood in the context of late nineteenth-century imperialism, often known as 'New Imperialism'. I will explore this form of imperialism primarily in the context of the British colonization of Africa. This was the period when the British Empire was expanding in Africa and consolidating its position in India with large military and civilian establishments. The British imperialist expansion in Africa was defined as 'constructive imperialism', which was based on the idea that imperialism was for the benefit of the colonized people and for the rational utilization of colonial resources. To understand how constructive imperialism became an important component of African colonialism, it is necessary to study the history and nature of British colonial expansion in late nineteenth-century Africa.

One significant mode of expanding imperial domains and establishing colonial rule in Africa in this period was through the protectorate system. The idea of the colonial protectorate emerged from a complex history of colonial dominions that developed in the nineteenth century. The main concept was indirect and informal rule, practised differently at various colonial locations. In India it emerged in the context of indirect rule of the princely states. In Latin America it developed in the context of control of the market, and modes of production and the 'peripheral economy', without political control, which is otherwise known as informal empire. In India one of the main routes of expansion of territorial power in the nineteenth century was by incorporating existing indigenous polities – the so-called 'princely states'. Often the EEIC would retain a native ruler as the figurative administrative head and appoint a British resident or political agent, who controlled the matters of the state through pointed 'advice'. After 1857, when direct territorial expansion ceased, the British used the concept of indirect rule largely to maintain the loyalty of the princes and through them the sizable populations of their states.

The British drew heavily from this Indian system as they sought to expand their territories in Africa in the second half of the nineteenth century.[33] Frederick Lugard (1858–1945), governor general of Nigeria, was responsible for applying the Indian method to Nigeria and Uganda in the 1880s. His colleague, G.T. Goldie, explained that the expansion of his Royal Niger Company was based on the British imperial experiences in India: 'to rule indirectly through the native feudatory princes'.[34] Thus the British controlled economic and military matters, while the African chiefs enjoyed (limited) administrative autonomy.

The economic basis of the protectorate system that developed in Africa emulated the informal empire in Latin America. According to John Gallagher and Ronald Robinson, private business interests in the Victorian period marked a shift in British imperialism – from formal and direct political control to informal modes of controlling a crucial sector, such as the economy, of other countries in order to strengthen their own commercial and economic interest, without assuming direct political control. Private entrepreneurs sought to convert independent peripheral regions into an extension of Victorian Britain, which constituted the 'informal empire'.[35] This form of imperial control was most visible in British economic investments in Latin America (particularly Argentina, Brazil and Peru) but was also evident, for instance, in Turkey and China.

The British territorial expansion in Africa took place at the time of the growth in indirect rule and informal empires in these different parts of the world. This led to the establishment of the protectorate system. This coexisted with the imperial policy followed more generally by all of the European colonial powers – that of 'sphere of influence'. This was a system similar to the protectorate system in which European powers controlled the economic, legal and jurisdictional sectors while the native chiefs were provided with the relative freedom to practise their cultural and ritual customs and administration in the tribal areas.

The protectorate and sphere of influence systems developed in the context of great rivalry among European powers such as Britain, the Netherlands, France, Belgium, Germany and Italy in securing territories in the African continent from the 1880s, which is commonly known as the 'Scramble for Africa'. While the British and the French slowly expanded their territories in Africa from the 1840s, Belgium (in the Congo from 1876) and Germany (in East and West Africa in the 1880s) gained some political and economic influence in Africa. Consequently, rapid changes occurred on the political map of Africa. In 1880 the region to the north of the River Congo became a French protectorate. In 1881, Tunisia became a French protectorate. In 1882, Britain occupied Egypt, and Italy began the colonization of Eritrea. In 1884, British and French Somaliland was created. In 1884, German South West Africa, Cameroon, German East Africa (Kenya) and Togo were created, and Río de Oro was claimed by Spain. Tensions between the European powers seeking African spheres of influence increased. In response, Chancellor Otto von Bismarck of Germany convened the Berlin Conference of 1884–5 in which 14 European nations participated.

In the Berlin Conference, European nations set the basic rules for dividing the African continent among themselves. Navigation on the main rivers such as the Niger and the Congo was to be free to all European nations. The conference also set the rule that to declare a protectorate over a region the European colonizer must show effective occupancy and develop a 'sphere of influence'. The conference formalized the hitherto ad hoc protectorate and sphere of influence systems as the mainstay of African colonial expansion and consolidation. The Berlin agreement stipulated that the colonial powers could declare Sub-Saharan Africa protectorates that could be established by diplomatic notification, even without actual possession on the ground. Therefore while it is true that territorial expansion in the Victorian period was shaped by local circumstances in Africa as argued by Gallagher and Robinson, it is also important to note that despite these local contingencies, there were broader overarching imperial systems of shared interests and structures across empires in Asia and Africa.

Following the Berlin Conference the term protectorate was widely used by European powers in securing control over territories which were considered to lack sufficient political organization. Under the now formalized protectorate system, European powers allowed nominal political power in the hands of the local African chiefs and rulers, who could retain indigenous political systems and customs while European powers and settlers controlled the essential economic interests and activities of the region.

Following the conference, European powers made bilateral treaties with each other to define the different European spheres of influence. In East Africa the German sphere was demarcated from the British and the Portuguese. In West Africa, British spheres of influence were distinguished from those of the German and the French. In equatorial Africa the French sphere was defined between the Congo Free State and the Spanish sphere. In north-east Africa, lines were drawn separating the British, French and Italian spheres. These not only formalized political boundaries (which often determined the borders of future African nation states) but also established the sphere of influence system as the main mode of maintaining colonial control.[36]

Both systems were premised on the belief among European nations that by colonizing Africa they were spreading civilization and modernity there through a mediated process. At the same time it enabled them to exploit the economic resources of the region. The European colonization of Africa, as we saw in Chapter 7, was shaped by the strong moral obligation that Europeans felt to introduce modern civilization (including modern scientific medicine), commercial use of natural resources and large-scale international trade. This approach also included a strong Christian and missionary component, as Europeans believed that the main reason for African backwardness and of African slavery was in the influence of Arab/Islamic trade and culture. It was within this idea of sphere of influence that constructive imperialism developed, as a moral justification for territorial and commercial expansion in the Age of Empire.

Tropical medicine and 'constructive imperialism'

In the Age of Empire, Western medicine was seen as an essential component of the beneficial aspects of colonialism, which also included formal (Western) education, modern economy, commerce and industrialization. This led to the convergence between imperial policies and medical research, which gave birth to tropical medicine as a specialization. In 1895, Joseph Chamberlain became the colonial secretary in Britain. He was convinced that disease control and medical intervention in the tropics was an indispensable part of the imperial mission. Through his efforts, political and economic power propelled the institutionalization of tropical medicine. He appointed Patrick Manson as medical advisor to the Colonial Office and put government opinion and resources behind Manson's efforts to establish the School of Tropical Medicine. Manson himself aspired to this: in many of his lectures in different medical colleges, he pointed out that the doctors who graduated from medical schools did not get to experience patients with specifically tropical diseases. Chamberlain raised private as well as public money to found the London School of Tropical Medicine and he fostered the appointment of committees of British researchers to go to the tropics to undertake field research. In his presidential speech at the Annual Dinner of the London School of Tropical Medicine, Chamberlain elaborated the 'constructive' aims of tropical medicine:

> I cannot myself think of any subject of scientific research and philanthropic enterprise which is more interesting, and the duty of supporting is one, which we owe to the Empire, and from which we cannot divest ourselves whatever our political opinions may be... This duty to which I refer has increased in recent years with the continual extension of our territory, with the increase of our scientific knowledge, and our opportunities, and also with what I may call the awakening of our Imperial conscience. We owe this duty to the vast population for which we have gradually made ourselves responsible and we owe it still more to those of our own race who are daily risking health and life to maintain the honour and interests of this country... In the last half century Africa has been disclosed to the world, and in what has been called 'the race for Africa' it was impossible that we should not have had our share.[37]

Patrick Manson played a vital role in directing the attention of the Colonial Office under Chamberlain towards tropical diseases such as sleeping sickness and malaria.[38] The London and Liverpool schools sent several expeditions to Africa to conduct medical surveys. By 1902 the control of the sleeping sickness epidemic formed an intrinsic part of colonial administrative policy in Uganda, and over the next two decades, field surveys played a major part in it.[39] British pathologist Joseph E. Dutton discovered the trypanosome in human blood in 1901 during the Liverpool school's expedition in Gambia.

There were differences in tropical medicine as practised in Asia and Africa. This was mainly due to the differences in how European colonialism and medicine advanced in these two regions. While in the Asian colonies, Western medicine made a much more gradual and negotiated intervention from the eighteenth century, in Africa, late nineteenth-century tropical medicine appeared as a more revolutionary and invasive medical tradition. This difference can be seen particularly in the history of the colonial medical services and the role that they played in promoting tropical medicine. Colonial medical services in

South Asia and Africa followed two very different models.[40] In South Asia, as we saw in Chapter 6, the IMS emerged slowly from the eighteenth-century military and commercial traditions of the EEIC and controlled both civilian and military medical services.

In Africa, on the other hand, colonial medical services developed more rapidly along with the establishment of colonial rule in the late nineteenth century. Colonial services remained disorganized and retained an ad hoc nature until the end of the nineteenth century, and catered mostly to European settlers and officers. In the initial years, doctors of the Imperial British East Africa Company provided healthcare exclusively for European families. It was only at the turn of the century that Chamberlain, with his emphasis on 'constructive imperialism', made the Colonial Office an important component of the British colonial governance and extended colonial medicine for healthcare to the natives as well. Colonial medical services in Africa created the link between constructive imperialism and tropical medicine. From 1900, Chamberlain reorganized the office of the Colonial Service, and, together with Manson, linked the activities of the London School of Tropical Medicine with those of the Colonial Service.[41] There were two major British colonial medical services in Africa: the West African Medical Service (WAMS, serving Nigeria, the Gold Coast, Sierra Leone and Gambia, 1902) and the East African Medical Staff (EAMS, in Kenya, Uganda and Tanzania, 1921). Both were administered closely from the Colonial Office in London, which controlled recruitment, pay, promotions and postings. The governor of each colony also played an important role. The EAMS was created in 1921. The WAMS received particular attention under Chamberlain, and it remained racial in its recruitment pattern and metropolitan in its administration.[42] The Colonial Service provided lucrative opportunities to the British/European doctors who wished to travel to and work in Africa but were not interested in medical missionary activities. Although the pay was modest and the work often arduous, the officials were driven by the belief that they were, through their medicine, bringing the benefits of Western civilization to Africa.

In East Africa the EAMS organized the sleeping sickness surveys in Uganda (1901–4). The real expansion of the African colonial services took place in the 1930s when the Colonial Administrative Service was created (1932) followed by the Colonial Medical Service (1934). As a result of these reorganizations and heavy investments made in tropical medicine in African surveys, by the 1930s the African medical services had become more lucrative in terms of pay and research facilities than their Indian counterpart, the IMS. Further specialized departments opened in the 1940s.[43] As Africa became the main area of operation of tropical medicine, Leonard Rogers, a member of the IMS, declared: 'The vast tropical areas of Africa under British administration are far more in need of medical research workers than India.'[44]

Several malaria and trypanosomiasis surveys were undertaken by experts in Africa and Asia from the 1890s. Historians have shown that the two diseases attracted the attention of Europeans not only because they were the most

Figure 8.1 Members of the Sleeping Sickness Commission, Uganda and Nyasaland, 1908–1913. Courtesy of the Wellcome Library, London.

devastating epidemics but also because they threatened metropolitan and economic interests in the colonies.[45]

The surveys enlisted a heterogeneous body of frontier men, medical missionaries, explorers and hunters to explore, survey and map the vast lands of Africa that had been allocated to various European nation states. These surveys provided opportunities for conducting field research, which drove European medical scientists 'far from their urban laboratories to "exotic" regions of the globe and immersed them in the adventure of the safari "bush"' (see Figure 8.1).[46]

Europeans first encountered sleeping sickness or trypanosomiasis in Africa in the 1860s. It is caused by the parasitic protozoan *Trypanosoma* and affects the central nervous system, ultimately leading to coma and death. Unfamiliar with the aetiology of this neurological affliction, European physicians described it in racial terms as characteristic of African idleness, referring to it as 'Negro lethargy'.[47] The spread of sleeping sickness in the early 1900s from northern Zaire (also known as the Belgian Congo) in a northerly direction, through the River Nile towards the Suez Canal, threatened critical imperial economic concerns, and efforts were made in the scientific investigation of the disease. Belgian, British and German authorities undertook several sleeping sickness surveys, sometimes jointly. The British Colonial Office, the Foreign Office and the schools of tropical medicine organized international conferences on sleeping sickness and published research materials. In 1902, during the Ugandan sleeping sickness epidemic of 1899–1905, a three-man team consisting of Castellani, George Carmichael Low and Cuthbert Christy conducted research in the region. Castellani discovered the *Trypanosoma* parasite during this expedition.[48] King Leopold of Belgium, for whom the Congo was an imperial adventure, grabbed the opportunity to invite experts from the Liverpool school to conduct experiments in the Congo. He also used the school's investigation there to propagate the benevolent nature of his colonialism internationally. In 1905, in the Congo Free State, British medical scientists, including Dutton, J.L. Todd and Christy, explored the entire length of the Congo River, and most of the Lualaba, examining and investigating the disease. Leopold used this opportunity to establish an important aspect of tropical medicine in Africa: medical research, particularly field research, was the essential component of and justification for colonialism.[49]

Malaria surveys were similarly carried out in Asia and Africa by British, French and German authorities. Following Ross' discovery of the *Anopheles* mosquito as the vector of the parasite, *Anopheles* surveys were carried out in the Dutch East Indies, throughout the Malayan archipelago. Similar surveys were undertaken by the LNHO in the Balkans, Italy, India, South Africa and Swaziland. Racial studies were also carried out to investigate how the parasite behaved among the Sub-Saharan Africans, the Malayans and the North Africans.[50]

Tropical medicine and the colonial burden

These field surveys in tropical medicine were undertaken at the same time that modern agriculture and plantations were being expanded, mines were being dug, and roads and railways were being built in Africa. Both followed the same rhetoric of colonial burden, ushering in European modernity and civilization in the Dark Continent. Tropical medicine appeared to them as their gift to Africa, along with economic and cultural modernity.

Yet the history of trypanosomiasis, malaria and other diseases prevalent in the tropics helps us to understand the social and economic history of colonialism itself. By linking the two historical processes of epidemics and economic developments, historians have shown that the relationship between disease and modernity was complex. They have challenged the imperial presumption that European civilization and economic impetus rid Africa and Asia of its tropical malaises. They have also pointed out the fallacies of the assumption that 'malaria blocks development' (the theory that suggests that epidemics of malaria create obstacles in the path of economic growth), which has continued to shape health policies in developing countries even in the postcolonial period.[51] It was often European colonialism that led to the spread of diseases such as malaria and trypanosomiasis.[52]

In India the increased rate and the geographical spread of malaria were linked to rapid deforestation, expansion of the railways and ecological changes that took place in the nineteenth century.[53] Similarly, famines in nineteenth-century India, particularly the Madras famine of 1876–8, took place within the wider context of social change that involved a complex chain of food shortages, malnutrition, consumption of harmful 'famine foods', migration and the spread of epidemic diseases, such as cholera, dysentery, malaria and smallpox.[54]

Randall Packard has connected malaria in Swaziland in southern Africa with the general changes in agricultural policies and rural impoverishment. Although malaria had been present in the region for a long time, it had occurred seasonally and caused limited fatalities. The increased occurrence of malaria, in terms of both frequency and severity in the colonial period, can be attributed to the demographic changes and the political economy of the region. There was a general movement of the Swazi people to the rocky and drought-prone regions (the Highveld and Lowveld) in the colonial period, due to a rise in population as well as restrictions placed by colonial regimes on prime agricultural (the Middleveld) lands. The growing inability of the Swazi people to feed themselves from their land and the subsequent growth of wage labour was also linked to the growth of mines and the introduction of maize cultivation in the region. The latter was in turn connected to global trade and depressions, exposing them to cycles of drought, loss of crops, declining economic conditions, rising food prices, famine and the major malaria epidemics of 1923, 1932 1939, 1942 and 1946.[55] It was not just malaria: in *White Plague, Black Labor*, Packard shows that the wider social and economic transformations in South Africa led to the spread of tuberculosis among the African population. The book studies the rise of tuberculosis in South Africa from the mid-nineteenth century, which developed along with industrialization and labour migration. The growth of gold mines and subsequent agricultural expansion drew a large numbers of labourers from various regions into the cities, which in turn led to overcrowded houses, low wages, inadequate diet and lack of sanitation. The conditions were ideal for the sharp rise in tuberculosis outbreaks. Packard shows that black workers suffered much more than white workers because of a lack of adequate medical care and diet, and due to the terrible living conditions.[56] The spread of the modern epidemic of HIV/AIDS in Africa has also been linked to the economic changes in Africa, the growth of plantations, the migration of labour and the general dislocation of social systems.[57]

Mariynez Lyons has shown that the exploitative and aggressive rubber-gathering practices in the rubber plantations of the Congo, which developed as part of King Leopold's 'civilizing mission', exposed labourers to sleeping sickness. The plantations put increasing pressure on the labourers to procure large quantities of rubber and forced them to spend several days in the forests away from their homes, collecting wild rubber. This often exposed them to the tsetse fly, the vector of sleeping sickness. On the other hand, high taxes and labour exploitation encouraged large-scale movement of labourers. This uncontrolled population migration led to the spread of sleeping sickness, particularly north of the Nile basin.[58]

These works draw from the historiographical trend of the 1970s and 1980s, which located diseases within larger and long-term structures of economic, social and ecological transformations. These also provided important insights into precolonial African ecological systems. John Ford suggested that British imperial policies transformed the ecology of vast parts of Africa, which was a critical factor in the spread of the trypanosomiasis epidemic. It was the first major work that connected ecological history with the history of epidemics. His research explored the relationships between vectors, hosts and parasites in diverse parts of eastern Africa, Rhodesia (now Zimbabwe) and Nigeria. Ford argued that in precolonial Africa the disease was successfully contained within smaller populations. Most importantly, he studies the role of precolonial African societies in creating barriers (such as 'no-man's land') to the spread of disease. Colonial agricultural policies disturbed the ecological balance, leading to the spread of the disease.[59] Later studies of precolonial Tanzanian ecological history support Ford's view of the ecological balance that existed in these regions before colonialism. Both human and bovine populations had acquired resistance to the disease through slow contact with the vector over the years, which restricted the spread of sleeping sickness.[60] This was disturbed by the large-scale migrations, which exposed virgin populations to the disease, leading it to take on epidemic proportions.

The other influential work was Helge Kjekshus' *Ecology, Control and Economic Development in East African History* (1977). He argued that colonialism in Tanzania from the 1890s led to 'ecological catastrophe' and a series of environmental and health disasters. Diseases such as rinderpest affecting both cattle and wild animals were introduced. The clearing of forests also led to droughts, and old pastoral systems and lifestyles were destroyed.[61]

These works have creatively and incisively used the ecological premises of tropical medicine (which suggested that tropical diseases were caused by tropical climate and environment) and the economic promises of 'constructive imperialism' to return tropical diseases to their ecological and economic context – in the clearing of forests, rise of vector infestations, drilling of mines, setting up of plantations, building of roads, movement of labourers and changes in landscapes. In doing so they have shown that the growing occurrence of diseases in the tropics was in fact a product of colonialism. They have also successfully dispelled the late nineteenth- and early twentieth-and twenty-first centuries ideas of the tropics as being perennially ridden with diseases. Thus what was supposed to be the colonial burden was indeed often the product of colonialism.

At the same time, some of these works, particularly those of Ford and Kjekshus, pose a different burden of history for historians. Their arguments broadly fall within the idea of the imperial apocalypse and are based on the presumption of a precolonial cultural, economic and ecological stability. This was an idea to which postcolonial movements such as the Ujamaa sought to return.[62]

Conclusion: What is tropical about tropical medicine?

So what is tropical about tropical medicine? This is related to the issue of what exactly the tropics are. The tropics are those regions which had enchanted Europeans from the seventeenth century with their natural riches and where they had ventured in search of wealth and adventure. In the process, most of these tropical regions became the colonies of Europe. The term 'tropical diseases' can itself be a misnomer. Many infections and infestations that are classified as tropical diseases, such as malaria, cholera, leprosy and dengue, used to be endemic in Europe and North America as well. Most of these diseases have been controlled or even eliminated from the developed countries. Historical research has also shown that most of these diseases were, and are, caused by lack of food, poor nutrition, environmental factors, lack of clean water and lack of medicine, whether in Europe or in the colonies. They had disappeared from the West mainly due to the provision of an abundant food supply, removal of poverty and the institution of public health measures, which led to better housing and hygiene, improved sewerage and a clean water supply. It is thus safe to say that there was nothing inherently pathological about the tropical climate or environment.

At one level, tropical medicine was sustained by the idea that the tropics were particularly unhealthy. This idea was constructed along with the growth of European colonialism in the tropics from the eighteenth century. Yet there is also the reality that tropical regions continue to suffer from diseases and malnutrition. Why is that so? This is a difficult question to address but part of the answer lies in the divergent histories of Europe and its colonies during the colonial period. This divergence was in the gap in economic wealth between Europe and its colonies, in the dependence of the latter on industrialized European economies, and in the simultaneous disappearance of epidemics in Europe and their resurgence in the tropics.

Tropical medicine developed as a research specialization based on the presumption that the tropics were different. However, it also developed to ensure that the tropics became more habitable for Europeans and thus sustainable as colonies. Along with it was the need felt for the economic and social transformations of these regions. This started the cycle of change, epidemics and preventive measures. Tropical medicine is no longer an imperial science and is undertaken by the international healthcare agencies in the twentieth and twenty-first centuries. With the end of empire, organizations such as the WHO took up tropical medicine as part of their attempts to eradicate malnutrition, establishing medical infrastructure, the control of epidemics and the spread of education.

Notes

1 For primary readings, see Charles Prestwood Lucas, *A Historical Geography of the British Colonies: parts 2 and 4* (Oxford, 1888–1901). A later version of it was

E. Benians, J. Holland Rose and A. Newton (eds), *The Cambridge History of the British Empire* (9 vols., Cambridge, 1929–59).

2 R. Hyam, *Britain's Imperial Century, 1815–1914: A Study of Empire and Expansion* (Batsford, 1976), p. 104.

3 Michael Worboys, 'The Emergence of Tropical Medicine: A Study in the Establishment of a Scientific Speciality', in G. Lemaine et. al. (eds), *Perspectives on the Emergence of Scientific Disciplines* (The Hague and Paris, Mouton, 1976) pp. 75, 76–98.

4 See, for example, P. Manson-Bahr, *Patrick Manson: The Father of Tropical Medicine* (London, 1962), Douglas M. Haynes, *Imperial Medicine: Patrick Manson and the Conquest of Tropical Disease, 1844–1923* (Philadelphia, 2001), W.F. Bynum and Caroline Overy (eds) *The Beast in the Mosquito: The Correspondence of Ronald Ross and Patrick Manson* (Amsterdam, 1998), Helen Power, 'Sir Leonard Rogers FRS (1868–1962), Tropical Medicine in the Indian Medical Service', thesis submitted to the University of London for the degree of Doctor of Philosophy, 1993.

5 W.F. Bynum, *Science and the Practice of Medicine in the Nineteenth Century* (Cambridge, 1994), pp. 149–50.

6 Ronald Ross, 'The Progress of Tropical Medicine', *Journal of the Royal African Society*, 4 (1905), 271–89, p. 271.

7 Ibid, p. 272.

8 Ross, 'Tropical Medicine – A Crisis', *BMJ*, 2771 (7 February 1914), 319–21, p. 319.

9 Charles Cameron, 'An Address on Micro-Organisms and Disease', *BMJ*, 1084 (8 October 1881), 583–86, p 584; Vandyke H. Carter, 'Notes on the Spirillum Fever of Bombay, 1877', *Medical and Chirurgical Transactions*, 61 (1878), 273–300; Charlton H. Bastian, 'The Bearing of Experimental Evidence upon the Germ-Theory of Disease', *BMJ*, 889 (12 January 1878), 49–52; W.M. Crowfoot, 'An Address on the Germ-Theory of Disease', *BMJ*, 1134 (23 September 1882), 551–4.

10 Eliza Priestley, 'The Realm of the Microbe', *The Nineteenth Century*, 29 (1891), 811–31, pp. 814–15.

11 William Roberts, 'Address in Medicine', *BMJ*, 867 (11 August 1877), 168–73.

12 'Professor Koch's Investigations on Malaria: Second Report to the German Colonial Office', *BMJ*, 2038 (10 February 1900), 325–27. For an analysis of Koch's research in tropical diseases, see Christoph Gradmann, 'Robert Koch and the Invention of the Carrier State: Tropical Medicine, Veterinary Infections and Epidemiology around 1900', *Studies in History and Philosophy of Biological and Biomedical Sciences*, 41 (2010), 232–40.

13 Quoted in R. Münch, 'Robert Koch', *Microbes and Infection*, 5 (2003), 69–74, p. 69.

14 Patrick Manson, 'The Life-History of the Malaria Germ Outside the Human Body', *BMJ*, 1838 (21 March 1896), 712–17; see also *BMJ* 1880 (January 9, 1897), 93–100, p. 94. See also Crowfoot, 'An Address on the Germ-Theory of Disease'.

15 'Introduction: Tropical Medicine Before Manson', in Arnold (ed.) *Warm Climates and Western Medicine*, pp. 1–19.

16 Harrison, *Public Health in British India*, pp. 36–59. See also his 'Tropical Medicine in Nineteenth-Century India', *British Journal for the History of Science*, 25 (1992), 299–318.

17 Mark Harrison, 'A Question of Locality: The Identification of Cholera in British India, 1860–1890', in Arnold (ed.), *Warm Climates and Western Medicine*, pp. 133–59.

18 Jeremy D. Isaacs, 'D D Cunningham and the Aetiology of Cholera in British India, 1889–97', *Medical History*, 42 (1998), 279–305, pp. 281–2.

19 Manson 'On the Development of *Filaria sanguinis hominis*, and on the Mosquito Considered as a Nurse', *Journal of the Linnean Society of London, Zoology*, 14 (1878), 304–11.

20 Ross, 'On Some Peculiar Pigmented Cells Found in Two Mosquitos Fed on Malarial Blood', *BMJ*, 1929 (18 December 1897), 1786–1788; Ross, 'Observations on a Condition Necessary to the Transformation of the Malaria Crescent', *BMJ*, 1883 (30 January 1897), 251–55.

21 Contradictory because, as we will see in the next chapter, French Pasteurists believed in the ubiquitous nature of germs – that the vitality of germs was not determined by specific regions or climates.

22 Helen J. Power, 'The Calcutta School of Tropical Medicine: Institutionalizing Medical Research at the Periphery', *Medical History*, 40 (1996), 197–214. See also Power, 'Sir Leonard Rogers FRS (1868–1962), Tropical Medicine in the Indian Medical Service'.

23 For a detailed account of the role of bacteriology in colonial India, see Chakrabarti, *Bacteriology in British India*, particularly chapters 1 and 2.

24 Worboys, 'Almroth Wright at Netley: Modern Medicine and the Military in Britain, 1892–1902', in Cooter, Harrison and Sturdy (eds), *Medicine and Modern Warfare*, pp. 77–97.

25 Jane Buckingham, *Leprosy in Colonial South India: Medicine and Confinement* (New York, 2002), pp. 107–33.

26 Bhattacharya, Harrison and Worboys, *Fractured States*, pp. 146–230.

27 Harrison, *Public Health in British India*, p. 99.

28 W.J. Moore, 'The Causes of Cholera', *Indian Medical Gazette*, 20 (1885), 270–3.

29 Harrison, 'A Question of Locality', pp. 133–59.

30 W. M. Haflkine, 'Le cholera asiatique chez la cobbaye', *Comptes Rendus des Séances et Mémoires de la Société de Biologie*, 44 (1892), 635–37.

31 Ilana Löwy, 'From Guinea Pigs to Man: The Development of Haffkine's Anticholera Vaccine', *JHMAS*, 47 (1992), 270–309.

32 N. Bhattacharya, *Contagion and Enclaves*.

33 For a detailed analysis of how the Indian experience of indirect rule influenced the protectorate system in Africa, see Michael H. Fisher, 'Indirect Rule in the British Empire: The Foundations of the Residency System in India (1764–1858)', *Modern Asian Studies*, 18 (1984), 393–428.

34 As quoted in ibid, p. 426.

35 J. Gallagher and R. Robinson, 'The Imperialism of Free Trade', *Economic History Review*, 6 (1953), 1–15.

36 Leuan Griffiths, 'The Scramble for Africa: Inherited Political Boundaries', *The Geographical Journal*, 152 (1986), 204–216, pp. 204–5.

37 Quoted in 'Tropical Medicine', *Amrita Bazar Patrika* (30 May 1905), 7.

38 Haynes, *Imperial Medicine*.

39 Maryinez Lyons, *The Colonial Disease; A Social History of Sleeping Sickness in Northern Zaire, 1900–1940* (Cambridge, 1992), p. 71.

40 For details of the emergence of the colonial medical service in India, see Chapter 6 of this volume.

41 Anna Crozier, *Practising Colonial Medicine: The Colonial Medical Service in British East Africa* (London & New York, 2007), pp. 3–4.

42 Ryan Johnson, ' "An All-white Institution": Defending Private Practice and the Formation of the West African Medical Staff', *Medical History*, 54 (2010), 237–54.

43 Crozier, *Practising Colonial Medicine*, p. 5.

44 Leonard Rogers, *Happy Toil: Fifty-Five Years of Tropical Medicine* (London, 1950), p. 256.

45 See, Lyons, 'Sleeping Sickness in the History of Northwest Congo (Zaire), *Canadian Journal of African Studies*, 19 (1985), 627–633, pp. 628–9.

46 Lyons, *The Colonial Disease*, p. 66.

47 Quoted in Ibid.

48 Ibid, p. 71.

49 Ibid, p. 74.

50 Edmond Sergent, 'Address Delivered by Dr. Sergent on the Occasion of the Award of the Darling Medal to Dr. Swellengrebel', Geneva (17 September 1938) http://whqlibdoc.who.int/malaria/CH_Malaria_266.pdf.

51 For a detailed analysis of the relationships between disease and economic development, see Packard, ' "Malaria Blocks Development" Revisited: The Role of Disease in the History of Agricultural Development in the Eastern and Northern Transvaal Lowveld, 1890–1960', *Journal of Southern African Studies*, 27 (2001), 591–612.

52 For an analysis of the broad historiographical trends in the study of trypano-somiasis, see James Giblin, 'Trypanosomiasis Control in African History: An Evaded Issue?', *The Journal of African History* , 31 (1990), 59–80.

53 Watts, 'British Development Policies and Malaria in India 1897–c.1929', *Past and Present*, 165 (1999), 141–81; Klein, 'Death in India: 1871–1921', *Journal of Asian Studies*, 32 (1973), 639–59.

54 Arnold, 'Social Crisis and Epidemic Disease in the Famines of Nineteenth-Century India', *Social History of Medicine*, 6 (1993), 385–404.

55 Packard, 'Maize, Cattle and Mosquitoes', pp. 189–212.

56 Packard, *White Plague, Black Labor; Tuberculosis and the Political Economy of Health and Disease in South Africa* (Berkeley & London, 1989).

57 Mike Mathambo Mtika, 'Political Economy, Labor Migration, and the AIDS Epidemic in Rural Malawi', *Social Science & Medicine*, 64 (2007), 2454–63.
58 Lyons, *The Colonial Disease*, pp. 33–4.
59 John Ford, *The Role of Trypanosomiases in African Ecology: A Study of the Tsetse Fly Problem* (Oxford, 1971).
60 Juhani Koponen, *People and Production in Late Precolonial Tanzania: History and Structures* (Helsinki, 1988).
61 Helge Kjekshus, *Ecology, Control and Economic Development in East African History* (London, 1977).
62 For Kjekshus's attachment to the villagization movement in Tanzania, see Kjekshus 'The Villagization Policy: Implementational Lessons and Ecological Dimension, *Canadian Journal of African Studies*, 11 (1977), 262–82.

9

Bacteriology and the civilizing mission

New Imperialism was marked by territorial expansion in the tropics and the simultaneous pathologization and identification of these regions as unhealthy. The word 'civilization' acquired diverse meanings in this context. As we have seen in the last two chapters, the 'civilizing mission' was a common theme of African colonialism in the late nineteenth century. This chapter will explore another aspect of the civilizing mission that developed with the rise of germ theory. I will focus primarily on the French Empire but will also explore other imperial contexts.

Civilization was an integral part of nineteenth-century French imperialism. In France the mission to civilize became part of the official doctrine when France began to expand its imperial possessions in Africa and Asia. French officials publicized the fact that the French had a special mission: *Mission civilisatrice*, to civilize the people whom they now ruled.[1] This particular mission had secular connotations. The essence was control and mastery – over nature, social and cultural habits, environment and reason over ignorance.[2] Pasteurism, through its rationale of attenuating bacteria and microbes both in pasteurization and in producing vaccines, represented primarily the scientific mastery over nature and diseases, and then subsequently the triumph of French rationality over the perceived ignorance in the colonies.

While in the British colonial context the civilizing mission was more closely linked to the activities of missionaries such as David Livingstone, who believed that modern civilization would be introduced to Africa by Christianity, in France it was linked more with secular republican principles. The French concept of 'citizenship of the republic', which is more an ideology than a geopolitical concept, developed from this late nineteenth-century mission to spread civilization and the ideals of the republic to the colonies.[3] French imperialism from the 1870s was driven by this scientific mission, in which hygiene and medicine, particularly bacteriology, played an important role.

There were indeed common features in the French and British civilizing missions, such as the strong moralistic tone, the sense of European cultural and racial superiority, the urge to spread European civilization – whether religion or science – in the colonies, and the mobilization of funds and investments that took place in Europe for this humanitarian cause. French Jesuit missionaries too settled in different parts of the French Empire from the seventeenth century. In the nineteenth century, French missionaries played roles very similar to those of the British missionaries: they ran schools, hospitals and orphanages and were in close contact with the people.[4]

Yet the Christian missionary activities and the secular missions remained two distinct aspects of the civilizing mission. I will look at the history of this secular civilizing mission in this chapter. First I will focus on the rise of germ theory and then the global expansion of French Pasteur Institutes. I will then explore how the idea of civilization featured in the introduction of germ theory in various fields in the colonies, in public health administration, in the vaccination campaigns and in commercial enterprises.

The second half of the eighteenth century was a period of relative decline of the French Empire, mostly following the losses sustained in the Seven Years War. There was a revival in the early nineteenth century with the colonization of Africa and parts of Asia. This phase started with Napoleon Bonaparte's invasion of Egypt in 1798, the first modern European invasion of the Arab world. According to Edward Said, the invasion of Egypt was a turning point in modern imperialism and Orientalism.[5] Why did Napoleon invade Egypt? First, with the English dominant presence in western parts, he had to turn east to expand his empire. Second, the Orient had attracted him from his childhood. As he became ruler of France he saw himself in the image of Alexander the Great, who had entered Egypt in 331 BC. The ancient past of Egypt had become a topic of intellectual interest in France in the late eighteenth century. As he invaded Egypt, Napoleon carried with him scientists, archaeologists, linguists and textual scholars, not the kind of people who usually accompany a military invasion. His invasion was also about the discovery of ancient Egypt. Soon after he arrived in Egypt, Napoleon made a declaration in Arabic that he had come to liberate Egyptians from the tyranny of the Ottoman Empire. He and the other French imperialists regarded the invasion of Egypt as a triumph not just of their military power but also of French Enlightenment over Oriental despotism. Thus the invasion of Egypt was an attempt at conquests of two kinds: of land and of minds. This marked the beginnings of the French civilizing mission.

After a gap during the Napoleonic Wars and the end of his era, the second phase of the French colonization of North Africa started in 1830 with the invasion of Algeria. This was followed by expansion in Asia and the occupation of Indochina in the 1860s. The establishment of the Third Republic (the republican government of France from 1870, which lasted until 1940) in 1870 started the most aggressive phase of French colonialism in Africa and Asia. The French occupied Vietnam between 1884 and 1885. They established a protectorate of Tunisia in 1881 and expanded in central and western Africa; Senegal, Guinea, Mali, the Ivory Coast, Benin, Niger, Chad and the Republic of Congo by the end of the century.

Away from the imperial battlefields, French science experienced a 'golden age' in the late nineteenth century, after a period of decline following the French

Revolution. The revival was mainly in laboratory research with the emergence
of Pasteurian science. Until the early part of the nineteenth century, physicians
understood disease in terms of miasmas and humours. By the 1830s, scientists
found evidences that yeast consisted of small spherules, which had the property of multiplying and were therefore living organisms.[6] Aided by new micro-
scopes, laboratories and experiments, scientists soon discovered a 'new world
of life'. By the 1860s, French microbiologist Louis Pasteur had established that
fermentation depended on living forms or bacteria and that there was vital and
dynamic character to this change. He proposed that subjecting milk, wine or
food to partial sterilization destroyed most of the microorganisms and enzymes
present in it, making it safe for consumption and improving its preservation. In
1864 he demonstrated the sterilization of milk by heating it to a high tempera-
ture and pressure before bottling. The action or process of subjecting milk, wine,
agricultural products and food to partial sterilization is called pasteurization and
is now in widespread use.

Pasteur then suggested that these bacteria also caused human and animal
diseases. This came to be known as germ theory, which developed predominantly
within French and German medical traditions from the 1880s. It suggested that
diseases were caused by germs or microorganisms, not miasma, the environment
or humours. German physician Robert Koch was first to devise a series of proofs
to verify the germ theory of disease. His postulates were first used in 1875 to
demonstrate that anthrax, a disease that affected cattle and was a major concern
for the farming and leather industry in Europe and the United States, was caused
by the bacterium *Bacillus anthracis*. These postulates are still used today to help to
determine whether a microorganism causes a newly discovered disease.

Pasteur, on the other hand, combined his work on the partial sterilization of
germs with the identification of germs to develop antibacterial vaccines. In the
1870s he applied this immunization method to anthrax. He successfully prepared
an anthrax vaccine in 1881. However, his most important breakthrough came
when in 1885 he developed a vaccine for the treatment of rabies, a dreaded
disease that was passed from dogs and wild animals to humans. Pasteur identified
the nervous system as the main target for the experimental reproduction of the
rabies virus. He and his collaborators attenuated the virus by repeated passages
through rabbits. Strips of fresh spinal cord material taken from rabbits that had
died from rabies were exposed to dry and sterile air for various lengths of time.
This tissue was then ground up and suspended in a sterilized broth. The solution
was used for the vaccine. The discovery of the vaccine for rabies led to the establishment of the first Pasteur Institute in Paris in 1888, where Pasteur vaccinated
people who came from different parts of Europe.

Vaccines and pasteurization became the two pillars of Pasteurian science from
the 1880s, which influenced French public health policies, veterinary medicine,
agriculture and the food industry. In Europe the Pasteur Institutes carried out
vaccinations for rabies, anthrax, tuberculosis and plague, and participated in
public health undertakings of food preservation, agricultural, dairy and meat

production, and the determination of diet and nutrition. Soon Pasteur Institutes spread to different parts of the world, particularly to the French colonies, and germ theory and vaccines became part of imperial medicine.

The historical significance of germ theory was in the break it marked from existing practices of medicine, which were based on humourology or miasmatic theories. There was now a greater specificity of causation of diseases, rather than multiple or pluralistic explanations. Germ theory suggested that germs were the real enemies, which could be identified and eradicated with the application of bacteriology. It also put laboratories at the heart of public health policies, and as new institutions and symbols of Western medicine and modernity in the colonies. This also led to a changing notion of the human body, where earlier the body was viewed as part of the environment. The body was no longer seen to be in either harmony or disharmony with the environment, which had particular significance for how Europeans viewed themselves in the tropical environment. Now the human body (some bodies more than others) appeared full of germs and needed to be vaccinated or isolated. This established the universality of germ theory: germs could be identified for any diseases in any part of the world and could be eradicated by vaccines. This also allowed for more intrusive public health measures whereby the state and doctors could inject antigens into the bodies of its citizens and others.

Germ theory introduced a new focus in colonial medicine as it challenged the main tenets of medicine, suggesting that it was germs that caused diseases in the tropics, not the heat or the miasma. It also brought about a new confidence in colonial medicine – there was no need to be afraid of the tropical climate or miasmas, or indeed the degeneration of white races in the tropical climate. I will now explore how effective and crucial that confidence was in imperial history.

Germs and civilization

I will start investigating the history of germs in imperialism by asking what the relationship was between germs and civilization. The discovery of germ theory identified specific pathogens as causal agents of specific diseases. This provided the new possibility of, and urgency for, the eradication of diseases. However, the propositions of germ theory were never fully accepted by doctors and health officials. Even after germ theory had been established, diseases prevalent in the tropics or among the poor (in Europe or in the United States) continued to be linked with filth as they had been in earlier centuries. For example, cholera in the 1890s was described as a 'filth disease carried by dirty people to dirty places'.[7]

The zymotic theory of the 1870s (which suggested that diseases were caused by decomposition and degeneration) created the link between the new germ theory and the earlier ideas of putrefaction, absorbing the same moral values associated with filth and decomposition. Both filth and germs were seen as contributing to disease.[8] Physicians believed that moral miasmas corresponded to physical ones; moral filth was as much a concern as physical filth.[9] Now germs represented filth in both sanitarian and moral terms. The theory of the 'human carrier' of germs, which suggested that even healthy individuals can carry germs in their bodies and infect others without themselves showing symptoms of the disease, reinstated medical segregation based on race and class. Thus the eradication of germs also became acts of cleansing: of filth, unclean habits and prejudices, and even segregating unwelcome races and ethnic groups. In other words, Pasteurian scientists, public health officers, vaccinators and governments saw the eradication of diseases in the tropics or among the poor not just as practices of the immunization of germs but as acts of social and cultural reform. These ideas of filth and germs found new meaning in the tropical colonies where imperial medical men represented themselves as being involved in a moral crusade against colonial and tropical germs, diseases and prejudices. Bacteriology in the colonies became the new symbol of scientific and industrial modernity. Pasteurization and vaccination promised commercial and industrial progress in the colonies. Thus the eradication of germs often symbolized the eradication of barbarism.

The word 'civilization' gathered different meanings in different colonial contexts in Asia, Africa, Australia and South America. It reflected the cultural supremacy imposed by one group or community on another in terms of class, caste and race. It allowed the former to determine the living conditions and economic activities of the latter. Germs provided scientific validity to political, economic and social discrimination and segregation. For example, in late nineteenth-century Rio de Janeiro in Brazil, the elites of the city used the terms 'civilization' and 'germs' to protect their own privileges and isolation from the poor of the city. This involved first identifying the poor as carriers of filth and disease, then starting a process of driving them out from the city centre, changing their daily habits and introducing large-scale enforced vaccination among them.[10] These growing divisions in the city of Rio had a broad imperial backdrop. Brazil was part of the 'informal empire' and was heavily dependent on foreign investments in its growing cocoa and rubber plantations, leading to the growth of the migratory labour force on the one hand and the urban elites on the other, and a growing divide between the rich and the poor. Rio grew rapidly due to the requirements of international capital, which financed, planned and oversaw the transformation of the economy not only of the city but of the whole of Brazil. The elites of the city used the term 'civilization' to 'reform' the lives of the poor. In these new measures of cleansing the city and introducing vaccination among the urban poor, Oswaldo Cruz, who was trained in the Paris Pasteur Institute, was appointed as director general of public health.[11] He established a bacteriological institute in Rio in 1902 and started a strict regime of vaccination in the city, as well as restricting the poor to certain areas, away from the sights of the rich.

In Australia, 'civilization' was often used to mark the dividing lines between the white settlers and Asian immigrants. In the twentieth century, the Australian government launched what a journalist described as a 'war on foreign germs', which was essentially a war against Chinese immigrants. In the government's imagination and plans for a 'White Australia', Chinese immigrants were seen as unclean reservoirs of dangerous germs. Such ideas became particularly evident following the outbreak of plague in China in the 1890s.[12] Quarantine systems in Australia became more rigorous and intrusive with the rise of theories of germs. To protect the Australian shores from the invasion of germs and immigrants, which were by now seen as synonymous, the Australian government even intervened in the public health matters of Melanesian and Polynesian islands in the 1920s. While doing so, Australian health officers viewed the islanders as 'primitive', easily susceptible to germs and in need of modern scientific health measures.[13] While establishing preventive health measures in the Pacific islands, Australian doctors believed that they were also bringing civilization to them.

In Palestine, germs and civilization connoted the lines between modernity and Orientalism. Dr Leo Böhm, a Zionist physician, launched a movement to establish a Pasteur Institute in the early twentieth century. This was part of the Zionist movement to 'civilize' and transform the wild land into a modern nation. His efforts were supported and encouraged by the World Zionist Organization, whose members shared his view that the establishment of Zionism in Palestine depended on the application of modern science and technology. He was also backed by the Association of Jewish Physicians and Natural Scientists for the Sanitary Interests in Palestine, which comprised several Jewish physicians and doctors based in Europe. Böhm launched his war against malaria in Palestine, which he saw as the bane of modern civilization. He also instructed people about healthy and sanitary habits, and about the merits of quarantine.[14]

Germs in the empire

Three important historical processes coincided with the spread of germ theory and the introduction of vaccination in the colonies. First, by the time bacteriology developed as a specialization, the tropics had been identified as unhealthy regions. Pasteur and Koch's ideas of germs and bacterial fermentation thereby assumed new potency there. Germ theory provided a new explanation and connotation to the phenomenon of putrefaction in the tropics, hitherto associated with heat. This provided both optimism about European colonization of and habitation in the tropics, as well as new apprehensions about germs in the colonies. On the positive side it helped to challenge the climatic determinism of tropical diseases, and there were now possibilities that, with vaccines, diseases could be successfully eradicated. At the same time, questions arose about whether they behaved differently in hot climates, whether germs were more virulent in the tropics and whether vaccines would lose their potency in the tropical heat.

Second, as a historical coincidence, precisely at the time, in the 1880s and 1890s, when bacteriology was taking significant strides, several epidemic outbreaks of cholera and plague took place in different parts of the tropical colonies, killing millions of people and often threatening European and North American trade and borders.[15] The 'Third Pandemic' of plague started in China in the 1880s and soon spread globally. From Pakhoi in China in 1882, plague spread to Canton (now Guangzhou, 1894) and Hong Kong (1894), reaching Bombay in India in 1896 where it raged until the end of the century. It spread to Madagascar (1898), Egypt (1899), South America (Paraguay, 1899), South Africa (1899–1902) and San Francisco, (1900). It also spread to Australia (1900–5) and the Soviet Union (1900–27). The 'Fifth Pandemic' of cholera originated in the Bengal region of India and swept through Asia, Africa, South America and parts of France and Germany. It claimed 200,000 lives in the Soviet Union between 1893 and 1894, and 90,000 in Japan between 1887 and 1889. The 'Sixth Pandemic' of cholera spread globally between 1899 and 1923. It killed more than 800,000 in India before moving into the Middle East, northern Africa, the Soviet Union and parts of Europe.

These pandemic outbreaks established the link between germs and the tropics, and imprinted in the popular and scientific discourse in Europe the need for bacteriological intervention in the tropics to protect primarily European lives and commercial interests there. The outbreaks of plague and cholera had seriously challenged the efficacies of existing preventive health measures in the colonies. Germ theory and bacteriology appeared as the new hope – a means of dispelling these fears and providing a new force in the expansion and consolidation of the empire. Pasteurism provided a critical moral and institutional imperative within colonial medicine in the fight against epidemics. Bacteriology became part of the urge to eradicate diseases from the tropics, to make these places habitable for Europeans and to introduce modern agriculture and industry to the colonies.

Rapid development in bacteriological investigations and the production of vaccines took place in the tropics, particularly for plague and cholera. Koch identified the *Comma bacillus* in 1883, following which Waldemar Haffkine developed the cholera vaccine in Paris in 1893. The history of the identification of the plague germ and its vaccines was more contested. In 1894, Alexandre Yersin (a Swiss physician) and Shibasaburo Kitasato (a Japanese bacteriologist) independently announced the isolation of the plague organism. They represented the two schools of bacteriology competing with each other in bacteriological research in Asia. Kitasato was trained under Koch in Germany. Yersin had joined Calmette at the French Pasteur Institute in Indochina. In 1895, Yersin opened a second Pasteur Institute at Nha Trang. Both investigators were previously in Hong Kong in June 1894 to study the epidemic of bubonic plague, which had spread through southern China and claimed over 40,000 lives. This created a competitive atmosphere and both claimed to isolate the plague bacteria simultaneously in 1894. Similarly, several different and competing vaccines were developed during the time of the plague pandemic in 1897 – by Alexandre Yersin, A. Lustig and Haffkine.

Ilana Löwy has described the search for germs of different diseases, particularly in the tropics in the late 1870s and the 1880s, as the triumph of the 'microbe hunters'. Apart from plague, cholera, rabies and anthrax, bacteriologists successfully isolated and cultivated a great number of pathogenic bacteria: typhus (1879), leprosy (1880), pneumonia (1882), tuberculosis (1882), diphtheria (1883–4), tetanus (1884), Malta fever (1886) and meningitis (1887). The isolation of these aetiological agents also opened the way to develop specific therapeutic antisera.[16] In the tropics the notion of 'hunting' for microbes developed in relation to colonial hunting sports as well. Koch became a game hunter when he went to Africa and hunting was an integral part of his sleeping sickness expedition of 1906–7 in Africa. He shot and autopsied several animals during the expedition, ostensibly to identify the animal hosts of *Trypanosoma*, but Koch in reality hunted any form of animal that he came across, such as herons, eagles, crocodiles and hippopotami.[17] The German bacteriologist's enthusiasm for conducting research in Africa and his fascination and fear of its natural world reflected the aggressive adventurism of European colonialism in that continent. Both wild animals and pathogens were seen as part of Africa's repulsive and dangerous wilderness and needed to be eliminated.

Such microbe hunting expanded the field of bacteriology considerably. Bacteriologists searched for and identified new germs in water, soil, and in animal and human bodies. In 1902, during his typhoid research in Trier, Koch proposed the theory of the 'carrier state'. He suggested that healthy individuals can show no signs of the disease but still carry in their gall bladders or intestines the bacteria of typhoid with which others may be infected. He used this to explain the endemic nature of the disease among certain populations who could infect others. Bruno Latour argues that the identification of humans as carriers of germs helped in the spread of Pasteurian ideas, as any individual could be a carrier of germs and thereby a subject of Pasteurian analysis.[18] The Pasteurian population, according to him, consisted of 'sick contagious people, healthy but dangerous carriers of microbes, immunized people, vaccinated people, and so on'.[19] This marked two important shifts in bacteriological research. On the one hand, the human body became the focus of research as the site of germs. On the other, this had important implications for twentieth-century racial pathology and tropical medicine. Koch forwarded this thesis in his research to tropical diseases such as trypanosomiasis, which led to the identification of Africans as the carriers of the disease.[20] In Eastern Europe, Jews were considered to be the carriers of typhoid and were subject to brutal sanitary measures.[21]

The third important feature in the introduction of bacteriological research in the tropics was the political context. Bacteriology developed at the same time as imperialism reached a new and critical phase: the period of New Imperialism.[22] The Berlin agreement (1885) stipulated that henceforth the colonization of Africa would be without war or bloodshed, but through the display of influence in economic and cultural spheres. The colonial Pasteur Institutes served this purpose effectively in the French Empire as they participated in both protecting the economic interests in the colonies through their veterinary research and pasteurization, and in demonstrating their humanitarian credentials by vaccinating locals against diseases. Pasteur Institutes became an important adjunct of French colonial hegemony and influence.

Expansion of French Pasteur Institutes in the colonies

Soon after the establishment of the Pasteur Institute in Paris, the French established similar institutes in their colonies in South East Asia, South Asia, Africa
and Cuba (e.g., Saigon [1891], Nha trang [1895], Hanoi [1925], Dalat [1931], Hue
in Vietnam, Vientiane in Laos and Phnom Penh [1953], Tunisia [1893], Algeria
[1900], Tangier in Morocco [1910], Casablanca [1929], Saint Louis in Senegal
[1896], Brazzaville in the Congo [1908], Kindia in Guinea [1922], Madagascar
[1898] and Havana [1885]). Following this trend, Pasteur Institutes were established by other colonial nations as well. In the British colonies, such as India, the
British established several bacteriological laboratories, starting with the Imperial
Bacteriological Laboratory in Poona (1890), the Bacteriological Laboratory in
Agra (1892), the Plague Research Laboratory in Bombay (1896), and the Pasteur
Institutes of India in Kasauli (1900), Coonoor (1907), Rangoon (1916), Shillong
(1917) and Calcutta (1924). There was also a Pasteur Institute in Palestine. They
were also introduced in peripheral regions, such as Brazil, Cuba and Argentina,
and in the settler colony of Australia.[23]
     The French Pasteur Institutes had a clear political mandate and worked closely
with a network of colonial lobbies, institutions and agents based in France to
spread Pasteurian science across the colonies. As the French colonial expansion
took place, an influential colonial lobby (*Parti colonial*) was formed in Paris in the
1870s. This comprised individuals from various fields and organizations. It was
the main colonial interest group in France that pushed for territorial expansion
and for greater French economic investments in the colonies. By the 1890s the
lobby was more formally organized, particularly under Eugéne Etienne.
     He was influential in the expansion of French science in the colonies. He
connected the colonial lobby with the other colonial institution based in France,
the Ecole Coloniale (1889), which held training classes in science and engineering
for colonial officials. The Ecole Coloniale played a significant role in promoting
French science, language, hygiene and technology in the colonies by training
students from the colonies in these subjects.[24]
     Under Etienne, the colonial lobby formed committees for each major
colony or group of colonies, such as the Committee for Madagascar (1895), the
Committee for French Asia (1901) and the Committee for French Africa (1890).
These promoted the cause of commercial opportunities in the colonies, highlighted the achievements of French colonial governors, doctors, engineers and
missionaries, and pleaded for investments in colonial infrastructure.[25] The French
Pasteur Institutes in Saigon, Tunis and Morocco were created out of a close association between Pasteurian scientists and the colonial lobby. The colonial Pasteur
Institutes sought to promote the interests of the lobby in agriculture, veterinary
medicine and public health in the French Empire
     The Pasteur Institutes also worked closely with the French Colonial Health
Service (CHS), which was formed in 1890 soon after the Pasteur Institute was
established in Paris. This institute often trained the staff of the CHS, who served

in Africa and Indochina, in microbiology. Albert Calmette went to Indochina as a member of staff of the CHS and was trained in the Pasteur Institute in Paris. The head of the CHS consulted Pasteur when establishing the Pasteur Institute in Saigon. Yersin travelled to Saigon from Paris on the request of the CHS to conduct bacteriological investigations on plague in 1894.

For these significant connections with committees and networks in France, the Pasteur Institutes in the colonies had strong metropolitan characteristics. Firmly controlled from Paris and Lille, these institutes represented both the French civilizing mission and the centralizing tendencies of French imperialism under the Third Republic.[26] The Paris Pasteur Institute saw its colonial institutes as peripheral, devoted to applied research based on the 'pure' research done in Paris and Lille, and carrying out the civilizing mandate of the home nation.

The Pasteur Institutes and the French civilizing mission

French colonial Pasteurism and the establishment of overseas Pasteur Institutes was a dramatic episode in the spread of Pasteurian science. It evoked ideas of the transformation of colonial economy, culture and health. It proposed the introduction of vaccines and pasteurization as features of the French imperial benevolence and civilizing mission.[27] The French civilizing mission derived its meaning from a faith in the superiority of French culture and of the French idea of the republic. France's revolutionary past (1789–1848) had led to the overthrow of its monarchy, and power being nominally bestowed in the hands of its citizens. It had also engrained the idea of racial and civilizational superiority and progressiveness (which supposedly led to the formation of its republic) within French colonialism. At the end of the nineteenth century, armed with the benefits of yet another 'revolution' – bacteriology – French officials, scientists and imperial governors descended on the colonies with a mission to civilize and to spread the messages and gifts of France to its colonies. The civilizing mission became a movement to introduce French modernity to the colonies.

The French Pasteur Institutes adopted the term 'mission' for their overseas expansion to suggest that they were undertaking humanitarian activities there. There was continuity with earlier medical missionary activities in the colonies that started in the 1840s. Now scientists posed as the new missionaries. On the other hand, the colonies provided a new field for French bacteriology to operate as a mission; new diseases provided bacteriologists with new scope for the application of their science. This new mission was to eradicate disease and prejudice from the colonies. The Pasteur Institutes there participated in launching the civilizing mission on three fronts: the human and animal body, the land and culture.

Pasteur Institutes started mass vaccination in the colonies in Africa and Asia, initially for rabies, but soon expanding to tuberculosis, typhoid, plague and snakebites, among others.

Several prominent Pasteurists (disciples or followers of Pasteur) ran the colonial institutes. Charles Nicolle ran the institute at Morocco, Albert Calmette was in the charge of the institute in Indochina, and Yersin was in charge of the Nha trang institute. They conducted investigations, and discovered new bacteria and vaccines in the colonies. They also conducted vaccinations among the indigenous population. Calmette discovered the Bacillus Calmette-Guérin (BCG) vaccine and antivenom serotherapy, Yersin discovered the plague bacillus and Nicolle introduced vaccination for Malta fever (also known as Mediterranean fever) in North Africa.

The introduction of the BCG vaccination against tuberculosis in Indochina in the 1920s was part of the French imperial *Mission civilisatrice* in the region.[28] Historically, tuberculosis, since it attacks the lungs, was associated with vitiated air, unhygienic living conditions, filth and poverty, and as a general threat to civilization.[29] The BCG vaccine was employed in the city of Cholon (Saigon) by the French authorities to usher in the 'medical civilization' of Indochina.[30] Tuberculosis became the target of, and the BCG vaccine provided a focal point to, the broad spectrum of the French civilizing mission in Indochina.

From the 1870s, as the French occupied the region, they introduced hygiene courses into the school curriculum, and teachers were instructed to check the cleanliness of each student and that of their families. In the 1920s, soon after the BCG vaccine was discovered, a more intensive social movement was launched. First, the governor of Indochina, Dr Cognacq, started a comprehensive study of the lifestyle and living conditions of the local people. Along with that, the scientists belonging to the local Pasteur Institute searched for the source of the infection in a systematic study of the habits of local children. The children of the Municipal Boys School of Cholon were studied closely. The health details of their families were also recorded. Soon local people joined in the fight against the disease. French bacteriologists introduced BCG vaccination in the city in 1924 among infants. The vaccinations were accompanied by a campaign by the government to convince local people of its benefits. Through a concerted effort of scientists, doctors and administrators, the anti-tuberculosis movement and the BCG vaccination campaign in Cholon city became a movement for the inculcation of 'civilized' habits and modernity.

In other French colonies, particularly in North Africa, vaccination was made compulsory primarily to safeguard the lives of the French settlers. Vaccinations were carried out for rabies, tuberculosis, plague and smallpox among the local population. The added factor in North Africa was the arrival of the Haj pilgrims, which French colonial governments viewed as a particular threat, so they introduced vaccination of pilgrims. Indigenous physicians in North Africa incorporated vaccination and other French medicines into their own medicine. This helped in the wider absorption of French medicine within indigenous culture and society.[31]

The project of modernizing the colonial landscape took place in the large-scale changes in colonial agriculture and industrial practices. One of the driving forces behind late nineteenth-century imperialism was the identification and exploitation of the economic resources of the colonies, in agriculture, farming, mining and industrial sectors. Bacteriology and pasteurization were integral to this agenda. Apart from vaccinating labourers in mines, factories and planta-tions against diseases such as typhoid, cholera, influenza and plague in India, Sri Lanka, Batavia, Brazil, Argentina, Cuba and southern Africa, bacteriology was employed in promoting colonial economic interests. The French used their Pasteur Institutes in South East Asia to protect French monopolies in the manu-facture of alcohol, tobacco, opium, silk and rice by devising pasteurization processes for these products. Since the alcohol and opium trade was an impor-tant revenue earner in the French colonial economy in Indochina, Pasteurian investigations into rice and opium fermentation benefited French colonial industries.[32]

In Africa the French Pasteur Institutes made major investments in veterinary and agricultural research to protect the colonial farming economy. This went along with efforts to protect the lives of French settlers and officials serving the colony through vaccination. In Tunisia (a French protectorate), the Pasteur Institute of Tunis was opened in 1893 with Charles Nicolle (1866–1936), who had trained at the Pasteur Institute of Paris, as the director, primarily to assist the Department of Agriculture in analysing diseases of wine. Apart from that, the laboratory produced and administered vaccines and sera against rabies, smallpox, diphtheria and typhus, and conducted laboratory tests for hospitals and dispen-saries.[33] The colonial Pasteur Institutes in Tunis and Morocco played an important role in the economic activities of the entire Mediterranean region by promoting vaccination as well as pasteurization in the farming industries. French Pasteurists sought to form a coalition of Pasteur Institutes around the Mediterranean to conduct research in a more coordinated and cooperative manner, by developing alliances with businesses and industries in the entire region. Nicolle, who ran the entire operation from his institute in Tunis, was once even described as the 'Emperor of the Mediterranean Sea'.[34]

The economic and medical influence of the Pasteur Institutes extended beyond the formal French Empire as similar institutes developed in Athens in Greece, under the direction of French bacteriologists or local physicians trained at the French institutes. The rulers of the Ottoman Empire consulted the Pasteur Institute of Paris on its efforts to industrialize Turkey. Even in India, southern America and Central America, bacteriology played important economic roles both in tackling diseases that affected labouring populations, such as yellow fever, cholera and influenza, and in promoting veterinary medical research and farming.[35]

Apart from assisting and protecting the economic interests in the colonies and thereby transforming the colonial landscape, bacteriology also helped in the establishment of European cultural superiority in the colonies. Three processes shaped the hegemonic role that bacteriology and germ theory played there. First was the emergence of the idea that bacteriology was 'scientific

medicine', as it had marked a break from traditional notions of humourology
and climate, and used extensive laboratory methods and experiments in the
study of diseases as well as the production of vaccines and sera. Second was the
emergence of the idea of the 'protectorate' in Sub-Saharan African colonialism
in the 1880s. This took place, as we saw in Chapter 8, particularly in Africa, in
the context of the so-called 'Race/Scramble for Africa'. The third was the hegemonic role of Pasteurism and of European colonialism in the late nineteenth
century.

As we have seen, this Pasteurian 'mission' was distinct from the Christian
mission, which developed alongside and operated with an urge to spread the word
of the Gospel in Africa, and to root out slavery from that continent. Yet the two
missions (the Pasteur Institutes and the French religious missions) often shared
common elements. In the French Empire, Pasteur Institutes often adopted strong
moral connotations and Catholic symbolism, despite the overt secular nature of
their science and of their civilizing mission. Pasteur's disciples represented him as
the prophet or 'pilgrim father'. Those, like Emile Roux, who were trained directly
under him and headed later institutions in the French colonies 'displayed the
proofs of his lineage: he [the French colonial Pasteurist] was trained by one of the
early companions of the prophet'.[36]

There was also a strong metaphor of conquest in colonial Pasteurian science.
Colonial Pasteurian scientists launched their institutes and their acitivities as a
movement to conquer the germs of the tropics at a time when these were equated
with colonial filth and backwardness. The colonial Pasteur Institutes were often
located in the middle of busy colonial cities from where the French bacteriologists
and military doctors boldly waged war against tropical diseases and even tropical
realities. Once satisfied with their activities, Pasteurists such as Ernest Conseil
and Emile Sergent declared Tunis and Algiers to be 'hygienized' and 'pasteur-
ized' cities, respectively.[37] The French bacteriologists informed the local population how they were saving lives by vaccinating people against deadly diseases,
and thus the benefits of French imperialism. Pasteurists such as Charles Nicolle
portrayed themselves as missionaries in North Africa, taking a backward country
to modern civilization.

Thus bacteriology in the colonies had a hegemonic character, which corresponded perfectly with the moral 'constructive' tone of New Imperialism. French
statesmen and imperial administrators recognized the role played by Pasteur in
French colonialism. Late nineteenth-century French statesman and historian
Gabriel Hanotaux described Pasteur as the 'grand master of modern colonization' and proclaimed that 'scientific medical colonization has become one of the
noblest adjuvant of French science'.[38]

The interplay of the three modes of the civilizing mission (of bodies, minds
and the land) of French imperial Pasteurism was most evident in Algeria. Even
though Algeria was not strictly a 'tropical' country (it is located above the Tropic
of Cancer, in the temperate climatic zone), French scientists and bacteriologists often saw it as belonging to the cultural and epidemiological domain of

tropical medicine. This idea of tropicality was infused with French ideas of Arab Orientalism, and these scientists increasingly described Arab culture, social values and practices as antithetical to civilization. In the last decades of the nineteenth century, French scientists referred to Algeria, Tunisia and Morocco as 'exotic pathologies'. French scientists working in the Pasteur Institutes of North Africa saw the indigenous population as 'virus reservoirs' which needed the modernizing imprint of Pasteurian bacteriology.[39]

By the late nineteenth century the Napoleonic vision of revitalizing Arab culture in North Africa through French enlightenment and 'assimilation' had largely disappeared. The local population in Algeria lived in segregated quarters and very few of them went to French/European schools. The Pasteur Institute that opened in Algiers offered new hope by introducing European modernity to the Arab world. A new civilizing mission was launched through the Pasteur Institute, with doctors and scientists at the forefront. As in Indochina, French scientists in Algeria set out to investigate the causes of diseases, inspired by the universality of Pasteur's ideas of germs and vaccination. They rejected the eighteenth-century ideas of medicine of hot climates, which suggested regional specificities and a climatic fatalism, and set out to expand the universality of Pasteurian science into the new paradigm of bacteriology in hot climates.

Two brothers, Edmond and Etienne Sergent, both of whom were trained in French microbiology, joined the Pasteur Institute at Algiers in 1912. They came with a clear mandate: to continue Laveran's earlier work on the malaria parasite in Algeria and thus extend the benefits of French *bienfaisance* (beneficence) to the country.[40] They started a malaria research programme along the Algerian railway network in the early twentieth century, which involved eclectic methods of weed removal from canal banks, destruction of larvae and spreading of oil on stagnant water surfaces. They also undertook mass investigations among the local Algerian population, who were regarded as 'germ carriers' and, as Moulin describes, 'came to be identified with the germs themselves'.[41] They also sought to popularize the bitter quinine drug by lacing it with chocolate and distributing these among local children. In their view, the eradication of malaria parasites and the large-scale modernization of the Algerian landscape through French science and industry required the intervention of Pasteurian science within Algerian bodies as well.

In 1927 the French colonial authority granted the Sergent brothers a 360-acre malarial marshland along the railway line as a field laboratory to conduct their anti-malarial work. Here the brothers started research to demonstrate how Pasteurian anti-malarial work could transform the Algerian landscape and cultural habits. They dug an extensive system of drainage canals with the help of local labour, planted trees and introduced larvae-eating fish in stagnant water. They introduced cereal crops and livestock to replace the marshes, persuaded Algerian farmers to settle on the land, taught them modern farming methods and planted several acres of vineyards (the latter was of great economic interest to the French colonial government). While doing so, the Sergent brothers claimed that they had 'humanized' the once 'savage' marshland with the help of Pasteurian science.[42]

Conclusion

With the emergence of germ theory, laboratory research became an integral part of modern medicine. The diagnostic skills and methods of the physicians, which helped in the understanding of diseases, were reinforced or replaced by laboratory experiments conducted to identify germs, analyse their life cycles and develop effective vaccines. Pasteur and his followers played a particular role in disseminating this influence in various sectors of modern life. Within France the spread of Pasteurian science took place through a range of networks, including the public hygiene movement, the medical profession, the farming profession and industry, which shaped French public health, farming practices and the economy. Bruno Latour has described this hegemonic spread of Pasteur's influence in France as the 'pasteurization of France'.[43]

In the colonies the change from clinical to laboratory medicine was even more revolutionary and radical. Here the spread of Pasteurian science and bacteriology served two main purposes: the colonization of the body and the colonization of the mind, both of which were connected to the civilizing mission. Vaccination and pasteurization marked a deeper intervention of imperial medicine and science within the human and animal bodies and agricultural practices than before. They helped to establish immunization (often compulsory) as a fundamental element of global health.

On the other hand, laboratory medicine marked another moment when medicine became 'Western'. Unlike humourology or miasmatic theories, in Asia and Africa there were no equivalents in indigenous medicines to laboratory medicine. While several indigenous traditions were quick to respond and incorporate germ theory and vaccination within their medical practice and cosmology (see Chapter 10), from the late nineteenth century the laboratory defined what was modern and Western about medicine. This was a new phase of imperialism too, which was not just shaped by the activities of traders and the conquests of armies. This was an imperialism that established European science and medicine as a critical aspect of the modernity aspired to by countries and communities in various parts of the world. The influence was also far-reaching, as we have seen: it spread to South America and Australia, which were no longer controlled formally as empires.

Notes

1 A.L. Conklin, *A Mission to Civilize: The Republican Idea of an Empire in France and Africa* (Stanford, 1997) p. 1.

2 Ibid, pp. 5–6.

3 M. Ticktin, 'Medical Humanitarianism in and beyond France: Breaking Down or Patrolling Borders?' in Alison Bashford (ed.), *Medicine at the border: disease, globalization and security, 1950 to the present* (Basingstoke, 2006) pp. 116–135.

4 For a collection of articles on the role of missionaries in the French empire, see, Owen White and and J.P. Daughton (eds), *In God's Empire: French Missionaries in the Modern World* (Oxford, 2012).

5 Said, *Orientalism*, p. 80.

6 Stig Brorson, 'The Seeds and the Worms: Ludwik Fleck and the early history of germ theories', *Perspectives in Biology and Medicine*, 49 (2006), 64–76.

7 Ernest A. Hart, 'Cholera: Where it comes from and how it is propagated', *BMJ*, 1696 (1 July 1893), 1–4, p. 1.

8 Hamlin, 'Providence and putrefaction; Victorian sanitarians and the natural theology of health and disease', *Victorian Studies*, 28 (1985), 381–411, pp. 386–7.

9 Felix Driver, 'Moral Geographies: Social Science and the Urban Environment in Mid-Nineteenth Century England', *Transactions of the Institute of British Geographers*, 13 (1988), 275–87.

10 Teresa A. Mead, *'Civilizing' Rio: Reform and Resistance in a Brazilian City, 1889–1930* (Philadelphia, 1997).

11 Ibid, 89–90.

12 Alison Bashford, 'At the Border: Contagion, Immigration, Nation', *Australian Historical Studies*, 33 (2002), 344–58.

13 Alexander Cameron-Smith, 'Australian Imperialism and International Health in the Pacific Islands', *Australian Historical Studies*, 41 (2002), 57–74.

14 Nadav Davidovitch and Rakefet Zalashik, 'Pasteur in Palestine: The Politics of the Laboratory', *Science in Context*, 23 (2010), 401–25.

15 Samuel K Cohn, '4 Epidemiology of the Black Death and Successive Waves of Plague', *Medical History Supplement*, 27 (2008), 74–100; and R. Pollitzer, 'Cholera studies: 1. History of the disease', *Bulletin of the WHO*, 10 (1954), 421–461, p. 449.

16 Löwy, 'Yellow fever in Rio de Janeiro', p. 145.

17 Christoph Gradmann (translated by Elborg Forster), *Laboratory Disease: Robert Koch's medical bacteriology* (Baltimore, 2009), 222–4.

18 Bruno Latour, *Pasteurization of France* (Cambridge, Mass, 1988), 80–6.

19 Latour, *Science in Action: how to follow Scientists and Engineers through Society* (Milton Keynes, 1987), pp. 115–6.

20 Gradmann, 'Robert Koch and the Invention of the Carrier State: Tropical Medicine, Veterinary Infections and Epidemiology around 1900', *Studies in History and Philosophy of Biological and Biomedical Sciences*, 41 (2010), 232–40.

21 Paul Weindling, *Epidemics and Genocide in Eastern Europe, 1890–1945* (Oxford, Oxford University Press, 2000), p. 6.

22 J. Cain and A.G. Hopkins, 'Gentlemanly Capitalism and British Expansion Overseas II: New Imperialism, 1850–1945', *The Economic History Review*, 40 (1987), 1–26.

23 Löwy, 'Yellow Fever in Rio de Janeiro', 144–163; Bashford, ' "Is White Australia possible?" Race, colonialism and tropical medicine', *Ethnic and Racial Studies*, 23 (2000), 248–71; Steven Palmer, 'Beginnings of Cuban Bacteriology: Juan Santos Fernandez, Medical Research, and the Search for Scientific Sovereignty, 1880–1920', *Hispanic American Historical Review*, 91 (2011), 445–68.

24 Michael A. Osborne, 'Science and the French Empire', *Isis* , 96 (2005), 80–7.

25 Robert Aldrich, *Greater France; A History of French Overseas Expansion* (Basingstoke, 1996), 100–1.

26 For a detailed account of French imperialism under the Third Republic and its strong sense of the civilizing mission, see Conklin, *Mission to Civilize*, 59–72.

27 Anne Marie Moulin, 'Patriarchal Science: The Network of the Overseas Pasteur Institutes', in Patrick Petitjean, Catherine Jami and Moulin (eds), *Science and Empires; Historical Studies about Scientific Development and European Expansion* (Dordrecht, 1992), pp. 307–22; Moulin, 'Bacteriological Research and Medical Practice in and out of the Pasteurian School', in Ann La Berge, and Mordechai Feingold (eds) *French Medical Culture in the Nineteenth Century* (Amsterdam/Atlanta, 1994) pp. 327–49.

28 Laurence Monnais, 'Preventive Medicine and "Mission Civilisatrice"; Uses of the BCG Vaccine in French Colonial Vietnam between the Two World Wars', *The International Journal of Asia Pacific Studies*, 2 (2006), 40–66.

29 Harrison and Worboys, ' "A Disease of Civilization": Tuberculosis in Africa and India', in Lara Marks and Worboys (eds), *Migrants, Minorities and Health; historical and contemporary* studies (London, 1997) pp. 93–124.

30 Monnaies, 'Preventive Medicine and "Mission Civilisatrice"', p. 64.

31 Ellen Amster, 'The Many Deaths of Dr. Emile Mauchamp: Medicine, Technology, and Popular Politics in Pre-Protectorate Morocco, 1877–1912', *International Journal of Middle East Studies*, 36 (2004), 409–28.

32 Annick Guenel, 'The Creation of the First Overseas Pasteur Institute, or the Beginning of Albert-Calmette's Pastorian Career', *Medical History*, 43 (1999), 1–25.

33 Kim Pelis, 'Prophet for Profit in French North Africa: Charles Nicolle and Pasteur Institute of Tunis, 1903–1936', *Bulletin of the History of Medicine*, 71 (1997), 583–622.

34 Pelis, *Charles Nicolle, Pasteur's Imperial Missionary: Typhus and Tunisia* (Rochester, NY, 2006), p. 121.

35 Ilana Löwy, 'Yellow fever in Rio de Janeiro'; Palmer, 'Beginnings of Cuban Bacteriology'; Mariola Espinosa, *Epidemic Invasions*; Chakrabarti, *Bacteriology in British India*.

36 Moulin, 'Patriarchal Science', 310–11.

37 Moulin, 'Bacteriological Research', 342.

38 Quoted in Osborne, 'Science and the French Empire', p. 81.

39 Moulin, 'Tropical without Tropics: The turning point of Pastorian Medicine in North Africa', in D. Arnold (ed.), *Warm Climates and Western Medicine*, pp. 160–180, p. 161.

40 John Strachan, 'The Pasteurization of Algeria?', *French History*, 20 (2006), 260–275, p. 268.
41 Moulin, 'Tropical without Tropics, p. 172.
42 Strachan, 'The Pasteurization of Algeria?'.
43 Latour, *Pasteurization of France.*

10

Colonialism and traditional medicines

Colonial societies were not passive recipients of modern medicine. Indigenous physicians, medical assistants and patients in Asia, South America and Africa engaged creatively with modern medicine, often defining their application in unique ways, but also transforming their own therapeutics in the process. Traditional medical practices in the colonies negotiated with modern medicine and emerged as alternative forms of medicine in the twentieth century.

The expansion of European commercial and cultural dominance led to the pre-eminence of European medicine in the colonies from the end of the eighteenth century. European medical traditions and practices were part of colonial establishments. European hospitals, drugs such as quinine, vaccines, European medical colleges – which offered people Western medical degrees and recognized only Western medicine – became dominant with the spread of colonial influence and power. As Europeans collected medical specimens in the tropics from the seventeenth century, they often extracted the materials used in the local drugs for their own medicines, but discouraged the adoption of those traditional medicines. They instead introduced and encouraged among Europeans and the locals the use of their own medicines. European colonial authorities also outlawed traditional practices, such as obeah and voodoo, which they associated with witchcraft and sorcery. Along with that, European colonial authorities controlled medical universities, degrees and licensing. These often led to the marginalization of traditional forms of medicine.

However, this is only half of the story. Indigenous forms of medicine continued to survive and even thrive among the colonized population despite this marginalization. There are two significant historical processes to examine in the history of the modern evolution of traditional medicine. The first is the invention of traditions. The second is the role of indigenous agency.

While Europeans discarded several indigenous forms of medical tradition as quackery, they also took a keen interest in finding the contents of such drugs, the medical plants or substances used as their ingredients; they recorded, classified and codified them for their own interest in natural history and in exotic drugs, as we saw in Chapter 2. In India and North Africa, where indigenous medicine had a rich textual basis, they also read and translated classical medical (Sanskrit and Arabic) texts to understand the classical roots of these medical traditions.

This was an integral part of the contemporary European search for the roots of their own medicines in ancient Greek and Latin texts.[1] Local practitioners in the colonies also responded to the introduction and dominance of Western medicine by codifying and standardizing these, selecting certain drugs or practices, which corresponded with modern practices and prescriptions, introducing new medical substances and techniques from modern biomedicine and vaccines, and producing indigenous pharmacopoeias. Through this local agency, indigenous medical practices in Asia, South America and Africa were 'invented' as new forms of traditional medicine. To understand this complex historical process it is useful to study the historical methodology that sought to understand the 'invention of traditions' in various fields.

Invention of Traditions

The book *Invention of Tradition* (1983), edited by Eric Hobsbawm and Terence Ranger, introduced a new approach to the historical understanding of traditions.[2] It demonstrated how many practices which are considered, or appear, to be traditional and thus of ancient heritage are often of quite recent inventions, often deliberately constructed as such to serve particular ideological and political ends. Traditions, the book asserts, are invented at a later period, in different contexts and to serve different purposes than they did in the past. Inventions of ancient heritage for a relatively modern practice take place because it provides authenticity to the latter. Invented traditions occurred more frequently at times of rapid social and economic transformation when 'old' traditions were seen to be disappearing. Invented traditions attempt to establish continuity with a suitable historic past and to assert the timelessness of certain principles, practices, architecture and dress.[3] This took place in the past and continues to do so in our contemporary world in a variety of areas. The book provides examples of the use of Gothic-style architecture (a medieval architecture, originating in France in the twelfth century and identified with the Renaissance) for the nineteenth-century rebuilding of the British parliament. This helped to give the building a historical appearance. Similarly, several of the ceremonies of the British monarchy, which appear to be old traditions, are of relatively recent making. The main point to take from this approach is that despite their expression, form and appearance, traditions are often of quite modern origins.

Inventions of scientific and medical traditions

This approach is also useful to understand the history of medicine and science, to see that several scientific ideas and traditions, which appear ancient or eternal, are often recent inventions, which have been deliberately made to look classical and consequently timeless. Examples of this are the Hippocratic Oath that doctors take as part of their graduation ceremony, which was only created in the mid-twentieth century to serve modern purposes and ethical concerns, with very little similarities to any ancient Greek tradition. However, the act of taking such an oath marks the occasion with a sense of tradition and solemnity.

In fact, historians have shown that one of the most classical traditions of knowledge – that is, ancient Greek science – was in fact 'invented' several times through complex historical processes, which made it appear timeless. The twelfth century, when the Mediterranean world was thriving in cultural and economic terms, was when the transmission of scientific knowledge from the Islamic world and ancient Greece became crucial to Europe. The Italian economy was thriving in this period and although the Crusades were in full flow, Europe's contacts with the Arab world continued through the Mediterranean. European philosophers in this period absorbed various influences in the process of evolving a new idea of nature. As Umberto Eco suggests, Europe was searching for a culture that would reflect a political and economic plurality. This plurality had to revolve around a new sense of nature, of concrete reality and of human individuality and not only the Church.[4] This invention of the Greek heritage of Europe shaped the birth of the Renaissance in Europe.

The significance of understanding 'invented traditions' is that it enables us to understand that traditions are not eternal or timeless; they are in fact products of historical processes. They are often made to appear timeless. It also helps us to understand that the difference between 'traditional' and 'modern' is often not very clear. What appears to be traditional can in fact be quite modern. Such an approach helps us to understand the history of traditional medicine in different parts of the world as products of modern colonialism. It also helps us to understand that in this invention scholars, traders and common people participated actively. We will see the role of human agency in inventing traditional medicine.

History of traditional medicine in India

Most classical scriptural medicines in India, such as Ayurveda and Unani-tibb, were based on humourology. Ayurveda (the Sanskrit roots of which mean the 'science of life') is the term used for a collection of theories of disease and therapeutic practices, the earliest of which go back to 300 BC. Ayurveda considers that the universe is made up of combinations of the five elements (*pancha mahabhutas*). These are *akasha* (ether), *vayu* (air), *teja* (fire), *aap* (water) and *prithvi* (earth). The five elements can be seen to exist in the material universe on all scales of life, and in both organic and inorganic things. In biological systems, such as the human body, elements are coded into three forces (kapha, pitta and vata), which govern all life processes. These regulate every physiological and psychological process in the living organism. The interplay between them determines the health and wellbeing of the individual.

The principal texts of Ayurveda are the *Charaka Samhita* (the dates differ from the fourth to the second century BC) and the *Sushruta Samhita* (around the fourth century AD). The *Charaka* is believed to be one of the oldest ancient writings on Ayurveda. It is not known who Charaka (the person who is believed to have composed it) was. It could also have been composed by a group of scholars or followers of a person known as Charaka. The language of Charaka is Sanskrit and it is written in verse form. Poetry was known to serve as a memory aid. For example, *Charaka Samhita* contains over 8,400 metrical verses, which are often committed to memory by modern medical students of Ayurveda. The *Sushruta Samhita* is believed to have been composed by Sushruta, who introduced surgical skills and knowledge to Ayurveda. This branch of medicine arose in part from the exigencies of dealing with the effects of war. This work is also said to be a redaction of oral material passed down verbally from generation to generation.

Among the scriptural medicinal traditions of South Asia, the Unani-tibb is perhaps the most eclectic, derived from Greek, Arabic and Jewish therapeutic practices and a product of the strong cultural contact between Europe and the Arab world through the Mediterranean from the twelfth century. This too is a humoural medical system that presents causes, explanations and treatments of disease based on the balance or imbalance of the four humours (*akhalat*) in the body: blood (*khun*), mucus (*bulghum*) and bile (yellow: *safra*; black: *saufa*). These combine with the four basic environmental conditions of heat, cold, moisture and dryness. As the Unani system developed in the culturally diverse Islamic medieval world, its therapeutics became extremely sophisticated. The ancient practitioners of Unani were thus well known in both the Arabic- and the Latin-speaking worlds. Ibn Sina (known in Europe as Avicenna, 980–1007) and Muhammad bin Zakaria Rhazi (Rhazes in Europe, 865–923) collected and codified the central texts of Unani-tibb. The key text is Ibn Sina's *Qanun* (*Canon of Medicine*). Arabic medicine, which developed from an interaction with Greek medicine, was the precursor to the fully developed Unani-tibb. The later interaction in the Mediterranean world between Spanish, Moorish and Jewish therapeutic practices contributed to the corpus of Unani texts.

Both of these traditions had a rich history in precolonial South Asia, but here the issue is their interaction with colonialism. Orientalism and nationalism played the most significant roles in the history of Indian traditional medicine during the colonial period. The hospitals that the British built in India in the eighteenth century expanded in the nineteenth century and became the principal medical colleges in the presidency capitals of Bombay, Calcutta and Madras. Their graduates generally entered the government subordinate medical services, or set up private practice in the cities. These doctors who were trained in Western medicine secured the most lucrative jobs and became elite medical professionals. At the same time, the medical profession in India came to be regulated by the state. The Medical Acts in the various provinces between 1914 and 1919 restricted government employments to graduates of government medical colleges. More specifically, the Indian Medical Degrees Act of 1916 restricted the use of the term 'doctor' to those who practised Western medicine. Others, who belonged to the various medical traditions, were formally designated as 'quacks'. While this did not affect the great majority of hakims (those who practise Unani-tibb) and vaids (Ayurvedic practitioners) in rural practice, it contributed to the marginalization of the Indian medical systems in the urban centres. This trend combined with a few other features – the creation, after nearly a century of British rule, of a Westernized educated elite; the rise of middle-class professional Indians (lawyers, minor civil servants, teachers) in the cities and a change in their consumption practices – pushed the hakims and vaids into disrepute.

There was another trend in the European attitude towards indigenous Indian medicine. This emerged from British Orientalism. Orientalism was an eighteenth-century intellectual tradition that developed as a result of the cultural contact between Europe and Asia. Orientalists believed that all 'classical' civilizations, such as the Greek and the Vedic (ancient Indian civilization), had common roots, and they studied classical languages such as Sanskrit, Latin and Arabic to identify the common heritage between them. India was an important centre for this scholarship and, in Calcutta, William Jones (1746–94) established the Asiatic Society in 1784 for the study of Indian classical texts, along with the discovery of Indian natural history. The Orientalists started the translation of classical Sanskrit and Arabic texts and compared these with Greek and Latin ones. This included studying ancient Indian medical texts, such as the *Sushruta Samhita*, and the modern discovery of Ayurveda. In addition, German missionaries, particularly in southern India, had collected and translated Tamil medical texts. As a result, scholars such as J.F. Royle published works such as the *Antiquity of Hindoo Medicine*.[5] Although European scholars initiated the movement, Indians (particularly the Brahmin scholars who knew Sanskrit) participated and played important roles in defining the discovery of ancient Indian medical heritage.

This study of ancient texts for the authenticity of the sources of therapeutics was not unique; a similar process had taken place in Europe. In the seventeenth and eighteenth centuries, medicine in Europe entailed a dual journey: a study of the natural history of plants and a search in ancient texts for 'authentic ancient remedies'.[6] European medicine in the modern period too frequently used classical Latin language to provide it with a sense of authenticity. In the eighteenth century, European naturalists and scientists such as Carl Linnaeus (who translated his own name into Latin: Linne to Linnaeus) and Antoine Lavoisier adopted Latin to develop the new nomenclatures and classifications for plants and chemical elements, respectively.

The other factor that helped the study of Indian texts by British scholars was simple pragmatism. When confronted with unfamiliar (to them) diseases, such as cholera or abscesses of the liver in the eighteenth century, British physicians often had to rely on indigenous medicines and study the ancient texts to understand the rationale behind the practices. The British found it necessary to train lower-level medical personnel in the rudiments of Western as well as Ayurvedic and Unani medicine. The first medical college in India (in Bengal) was the Native

Medical Institution (1824–1934), where an eclectic syllabus was taught to Indian graduates. The translation of ancient classical texts led to compilations of Unani and Ayurvedic texts – for instance, *Taleef Shareef or the Indian Materia Medica* (1833) by George Playfair.[7] Whitelaw Ainslie, a British surgeon at the EEIC's establishment in Madras, compiled *Materia Indica* (1826).[8]

Another historical development took place in the nineteenth century. Beyond the classical texts, indigenous medicines had survived in India during the colonial period, particularly among a large section of the Indian citizenry. The great majority of the population depended on vaids and hakims for their day-to-day medical care. Besides them there were huge numbers of local herbalists who practised an amalgamation of Unani, Ayurveda or siddha as well. Faced with the pressures of licensing, codification and textualization, indigenous medical practitioners of the elite systems – that is, Ayurveda and Unani – sought to codify their therapeutic practices, invent new traditions of therapy and resort to marketing strategies that suited the consumer culture of late colonial urban India. They adopted some practices that they felt enhanced the 'scientificity' of their diagnoses. These included feeling the pulse of the patient, as well as a greater use of minor surgeries and the study of anatomy.

The most important developments, however, were in the calls for the professionalization of Unani and Ayurveda, the packaging of medicinal products in factories for sale similar to modern drugs, and the relationship of the profession with Indian nationalism. Previously, medical knowledge was acquired through the system of apprenticeship in which the closed 'genealogical' line was followed, from one disciple to another. Legitimacy was retained by virtue of apprenticeship. From the late nineteenth century, print and the use of vernacular language became an important part of retaining legitimacy and tradition. In the case of Unani, this represented the translation of Arabic and Galenic texts into Urdu – a more commonly used language in India. A second part of this institutionalization was the transformation of the requirements of the profession from scholastic knowledge to demonstrable knowledge. Hakim Ajmal Khan (1868–1927) championed the cause of the Unani system in India. Ayurvedic texts were similarly translated into local vernacular languages, such as Punjabi, Bengali and Hindustani.

In both systems, medical journals were published from the late nineteenth or early twentieth century, following the European tradition of medical journals as the source of medical knowledge and interaction. The *Tibb-i-Ayina* ('Mirror of Medicine') was the most famous Unani monthly journal published in the nineteenth century, which attempted to bring Hindustani cures to the attention of professionals who were trained in Western ways, and to familiarize hakims with the ways and practices of Western medicine. This was edited by Imad al-Din Ahmad, curator of the Government Medical School in Agra. The journal *Oudh Akhbar*, published in Lucknow, attempted to establish Unani as being on a par with Western medicine and at the same time reiterated the fact that Western medicine, whatever its proof of progress, was incompatible with the Indian temperament because it was not rooted in the Indian locale.[9] Ayurveda followed a similar trajectory. There was a similar creation of public space where Ayurveda manuals and textbooks appeared in print in several regional languages as well as in Hindi. The first Ayurvedic periodical was *Ayurveda Sanjivani*, which was published in Bengali.

These movements gained sustenance from Indian nationalism, which developed from the late nineteenth century. Indian nationalism was marked by a political struggle for power with the colonial state, as well as a search for the roots of Indian identity. Traditional medicine became a part of that search for Indian identity and the restoration of Indian epistemologies to their supposedly glorious past. In certain aspects, Ayurveda disseminated further and reinvented itself more robustly than Unani because in the course of the nineteenth century it became associated with Hindu nationalism, which supported a 'revival' of Hindu knowledge. Ayurveda dominated the nationalist conceptualization of indigenous Indian medicine and was utilized by Hindu nationalists in the formation of a particular Hindu 'national' identity.

Indigenous medical practitioners used the political platform of Indian nationalism to organize several countrywide Ayurveda and Unani-Tibb conferences between 1904 and 1920. These had a strong nationalist bias. In 1920 the INC (the mainstream nationalist party) convention at Nagpur recommended the acceptance of the Ayurvedic system of medicine as India's national healthcare system. In 1921 the nationalist leader, M.K. Gandhi, inaugurated Ayurvedic and Unani Tibbia College in Delhi, which was intended to train practitioners of Indian systems of medicine.

Along with these textual developments and political events, the indigenous practitioners established pharmaceutical industries based on indigenous medical products. The Hamdard Laboratories were founded in 1906 in Delhi by Hakeem Hafiz Abdul Majeed, a well-known Unani practitioner, to mass-produce Unani products and to package and market them on a large scale. This was a fundamental departure from Unani practice, where the physician himself would mix prescriptions, with the amounts differing according to the needs of individual patients. The mass commercialization of Indian pharmaceutical products also occurred through the establishment of Ayurvedic companies, such as Dabur and Zandu. These trends were helped by nationalism, a rise in consumption by the middle classes in the cities and the reinvention of indigenous therapeutics in response to Western medicine.

In the interwar years, even Western-trained Indian doctors, such as R.N. Chopra and S.S. Sokhey, participated in the search for indigenous alternatives to European pharmaceuticals.[10] Chopra wanted to establish what he referred to as 'Indian pharmacology' – a modern medical tradition combining laboratory research with classical Indian *materia medica*. This had important financial connotations at a time when modern pharmaceutical drugs were very popular in India and the Indian pharmaceutical market was dominated by multinational companies. Chopra's quest was to find cheaper indigenous alternatives. There was, indeed, a wider trend among Indian physicians and chemists to identify cheaper and easily accessible indigenous drugs.

The problem with this revival of classics was that this was an overwhelmingly textual one based on classical languages, and the epistemological contribution and cultural accretions of several hundred years, which took place outside such texts, were erased. Moreover, only the elite interpretations of such texts (Brahminical interpretations) were acknowledged as pure and authentic and the various non-Brahmanical medical traditions were overlooked. At the same time, it needs to be added that throughout this period, Western medicine remained dominant and mainstream in colonial India. Particularly with the repeated epidemics of cholera and plague, Western drugs, vaccines, laboratories and hospitals became firmly established as vital to colonial medicine and public health. The traditional medicines, which received no governmental patronage, became classified as 'alternative'. State medicine in colonial India was clearly Western medicine.

Historians have analysed various aspects of this rise of traditional medicine in colonial India. One view has been to see it as a form of resistance. Indigenous practitioners, such as Ajmal Khan and P.S. Varier, reacted to the onslaught of new discoveries in Western medicine in the nineteenth century by synthesizing traditional medicines and organizing their commercial distribution.[11] On the other hand, it also led to revitalization and hegemony as key to the emergence of traditional medicine in colonial India. The popular movement, while opposed to colonial medicine, also incorporated elements of modern medicine, leading to the emergence of a new, vigorous, traditional medical practice that was textualized and modernized. This established a cultural hegemony, in which only selective textual traditions, which could claim and construct an ancient lineage, became dominant and came to be known as Indian traditional medicine.[12] It is important to remember that this did not include a whole range of local or customary medical practices, which are not based on classical scriptures but are practised everyday on the roadsides of Asia, catering to a large to section of the poor. These have neither been codified nor received government support or recognition, even in the postcolonial period. The more recent historiographical trend, however, has been to highlight the culturally pluralistic (and regenerative) rather than the hegemonic aspects of the reinvention of Ayurveda.[13]

Both Unani-tibb and Ayurveda are known as traditional Indian medicine. They are also described as ancient and ageless practices of Indian heritage. It is evident from this history that this idea of the ancient tradition was created from the eighteenth century with the fresh study of classical Indian texts and interaction between European and indigenous medical practitioners. Many of the medical practices in India in the eighteenth and nineteenth centuries were flexible and hybrid with little direct link to any particular classical text.

With the Orientalist and nationalist search for authentic Indian classical traditions, the link between classical texts and everyday medical practices was enforced and fixed. This link helped to establish the authenticity of these medicines. To be authentic, traditional medicine needed to have a classical heritage.

In the process, others, which had no such linkage, were discarded or remained marginalized. At the same time as establishing the links with antiquity, Indian physicians also modernized these medicines by adopting the new cultures of print, medical colleges and markets.

Colonialism and traditional medicine in Africa

On the one hand, traditional medicine in Africa emerged out of a long interaction between African healing practices and Christianity, an outcome of the activities of the medical missionaries from the late nineteenth century. On the other, these processes were a consequence of the colonial transformation of African society, economy and culture, and the introduction of modern biomedicine. As we have seen, the colonization of Africa took place at a time of great assertion of European superiority – the so-called 'civilizing' mission – as well as at the time of the growing influence of European biomedicine, vaccines and laboratories. Europeans felt that they clearly had something to offer to Africa and at the same time remove some of their traditional practices. Colonial Africa thus experienced much more drastic intervention of European medicine and modernity in general, than other parts of the world in the nineteenth century.

Four historical factors shaped the history of African traditional medicine during the colonial period: the colonial state and its legal and administrative machinery; the introduction of biomedicine and colonial bioprospection; the presence of medical missionaries; and finally the protectorate system, which provided relative autonomy to African cultures and practices while at the same time destroying traditional links between healing and sovereignty. Within the protectorate system, new forms of traditions and authorities were established. Once the traditional links between land, sovereignty and healing were severed, the colonial authorities, particularly the plantation managers in order to prevent peasant unrest, established the tribal headman as the new political authority. However, the chief did not enjoy traditional medical authority. In order to report sick and be certified legitimately, the African labourer needed to be checked by a Western doctor.[14]

Historians have widely debated the nature of the impact of colonialism on African traditional healing practices. These debates have been important because on these the issues the general nature of the impact of colonialism and of African agency are also hinged. Steven Feierman's work falls within the tradition of the social history of health and medicine, and he locates African traditional healing practices within African social and political history.[15] He argues that traditional modes of healing were deeply connected with political power and control over resources. For example, in precolonial Nigeria, healers were also responsible for maintaining public order, administration of the law and explaining any misfortune or accident. In the Shambaa kingdom in Tanzania, on the other hand, those who administered medicine were also believed to be responsible for the fertility of the land. Essentially the point that Feierman makes is that precolonial African 'healing was bound up with political and economic processes'.[16] He then demonstrates how with colonialism these relationships were often severed. Colonialism destroyed the traditional ties between healing and public authority by wresting control of the modes of economic production and political sovereignty, and by divorcing traditional medical links with these. The colonial powers were sometimes aware of these linkages and deliberately undermined the powers of the healers in order to withdraw political authority from them. The attack on traditional healers by colonial forces, such as in the suppression of Shona spirit mediums in Zimbabwe, the British attack on the Aro oracle in Nigeria and the German persecution of healers in Tanganyika, were not just attempts to introduce European scientific ideas and practices but were also designed to destroy traditional political authority.

The destruction of traditional forms of medicine has also been part of more violent processes of colonial governance, which negated various aspects of traditional culture. In the Belgian Congo, a brutal colonial regime destroyed local therapeutic systems and rationales, leading to the emergence of a new and hybrid Congolese therapeutic system and a new hybrid therapeutic language.[17] European colonial officials often misunderstood African medical practices as signs of primitive African culture and tended to lump them together with witchcraft, magic and superstition. They often instituted laws that banned these practices, forcing people to act in covert and subversive ways. At the same time there were dramatic transformations of traditional medicine during the colonial period.[18] There were several instances when African healers incorporated modern biomedical approaches within their own traditional practices.[19]

Traditional methods coexisted with the introduction of European biomedicine and vaccines, and the simultaneous bioprospecting for African plants and indigenous drugs by European scientists and pharmaceutical companies. The European collection of medicinal plants in Africa, which started from the late nineteenth century, led to the transformation of traditional plant-based therapies. The imperial expansion in Africa took place at the time of the rise of pharmaceutical chemistry in Europe. Europeans were simultaneously introducing new pharmaceuticals in African markets and looking for new ingredients for them. This led to European interest in African herbal preparation, as well as the translation of ethnobotanical knowledge in modern Africa.[20] *Strophanthus*, a plant which was used to produce the drug strophanthin in Europe, was traditionally used by western African healers both as a poison and as a medicine. It was 'discovered' by David Livingstone during his Zambezi Expedition. The drug, which Africans used in their poisoned arrows against the British on the Gold Coast, entered the British Pharmacopoeia in 1898 following a complex history of European distrust and anxiety, as well as an interest in African medicinal plants and healing practices. By the early twentieth century, with the establishment of a British military presence in West Africa, the British outlawed African use of the plant in poisoned arrows. Laboratory experiments on the drug in Edinburgh under Scottish pharmacologist

Thomas Fraser led to the discovery of its 'active ingredient', strophanthin, in 1873, which was found to be a particularly effective cardiac drug. Following this discovery, British pharmaceutical companies, such as Burroughs Wellcome & Co, procured the plant in large quantities to produce the drug on an industrial scale.[21] Growing international pharmaceutical demand for *Strophanthus* seeds led to an export scheme from the Gold Coast during the First World War. This coincided with the marginalization of its use in African medicine.[22]

The Christian missions were the sites where a more hybrid medical tradition emerged from the late nineteenth century. The missionaries sometimes opposed the attempted marginalization and destruction of traditional medicine and cultural practices by colonial regimes. The hybrid medical tradition was a product of this new social and institutional framework within which Christianity took root in Africa. While seeking to introduce a new moral order, the Church still depended on traditional customs. The missionaries often found African healing practices incompatible with their Christian doctrine and attempted to create new African Christian communities that were wholly dependent on mission medicine built on rational principles and foundations. Despite attempts by the medical missionaries to establish medical hegemony, the African Christian elite exercised a certain level of autonomy with respect to medicine and other cultural practices within the churches. African churches also adopted a more lenient attitude towards their congregations' use of African medicine. Meanwhile some African medicines and practices continued to be used because missionaries quietly accepted their efficacy.

Often discarded as witchcraft, evil practices and superstitions, African healing practices therefore survived. Sometimes this was because Western medicine was expensive and inaccessible, or failed to treat medical conditions and diseases in Africa. The lack of investment in medical infrastructure and public health by colonial governments allowed these practices to continue and even thrive as new forms of epidemic swept through the continent. They survived also because they were deeply linked with African social and cultural life. At the same time, within the protectorate system, there was relative cultural autonomy for African practices. While Africans converted to Christianity, they also retained and then imbued their religion and everyday life with traditional practices. Some British physicians even claimed to believe in, or to be adept at, local medicine in order to promote their own therapies. The colonial powers also encouraged the commercialization of African 'native medicine', which competed with Western biomedicine from the 1920s for the rapidly growing African population, particularly in the cities.[23]

This native medicine was a product of historical selection. Certain African practices, which Europeans were more comfortable and familiar with and which appeared to be free from 'primitive' practices, were encouraged more than others. There was also the tendency to accept only those medicines that seemed consistent with scientific and biomedical principles or with Judeo-Christian morality.[24] At the same time the continent of Africa was itself undergoing major transformations as a product of colonialism. In the first half of the twentieth century there were major demographic changes there, with the establishment of industrial systems such as mines and plantations, along with markets and cities. This led to the movement of labour, leading to epidemics, breakdown of traditional social customs and the birth of new ones, and the transformation of traditional belief systems, which were increasingly infused with a Christian ethos. Modern methods, such as injections, became popular in traditional African medicine. Such fusions of African healing and biomedicine have continued in the contemporary period, particularly in response to modern epidemics, such as HIV/AIDS.[25] In the process, a new and hybrid traditional African medicine was born in the twentieth century, imitating modern biomedicine, reinventing African culture, practised in Western-style hospitals and dispensaries,[26] standardizing medical dosages and serving the new social, cultural and economic realities. In Africa a different kind of invention of tradition took place. This was based not on the reading of classical texts, as in India, but on the establishment of new social, economic and cultural systems during colonialism within which the very meaning of 'African tradition' came to be defined.

The modern invention of traditional Chinese medicine

The evolution of traditional medicine in China is one of the most significant developments in the medical history of the twentieth century. Traditional medicine developed in China as part of the country's search for national identity during the Cultural Revolution (1966–78). Chinese medicine is also one of the most widely used and globally established forms of traditional medicine, which set the trend for the development of other forms of traditional and alternative medicine.

Although not all of China was colonized, several parts of southern China were, or experienced colonial influence and control by different European nations from the eighteenth century. China had a long history of contact with Europe from the medieval period through trading routes, particularly the Silk Route, which ran through Central Asia and southern Europe to the eastern Mediterranean ports. It was during these trading connections that the Venetian merchant Marco Polo (1254–1324) travelled to China across Central Asia in the thirteenth century. During the Age of Commerce, in the seventeenth century, the Portuguese arrived at and traded in the ports of southern China in search of spices, and as part of their maritime expansion in South East Asia.

It was in the nineteenth century that European nations sought to establish colonial control in China. Until the 1830s, European trade and colonial control in China was restricted to the port of Canton. Britain came into conflict with the Chinese rulers in the 1840s in its effort to expand its colonial control beyond Canton. This eventually unfolded in a military conflict known as the First Opium War (1839–42). Opium was not native to China, although the Chinese had a long

tradition of using it for medicinal purposes. The Dutch and English traders introduced it as a recreational drug in southern China from the seventeenth century.
The British encouraged large-scale cultivation of opium in India in the eighteenth century, which was then sold in the Chinese markets. This allowed them
to conduct trade with China, which otherwise had little interest in European
products. Alarmed by the growing opium addiction in China as well as the
expansion of British colonial power, the Chinese emperor, Daoguang (1821–50),
sent forces to Canton in 1839 to destroy the British opium-trading establishment there. In response, Britain sent a naval fleet to China and the war began.
The British overpowered the Chinese forces and this resulted in the Treaty of
Nanjing of 1842. This granted the opening of five 'treaty ports' along the southeast coast and the abolition of the trading monopoly that the Chinese rulers had
imposed on foreign traders in Canton. The Chinese kingdom was reduced to a
semicolony. The construction of the railway networks by the French in Yunnan
province in southern China opened up interior parts of China to colonial trade
and influence.

One significant development of this colonial contact between China and
Europe from the seventeenth century was the presence of Christian missionaries.
Both Jesuits and Protestant missionaries had come to China with the Portuguese,
French and British traders, but they often maintained their independent presence and activities. Jesuit missionaries introduced Western medicine in southern
parts of China and Hong Kong from the eighteenth century. The novel herbal
medicines that they brought with them (e.g., cinchona) were soon incorporated
within Chinese pharmacopoeia. Jennerian vaccines rapidly replaced older treatments against smallpox. Medical missionaries in the nineteenth century established missionary hospitals and medical schools. The British missionaries used
Western medicine extensively in Yunnan province.[27]

The growth of colonial power in the nineteenth century led to the dominance
of Western medicine in parts of China. From the late nineteenth century, both
British and French colonial authorities introduced mass vaccination and sanitary
measures to protect their colonial interests from the spread of plague and cholera
epidemics.[28] The British also enacted contagious diseases legislation in Hong
Kong in the 1850s.[29] Along with introducing Western therapeutics and hospitals,
Europeans viewed Chinese medical practices as superstitious and unscientific,
which reflected their attitude towards Chinese culture and society generally as
being backward and regressive.

With the onset of the communist regime (1949) and during the Cultural
Revolution, the Chinese government invested heavily in traditional medicine in
an effort to develop affordable medical care and public health facilities. Modernity,
cultural identity and the socioeconomic reconstruction of China were the main
facets of the Cultural Revolution. The movement sought to define a new and
modern China against its colonial and feudal past.

As part of this search for a new national identity, the Chinese government
established primary healthcare systems and sought to revitalize traditional

medicine. During the Cultural Revolution, the Ministry of Health directed healthcare throughout China and established primary care units. Interestingly, the term 'doctor' was used quite widely in this period in China, not just for someone with a medical degree but also for anyone who helped the sick. At the same time the Chinese government revolutionized traditional medicine by establishing medical colleges and hospitals, standardizing treatments and drugs, and instituting training in traditional medicine. Chinese doctors who were trained in Western medicine were also taught traditional medicine, while traditional practitioners were trained in modern methods. The result was a dynamic incorporation of modern medical concepts and methods along with a revitalization of some selected and suitable aspects of traditional practices. Thus, traditional medical practices in China were reformulated in response to Western medicine during the Cultural Revolution.

The 1960s was also a period of renewed interest in Chinese science and a greater appreciation of the links between traditional Chinese social structures and technological practices among Western scholars such as Joseph Needham. To Needham, Chinese science had its own distinct orientations, shaped by its particular social and material culture.[30]

Through these processes, a new tradition of Chinese medicine, formally known by the acronym TCM (traditional Chinese medicine), was created in the 1950s.[31] This form of Chinese medicine was relatively modern and comparable to modern biomedicine in terms of its reliance on statistics, diagnostic tests and standardization. TCM is a hybrid and invented tradition of medicine that combines elements of folk medicine with that of Western therapeutics and allopathic diagnostics and pharmaceuticals. At the same, despite this relatively modern creation, practitioners and advocates of TCM often claim its ancient heritage. Volker Scheid has written a comprehensive history of Chinese traditional medicine and shows that indigeneity, self-reliance and affordability were the driving forces of the Cultural Revolution, which shaped the emergence of TCM. During the post-Cultural Revolution era, TCM was transformed when the socialist government of China embraced economic liberalization and a new global medical marketplace emerged. TCM became a global medical tradition and economic force.[32]

In a critical history of the emergence of traditional Chinese medicine, Kim Taylor has argued that Western ideas directly and indirectly played an important role in the making of traditional Chinese medicine. During and after the Cultural Revolution, Chinese doctors and the Chinese government 'invented' a new Chinese medicine in order to create a medicine that was China's own heritage and was also different from Western medicine.[33] On the one hand, they repeatedly advertised and highlighted the richness of China's medical heritage. On the other, they reinforced the Western idea of Chinese medicine and of Chinese civilization as being static, unchanging and monolithic. During the 1960s and 1970s, Taylor writes, China politically and culturally 'withdrew into herself'. Western scholars had little knowledge of the changes taking place in

Chinese medicine and society, which also reinforced the appearance of Chinese medicine as a monolithic one. It was during this period that the 'basic theory of TCM' based on unification and simplification of a variety of heterogeneous medical traditions that existed in China was compiled.[34] Therefore a new tradition of medicine now served the interests of the new nation and, in the process, reinforced the idea of a 'traditional' China. Others have shown that within China, TCM has dominated and even marginalized many folk practices of medicine.[35]

There was another development in China in the experiments with new modes of delivering medicine, which was often more important than the form of medicine being delivered. During the Cultural Revolution in 1968, the Chinese Communist Party endorsed a radical new system of healthcare delivery for the rural areas. Every village was allotted a barefoot doctor (a medical person trained with the basic skills and knowledge of modern medicine who could cope with minor diseases) who was responsible for providing basic medical care. The barefoot doctors became symbols of the Chinese Cultural Revolution as heroic, gallant medical men combining traditional Chinese values with modern scientific methods, travelling though the remote countryside delivering health and medicine to poor peasants. This was indeed a new and revolutionary mode of health delivery in non-Western countries, beyond the existing ones of establishing hospitals, asylums and dispensaries, or conducting vaccination campaigns and sanitation programmes in remote places. The main impact of this was the introduction of modern Western medicine into villages that were hitherto served by traditional Chinese medicine.[36]

Today, TCM is a massive medical establishment in modern China and is integral to its primary healthcare system. By 2001 there were more than 2,000 hospitals in China practising TCM, and these were served by 80,000 traditional Chinese medical physicians. There are also several medical training institutions (the most famous is Beijing University of Chinese Medicine), which provide advanced training in TCM. They also accept foreign and overseas Chinese students.[37] At the same time, TCM also became a global brand. From the 1980s it became a translational medical tradition, in places such as Shanghai, Tanzania and California. It has also become the mainstream medicine in different parts of the United States. In the process, TCM catered to a range of requirements, and social and economic contexts. This transformation took place because of a particular form of appeal that TCM had for the white middle class of the West. At a time when modern biomedicine was increasingly seen as impersonal and being controlled by pharmaceutical giants with only profit motives, TCM appeared as the more organic, personal and kinder 'alternative'. In the process, paradoxically, TCM itself became globalized and corporatized.[38] What is now known as TCM is thus a relatively modern medicine, which emerged through a historical process from the 1950s when political leaders and doctors in China sought to define a new national culture, economy, education and health infrastructure.

Conclusion

The traditional medicines that are prevalent at present are not traditional in the true sense of the term. They are invented traditions and new medicines. It is possible to argue that even Western medicine is an invented tradition, in which various forms of practices and traditions from across the world were assimilated. The main issue here is of trust and authenticity. While Western medicine garnered its trust first from observation and empiricism (in the seventeenth and eighteenth centuries) and then from laboratory experiments (in the nineteenth and twentieth centuries), traditional medicines derive their authenticity and evoke trust by adhering to a certain tradition that is distinct from Western or European practices. Its distinction lies in it being the 'other'. This was in response to the consequences of colonialism and with the rise of Western medicine. To achieve that, traditional medicines created their particular lineages that appeared to be pure and uniform. At the same time, it is important to remember that there exists, or existed, numerous traditions and practices that have not been integrated into these and have either remained marginalized or been lost.

Traditional medicine today is a crucial part of global healthcare systems, particularly in the developing countries. In some Asian and African countries, 80 per cent of the population depend on traditional medicine for primary healthcare. In many developed countries, 70 per cent to 80 per cent of the population use some form of alternative or complementary medicine (e.g., acupuncture). With immigration and global movement of people, these medicines have also become global medicines. In many parts of the world, alternative medicines are highly popular.[39]

Notes

1 For a detailed account of the inventions of classical roots of Western science, see Chakrabarti, *Western Science in Modern India*, pp. 4–9. For the discovery of classical roots of Indian science, see Pratik Chakrabaorty, 'Science, Nationalism, and Colonial Contestations: P. C. Ray and his *Hindu Chemistry*', *Indian Economic and Social History Review*, 37 (2000), 185–213.
2 E.J. Hobsbawm and Ranger, *The Invention of Tradition* (Cambridge, 1983).
3 'Introduction: Inventing Traditions', ibid, pp. 1–14.
4 Umberto Eco, 'In Praise of St. Thomas' in his *Travels in Hyperreality: Essay* (London, 1987), pp. 257–68.
5 J.F. Royle, *An Essay on the Antiquity of the Hindoo Medicine* (London, 1837).
6 H.J. Cook, 'Physicians and Natural History', pp. 92–3.
7 G. Playfair, *Taleef Shareef or the Indian Materia Medica* (Calcutta, 1833).

8 Ainslie, *Materia Indica, Or, Some Account of Those Articles Which are Employed by the Hindoos, and Other Eastern Nations, in Their Medicine, Arts, and Agricultural* (London, 1826).
9 Seema Alavi, 'Unani Medicine in the Nineteenth-Century Public Sphere: Urdu Texts and the Oudh Akhbar', *Indian Economic and Social History Review*, 42 (2005), 101–29.
10 R.N. Chopra, *Pharmacopoeia of India* (Delhi, 1955). Sahib S. Sokhey, *The Indian Drug Industry and its Future* (New Delhi, 1959).
11 Kumar, 'Unequal Contenders', pp. 176–9.
12 K.N. Panikkar, 'Indigenous Medicine and Cultural Hegemony: A Study of the Revitalization Movement in Keralam', *Studies in History*, 8 (1992), 287–308.
13 See the articles in Dagmar Wujastyk (ed.), *Modern and Global Ayurveda: Pluralism and Paradigms* (Albany, 2008).
14 Feierman, 'Struggles for Control: The Social Roots of Health and Healing in Modern Africa', *African Studies Review*, 28 (1985), 73–147, pp. 118–9.
15 Ibid, p. 116.
16 Ibid.
17 Nancy Rose Hunt, *A Colonial Lexicon: Of Birth Ritual, Medicalization, and Mobility in the Congo* (Durham, NC, 1997).
18 Meredeth Turshen, *The Political Ecology of Disease in Tanzania* (Rutgers, 1984). For a detailed historiographical review of African medicine and colonization, read Kent Maynard, 'European Preoccupations and Indigenous Culture in Cameroon: British Rule and the Transformation of Kedjom Medicine', *Canadian Journal of African Studies*, 36 (2002), 79–117.
19 Karen Flint, *Healing Traditions: African Medicine, Cultural Exchange, and Competition in South Africa, 1820–1948* (Ohio, 2008).
20 Abena Osseo-Asare, 'Bioprospecting and Resistance: Transforming Poisoned Arrows into Strophantin Pills in Colonial Gold Coast, 1885–1922', *Social History of Medicine*, 21 (2008), 269–90.
21 Hokkanen, 'Imperial Networks, Colonial Bioprospecting'.
22 Osseo-Asare, 'Bioprospecting and Resistance'.
23 Flint, 'Competition, Race, and Professionalization: African Healers and White Medical Practitioners in Natal, South Africa in the Early Twentieth Century', *Social History of Medicine*, 14 (2001), 199–221.
24 Maynard, 'European Preoccupations and Indigenous Culture in Cameroon'.
25 Brooke Grundfest Schoepf, 'AIDS, Sex and Condoms: African Healers and the Reinvention of Tradition in Zaire', *Medical Anthropology*, 14 (1992), 225–242.
26 Anne Digby and Helen Sweet, 'Social Medicine and Medical Pluralism: the Valley Trust and Botha's Hill Health Centre, South Africa, 1940s to 2000s', *Social History of Medicine*, first published online 26 September 2011 doi:10.1093/shm/hkr114.
27 Elisabeth Hsü, 'The Reception of Western Medicine in China: Examples from Yunnan', in Patrick Petitjean, Cathérine Jami (eds), *Science and Empires: Historical*

Studies About Scientific Development and European Expansion (Dordrecht, 1992), pp. 89–102.

28 Francis F. Hong, 'History of Medicine in China; When Medicine Took an Alternative Path', *McGill Journal of Medicine*, 8 (2004), 79–84.

29 Philippa Levine, 'Modernity, Medicine, and Colonialism: The Contagious Diseases Ordinances in Hong Kong and the Straits Settlements', *Positions*, 6 (1998), 675–705.

30 J. Needham, *The Grand Titration: Science And Society In East And West* (London, 1969).

31 Kim Taylor, 'Divergent Interests and Cultivated Misunderstandings: The Influence of the West on Modern Chinese Medicine', *Social History of Medicine* 17 (2004), 93–111, see pp. 100–1 for details of how the term 'traditional Chinese medicine' came to be used and accepted in the 1950s.

32 Volker Scheid, *Chinese Medicine in Contemporary China: Plurality and Synthesis* (London & Durham, 2002).

33 Taylor, 'Divergent Interests and Cultivated Misunderstandings', pp. 93–111.

34 Ibid, p. 97.

35 F. Fruehauf, 'Chinese Medicine in Crisis: Science, Politics and the Making of "TCM" ', *Journal of Chinese Medicine – HOVE*, 61 (1999), 6–14.

36 Xiaoping Fang, *Barefoot Doctors and Western Medicine in China* (Rochester, NY, 2012).

37 Ruiping Fan, 'Modem Western Science as a Standard for Traditional Chinese Medicine: A Critical Appraisal', *The Journal of Law, Medicine & Ethics*, 31 (2003), 213–21.

38 Mei Zhan, *Other-Worldly: Making Chinese Medicine Through Transnational Frames* (Hastings, 2009).

39 'Traditional medicine', fact sheet no. 134, December 2008, WHO, http://www.who.int/mediacentre/factsheets/fs134/en/#.

Conclusion: The colonial legacies of global health

The history of colonial medicine unfolded in two main trajectories: assimilation and divergence. On the one hand, the history of medicine and empire was shaped by global interactions and assimilations from the sixteenth century to the twentieth. These were in the sharing of ideas, medical traditions, drugs and general interactions between diverse groups, such as African-Amerindian-Spanish, Dutch-Indonesian-Malabarian-Portuguese-Indian and African-French-Scottish. Such interactions led to the hybridization and plurality of medical practices, theories of diseases, uses of drugs and ultimately to the making of modern medicine. On the other hand, it is also a history of the rise of European imperial power, which led to growing differences between Europe and the rest of the world, and the divergent histories of hospitals, preventive medicine, epidemics and mortality rates between Europe and its colonies. Some of the divisions were of an economic and political nature, as European nations controlled and ruled over vast parts of the world; with the growth of the medical marketplace and pharmaceutical industries in Europe and the marginalization of indigenous traditions in Asia and the Americas; with the decline in epidemics in Europe and their rise in the colonies; and with demographic growth and mortality decline in Europe and depopulation in places such as the Polynesian islands. There were other divergences that were more imagined or invented. These were evident, as we have seen, in assertions of differences in the pathology and culture of Europe and the tropics, in racial characteristics and in the differences asserted between Western and traditional or alternative medicines. Assimilation and divergence have been the legacy of colonialism for twentieth-century global health. Global health agendas have been determined by attempts to bridge the gaps and thereby to square the circle, so as to provide equitable healthcare in an economically and socially asymmetrical world.

The late nineteenth century, when large-scale colonial expansion was taking place overseas, was a period of relative peace within Europe. The threat of epidemic diseases decreased drastically in Europe in this period. There were no major outbreaks of cholera in Western Europe after the 1870s. Malaria was restricted to parts of eastern and southern Europe. The expansion of public health infrastructure and improvements in general medical facilities was largely responsible for this. This was also the period of popular movements and awareness in Europe: ordinary people participated in and asserted their views about public health, ethics of laboratory research and living conditions. In the colonies, though, such measures were introduced at a time of aggressive colonization, growing authoritarianism of the state, missions of modernity and civilization, and large-scale military mobilization. They were also shaped by the general view of the tropics as reservoirs of disease. Thus the contexts within which preventive medicine and public healthcare measures, such as vaccination, sanitation and the collection of vital statistics of the population, were undertaken in Europe and in the colonies were quite different.

However, this period of peace in Europe was short-lived. The First World War and the subsequent pandemics of flu, typhoid and famines plunged Europe and other parts of the world into a humanitarian catastrophe. The war exposed the internal health crisis within Europe, which it had itself largely precipitated. It had resulted in the deaths of more than 15 million people, with a further 7 million permanently disabled and 15 million seriously injured. Apart from military casualities, disease – particularly malaria – caused havoc among troops, particularly those on the Eastern Front. In places such as Macedonia, East Africa, Mesopotamia and Palestine, malaria was a major health concern among British, French, American and German troops. At times almost half of the British forces in Palestine and Macedonia were immobilized by the disease. The French suffered similarly in Macedonia. Then there was the outbreak of pandemic influenza (commonly known as 'Spanish flu', as the Spanish King Alfonso XIII was the first prominent figure to suffer from the disease and thus received most media coverage in Spain) in 1918. It lasted until 1920 and was one of the deadliest epidemics in human history as it killed around 50 million people across the world. While the war did not directly cause the flu, the closely packed quarters of troops and massive troop movements increased transmission. This was followed by famine and typhoid epidemics in the Soviet Union between 1918 and 1922. The famine was caused by drought and the political turmoil following the war and the civil war during the Russian Revolution, which led to the dislocation of people, homelessness and a shortage of food and drink. The famine and typhoid led to massive migration from the Soviet Union to Central Europe.

Thus following the war, European nations were faced with their own health challenges and with the realization that these were connected with the conditions in and beyond their formal colonies. The experiences of colonial medicine proved vital at this time. The British government, for example, sought to utilize British colonial medical expertise in dealing with the malaria crises during the First World War. It appointed Ronald Ross to accompany British troops in Egypt, Greece and Gallipoli during the war as a consultant to undertake preventive action against malaria. The sanitary and quarantine measures and field studies similar to those conducted in the West Indies, Africa and Asia for yellow fever and malaria from the late nineteenth century were undertaken in parts of Europe and the United States in the post-war period.

The end of the war led to the formation of the LNHO in 1921. The medical and social crises that emerged in the wake of the war led to the realization of the need for new, broader international health collaborations. Isolated conferences and sporadic sanitary measures did not seem adequate to prevent diseases and epidemics on an international scale. The LNHO was also inspired by the ideal that better provision of international health and welfare would reduce social conflict and help to prevent future wars. The first focus of the LNHO was on the epidemics in post-war Europe. It took measures to stop the spread of typhoid to Europe. It also organized expert-led scientific surveys in different parts of the world against malaria. Medical experts under the LNHO also compiled and compared mortality rates, causes of deaths and levels of malnutrition to improve living conditions, particularly in the colonial countries.

However, the war had left European economies devastated and the LNHO faced the problem of funding these measures. An important event in the development of international cooperation and public health in general was the establishment of the Rockefeller Foundation in 1913 and its International Health Commission, later renamed the International Health Board and the International Health Division. In the interwar period, the International Health Division played a key role in the internationalization of health. The Rockefeller Foundation also generously helped the LNHO and the personnel of national health administrations. It initiated a new era in international health in which corporate philanthropy became part of global health. The Rockefeller Foundation insisted on philanthropy rather than charity. It defined philanthropy as an investment, to be offered to government institutions, not individuals, of limited duration in order to stimulate self-help and not lead to dependence. The main area of focus of the foundation was on the control and elimination of communicable diseases, so it made investments in the colonies in Asia, Africa and Latin America. Some of the major projects that it undertook were in medical research and eradication programmes for yellow fever in Africa and South America (1915–45), tuberculosis in France (1917–24), malaria in Asia, Africa and the United States (1915–35) and hookworm in India.

The LNHO similarly invested in malaria-eradication programmes in Europe, Asia and Africa. Several Malaria Commissions were organized by the LNHO in the 1920s, in India, Bulgaria and Greece. In 1925 the first International Malaria Congress was organized in Rome. In Europe the main malaria survey was launched along the River Danube. In England, Ross enlisted 'mosquito brigades' to eliminate mosquito larvae from stagnant pools and marshes. Medical surveys and larviciding operations were conducted in India – in Bombay, Jhansi, Poona, Meerut, Secunderabad and all other military posts. In 1917 the Bengal Nagpur Railway and the East India Railways formed a separate malaria-control organization specifically to control the disease in and around stations. The Rockefeller Foundation undertook similar larviciding and breeding-pool removal programmes during the 1920s in the tea plantations of Assam and in Mysore.

In the aftermath of the Second World War there was a long and uncertain period of decolonization when several of the colonial nations in Asia and Africa achieved independence. This was also the period of global collaboration in healthcare, particularly under the WHO. This highlighted the great contrasts in health provisions, mortality rates and living conditions between Europe and the United States on the one hand and the postcolonial nations on the other.

One of the greatest attempts at assimilation in global health took place in the 1930s and 1940s, in new ideas and visions of socialist public health and medicine. The main proponent of this was Henry Sigerist, a Swiss-born doctor and medical historian. He worked in the Johns Hopkins University School of Medicine in the 1930s and was deeply influenced by the socialist public health policies of the Soviet Union. In his book *Socialised Medicine in the Soviet Union* (1937) he promoted the Soviet structure of free and universal public health facilities and encouraged its adoption by other countries.[1] He stressed the need for a national health service and a socially equitable distribution of healthcare with funding from the state. In Britain a small group of radical socialist physicians who had been deeply influenced by the developments in the Soviet Union and Sigerist's 'socialist medicine' formed the Socialist Medical Association (SMA). This played a critical role in instituting the post-war National Health Service in Britain.[2] Sigerist's ideas also influenced health planning in Canada and in post-Independence India.

The Second World War rendered the LNHO dysfunctional and paved the way for the formation of the WHO, which formally began working in 1948 in Geneva, Switzerland. The WHO marked a new era in global health and epidemic control. Its main activities were in global vaccination campaigns, particularly for measles, polio and smallpox in children; addressing questions of poverty and health together; and ensuring basic medical infrastructure in different parts of the world. The challenge for the WHO was to extend the provisions of welfare on a global scale, to ensure that the citizens of poor countries had access to the basic provisions of healthcare and medicines. This was particularly difficult in the face of problems of epidemics, malnutrition and a lack of basic medical facilities in poor countries in Asia and Africa.

From the 1960s the WHO launched global disease eradication programmes and achieved great success in tackling smallpox. In 1967 it launched an intensified plan of action against smallpox, which threatened 60 per cent of the world's population. Through the success of the global campaign, smallpox was confined to the Horn of Africa and then to a single last natural case, which occurred in Somalia in 1977. The WHO achieved far less success with other infectious diseases, particularly malaria. It adopted a formal policy on the control and eradication of malaria in 1955, mainly using DDT and the distribution of quinine. Despite the successive programmes and investments, malaria remains a global health problem, presenting a serious infection risk to 2.7 billion people.

One of the main indicators of health in modern societies has been the decline in mortality rates. This has generally been accepted as a sign of better preventive medicine and improvements in general living conditions. Mortality rates also narrate the story of divergence in global health. In Europe there was a noticeable decline in mortality from the end of the eighteenth century. The greatest fall was in the late nineteenth century, despite the heavy causalities of the two world wars and the Spanish flu epidemic in 1918–19. This decline has been for both infants and adults, from infectious and other diseases, and life expectancy. In some countries, such as England, the decline started early – by the middle of the eighteenth century. Between 1730 and 1815, England's population doubled from 5.3 million to 10 million. It doubled again in the next 55 years to reach 21 million in 1871. The rate of growth has slowed only slightly since then, reaching 35.5 million by 1911. The mortality decline in the United States was slightly later, in the early twentieth century, when it declined by 40 per cent between 1900 and 1940.[3]

Historians have debated about the extent to which medical intervention helped in this decline. Thomas McKeown put forward the view that the growth in population in the industrialized world from the late 1700s to the present day was due not to advancements in the field of medicine or public health but to improvements in overall standards of living, especially diet and nutrition, resulting from better economic conditions.[4] This belief was contested by Simon Szreter, who highlighted not only the inconsistencies in McKeown's statistics but also the importance of sanitation and public health measures, such as supplies of clean water and milk, a nutritious diet, vaccination campaigns, and better medical facilities and diagnostics. Overall the decline in mortality in the West has been ascribed to better primary and public health, social welfare policies and economic growth.[5]

This link between public health, economic condition and mortality rates has assumed even greater significance in understanding the role of medicine in the developing and poor nations. The rapid decline in mortality rates in South America, South Asia and Africa almost invariably came in the postcolonial period, from the 1950s.[6] Sometimes the drop in mortality in the underdeveloped countries has been more rapid than that in Europe, and also less sustainable. One of the most noticeable mortality declines of the twentieth century, particularly in the second half of the century, was in Sub-Saharan Africa. By the end of the century, mortality among children under five had decreased from about 500 per 1,000 to about 150. Similarly, the average life expectancy, which was less than 30 years about 100 years ago, had increased to more than 50 years by the early 1990s. Much of the mortality decline happened in the second half of the twentieth century. This too has been explained in terms of both medical intervention in the shape of global and local health measures for the control of epidemic diseases, such as cholera, plague and malaria, and vaccination campaigns, as well as economic growth.[7]

However, the disturbing fact is that in Africa from the 1990s, mortality decline stalled for the region overall, with many countries experiencing reversals in the upward trend in life expectancy, largely because of the rise of HIV/AIDS mortality.[8] By 2000, AIDS had killed 1 million people every year, making it the world's biggest killer, and 95 per cent of these deaths took place in the developing countries, particularly Sub-Saharan Africa. There are a number of critical issues, such as disparate access to resources, political power, education, healthcare and legal services, which have contributed to this.

The mortality rates of both infants and children under the age of five have declined in postcolonial India. However, these rates are consistently higher among those infants born to illiterate mothers than those born to mothers who have received some education. Children born to mothers with at least eight years of schooling have a 32 per cent lesser chance of dying in the neonatal period and a 52 per cent lesser chance in the post-neonatal period, as compared with illiterate mothers.[9] In other words, this highlights the problem of the inequitable distribution of medical facilities in the country. India has a growing divergence in wealth and resources between the poor and the rich (where 30 per cent of the population are still under the poverty line and face starvation and malnutrition, while 71 per cent do not have public or private health cover[10]), which needs to be addressed. HIV/AIDS has exposed, more than any other disease, the great divergence in global health, which is not just along geographical or national lines but also in terms of class. Disease in general, whether in inner city New York or in rural Haiti, is a product of longstanding social and economic deprivation. Paul Farmer has shown that diseases such as HIV and tuberculosis spread particularly in those resgions and among those communities that were economically marginalized and suffered from 'structural violence', particularly in terms of the availability of drugs and other medical facilities.[11]

Notes

1 Henry E. Sigerist, *Socialised Medicine in the Soviet Union* (London, 1937).
2 John Stewart, *'The Battle for Health': A Political History of the Socialist Medical Association, 1930–51* (Aldershot, UK, 1999).
3 Szreter, 'Economic Growth, Disruption, Deprivation, Disease, and Death', p. 697.
4 T. McKeown, *The Modern Rise of Population* (New York, 1976).
5 Szreter, 'The Importance of Social Intervention in Britain's Mortality Decline, c. 1850–1914: A Reinterpretation of the Role of Public Health', *Social History of Medicine*, 1 (1988), 1–38.
6 Kingsley Davis, 'Amazing Decline of Mortality in Underdeveloped Areas', *The American Economic Review*, 46 (1956), 305–31.
7 Jacob Adetunji and Eduard R. Bos, 'Levels and Trends in Mortality in Sub-Saharan Africa: An Overview', in D.T. Jamison, R.G. Feachem, M.W. Makgoba, et al. (eds), *Disease and Mortality in Sub-Saharan Africa*, 2nd edition (Washington, DC, 2006), pp. 11–14.
8 Ibid.
9 *Report, The Infant and Child Mortality India; Levels, Trends and Determinants*, New Delhi, India, 2012, http://www.unicef.org/india/Report.pdf.
10 Ajay Mahal, et al (eds), *India Health Report 2010* (New Delhi, 2010), p. 83.
11 Paul Farmer, *Infections and Inequalities: The Modern Plagues* (Berkeley, 2001).

Bibliography

'An Account of Some Books', *Philosophical Transactions of the Royal Society* (*Philosophical Transactions*), 13 (1683), 100.

'An Account of the Cachexia Africana', *The Medical and Physical Journal*, 2 (1799), 171.

'Professor Koch's Investigations on Malaria: Second Report to the German Colonial Office', *British Medical Journal*, 2038 (10 February 1900), 325–7.

'The Madras Medical School', *Madras Journal of Literature and Science*, 7 (1838), 265

Abbri, Ferdinando. 'Alchemy and Chemistry: Chemical Discourses in the Seventeenth Century', *Early Science and Medicine*, 5 (2000), 214–26.

Adetunji, Jacob and Eduard R. Bos. 'Levels and Trends in Mortality in Sub-Saharan Africa: An Overview', in D.T. Jamison, R.G. Feachem, M.W. Makgoba, et al. (eds) *Disease and Mortality in Sub-Saharan Africa*, 2nd edition (Washington, DC, 2006), pp. 11–14.

Ainslie, Whitelaw. *Materia Medica of Hindoostan, and Artisan's and Agriculturalist's Nomenclature* (Madras, 1813).

——. *Materia Indica, Or, Some Account of Those Articles which are Employed by the Hindoos, and Other Eastern Nations, in their Medicine, Arts, and Agricultural* (London, 1826).

Alavi, Seema. 'Unani Medicine in the Nineteenth-century Public Sphere: Urdu Texts and the Oudh Akhbar', *Indian Economic and Social History Review*, 42 (2005), 101–29.

Aldrich, Robert. *Greater France; A History of French Overseas Expansion* (Basingstoke, 1996).

Ali, M. Athar. *The Mughal Nobility under Aurangzeb* (London, 1966).

Allen, Phyllis. 'The Royal Society and Latin America as Reflected in the *Philosophical Transactions* 1665–1730', *Isis*, 37 (1947), 132–8.

Allen, William and T.R.H. Thompson. *A Narrative of the Expedition Sent by Her Majesty's Government to the River Niger, in 1841*, vol. 1 (London, 1848).

Amrith, Sunil S. *Decolonizing International Health: India and Southeast Asia, 1930–65* (Basingstoke, 2006).

Amster, Ellen. 'The Many Deaths of Dr. Emile Mauchamp: Medicine, Technology, and Popular Politics in Pre-Protectorate Morocco, 1877–1912', *International Journal of Middle East Studies*, 36 (2004), 409–28.

An Account of the Religion, and Government, Learning, and Oeconomy, &c of the Malabarians: Sent by the Danish Missionaries to their Correspondents in Europe, Translated from the High-Dutch (London, 1717).

Anderson, M.S. *War and Society in Europe of the Old Regime 1618–1789* (London, 1988).

Anderson, Warwick P. 'Immunities of Empire: Race, Disease and the New Tropical Medicine, 1900–1920', *Bulletin of the History of Medicine*, 70 (1996), 94–118.

———. 'Geography, Race and Nation: Remapping "Tropical" Australia, 1890–1930', *Historical Records of Australian Science*, 11 (1996), 457–87.

———. *The Cultivation of Whiteness: Science, Health and Racial Destiny in Australia* (Carlton South, Victoria, 2002).

Anker, Peder. *Imperial Ecology: Environmental Order in the British Empire, 1895–1945* (Cambridge, MA, 2001).

Arasaratnam, Sinnappah. *Merchants, Companies and Commerce on the Coromandel Coast, 1650–1740* (Delhi, 1986).

Arnold, David (ed.), *Imperial Medicine and Indigenous Societies: Disease, Medicine, and Empire in the Nineteenth and Twentieth Centuries* (Manchester, 1988).

———. 'Social Crisis and Epidemic Disease in the Famines of Nineteenth-Century India', *Social History of Medicine*, 6 (1993), 385–404.

———. *Colonizing the Body: State Medicine and Epidemic Disease in Nineteenth-Century India* (Berkeley & Los Angeles, 1993).

———. (ed.), *Warm Climates, Western Medicine: The Emergence of Tropical Medicine, 1500–1900* (Amsterdam, 1996).

———. 'Introduction: Tropical Medicine before Manson', in Arnold (ed.), *Warm Climates and Western Medicine*, pp. 1–19.

———. *Science, Technology and Medicine in Colonial India* (Cambridge, 2000).

———. 'Race, Place and Bodily Difference in Early Nineteenth-Century India', *Historical Research*, 77 (2004), 254–73.

Attewell, Guy. *Refiguring Unani Tibb: Plural Healing in Late Colonial India* (Hyderabad, 2007).

Babar, Zaheer. *The Science of Empire: Scientific Knowledge, Civilization, and Colonial Rule in India* (Albany, NY, 1996).

———. 'Colonizing Nature: Scientific Knowledge, Colonial Power and the Incorporation of India into the Modern World-System', *British Journal of Sociology*, 52 (2001), 37–58.

Bakewell, Peter (ed.), *Mines of Silver and Gold in the Americas* (Aldershot, 1997).

Ballhatchet, Kenneth. *Race, Sex and Class under the Raj: Imperial Attitudes and Policies and their Critics* (London, 1980).

Banerji, Debabar. 'The Politics of Underdevelopment of Health: The People and Health Service Development in India: A Brief Overview', *International Journal of Health Services* 34 (2004), 123–42.

Banthia, Jayant and Tim Dyson. 'Smallpox in Nineteenth-Century India', *Population and Development Review*, 25 (1999), 649–80.

Barnes, Barry. *Interests and the Grounds of Knowledge* (London, 1977).

Barrett, T.P. Monath 'Epidemiology and Ecology of Yellow Fever Virus', *Advances in Virus Research*, 61 (2003), 291–315.

Basalla, George. 'The Spread of Western Science', *Science*, 156 (1967), 611–22.

Bashford, Alison. '"Is White Australia Possible?" Race, Colonialism and Tropical Medicine', *Ethnic and Racial Studies*, 23 (2000), 248–71.

——. 'At the Border: Contagion, Immigration, Nation', *Australian Historical Studies*, 33 (2002), 344–58.

——. *Imperial Hygiene: A Critical History of Colonialism, Nationalism and Public Health* (Basingstoke, 2004).

Bassett, D.K. 'British "Country" Trade and Local Trade Networks in the Thai and Malay States, c. 1680–1770', *Modern Asian Studies*, 23 (1989), 625–43.

Bastian, Charlton H. 'The Bearing of Experimental Evidence upon the Germ-Theory of Disease', *BMJ*, 889 (12 January 1878), 49–52.

Bayly, C.A. *Imperial Meridian: The British Empire and the World, 1780–1830* (London, 1989).

——. '"Archaic" and "Modern" Globalization in the Eurasian and African Arena', in Anthony G. Hopkins, ed., *Globalization in World History* (New York, 2002), pp. 47–73.

Beinart, William. 'Men, Science, Travel and Nature in the Eighteenth and Nineteenth-Century Cape', *Journal of Southern African Studies*, 24 (1998), 775–99.

Benians, E., J. Holland Rose and A. Newton (eds). *The Cambridge History of the British Empire* (9 vols, Cambridge, 1929–59).

Berg, Maxine. 'In Pursuit of Luxury: Global History and British Consumer Goods in the Eighteenth Century', *Past & Present*, 182 (2004), 85–142.

Bhattacharya, Nandini. 'The Logic of Location: Malaria Research in Colonial India, Darjeeling and Duars, 1900–30', *Medical History*, 55 (2011), 183–202.

——. *Contagion and Enclaves; Tropical Medicine in Colonial India* (Liverpool, 2012).

Bhattacharya, Sanjoy, Mark Harrison and Michael Worboys. *Fractured States: Smallpox, Public Health and Vaccination Policy in British India, 1800–1947* (Hyderabad, 2005).

Blanco, Richard L. 'Henry Marshall (1775–1851) and the Health of the British Army', *Medical History*, 14 (1970), 260–76.

——. 'The Development of British Military Medicine, 1793–1814', *Military Affairs*, 38 (1974), 4–10.

——. 'The Soldier's Friend Sir Jeremiah Fitzpatrick, Inspector of Health for Land Forces', *Medical History*, 20 (1976), 402–21.

Blane, Gilbert. *Select Dissertations on Several Subjects of Medical Science* (London, 1833).

Bloor, David. *Knowledge and Social Imagery* (London, 1976).

Bougerol, Christiane. 'Medical Practices in the French West Indies: Master and Slave in the 17th and 18th Centuries', *History and Anthropology*, 2 (1985), 125–43.

Boyd, H. Glenn. 'A Brief History of Medical Missions', *Gospel Advocate*, 132 (1990), 14–15.

Bradfield, E.W.C. *An Indian Medical Review* (Delhi, 1938).

Brandon, George. 'The Uses of Plants in Healing in Afro-Cuban Religion, Santería', *Journal of Black Studies*, 22 (1991), 55–76.

Brantlinger, Patrick, *Rule of Darkness: British Literature and Imperialism,1830–1914* (Ithaca & London, 1988).

Braudel, Fernand. *Civilization and Capitalism, 15th–18th Century: The Perspective of the World* (Berkeley, 1992).

Bravo, Michael T. 'Mission Gardens: Natural History and Global Expansion, 1720–1820', in Schiebinger and Claudia Swan (eds) *Colonial Botany: Science, Commerce, and Politics in the Early Modern World* (Philadelphia, 2005), pp. 49–65.

Brenner, Robert. *Merchants and Revolution: Commercial Change, Political Conflict, and London's Overseas Traders, 1550–1650* (Cambridge, 1993).

Brentjes, Sonja. 'Between Doubts and Certainties: On the Place of Science in Islamic Societies within the Field of History of Science', *NTM*, 11 (2003), 65–79.

Brewer, Anthony. *Marxist Theories of Imperialism: A Critical Survey*, 2nd edition (New York, 1990).

Brewer, John. *The Sinews of Power: War, Money and the English State 1688–1783* (London, 1994).

Brimnes, Niels. 'Variolation, Vaccination and Popular Resistance in Early Colonial South India', *Medical History*, 48 (2004), 199–228.

Brockway, Lucille. *Science and the Colonial Expansion: The Role of British Royal Botanic Gardens* (New York, 1979).

Brorson, Stig. 'The Seeds and the Worms: Ludwik Fleck and the Early History of Germ Theories', *Perspectives in Biology and Medicine*, 49 (2006), 64–76.

Bruijn, Iris. *Ship's Surgeons of the Dutch East India Company; Commerce and the Progress of Medicine in the Eighteenth Century* (Leiden, 2009).

Bryant, J.E., E.C. Holmes, A.D.T. Barrett. 'Out of Africa: A Molecular Perspective on the Introduction of Yellow Fever Virus into the Americas,' *PLoS Pathogens*, 3 (2007) doi:10.1371/journal.ppat.0030075.

Bryson, Alexander. 'Prophylactic Influence of Quinine', *Medical Times and Gazette*, 7 (1854), 6–7.

Buckingham, Jane. *Leprosy in Colonial South India: Medicine and Confinement* (New York, 2002).

Burmanni, Nicolai Laurentii. *Flora Indica: Cui Accedit Series Zoophytorum Indicorum, Necnon. Prodromus Florae Capensis* (Amsterdam, 1768).

Burnard, Trevor and Kenneth Morgan. 'The Dynamics of the Slave Market and Slave Purchasing Patterns in Jamaica, 1655–1788', *The William and Mary Quarterly* (2001), 205–28.

Bynum, W.F. *Science and the Practice of Medicine in the Nineteenth Century* (Cambridge, 1994).

Bynum, W.F. and Caroline Overy (eds). *The Beast in the Mosquito: The Correspondence of Ronald Ross and Patrick Manson* (Amsterdam, 1998).

Cain, J. and A.G. Hopkins, 'Gentlemanly Capitalism and British Expansion Overseas II: New Imperialism, 1850–1945', *The Economic History Review*, 40 (1987), 1–26.

Cameron, Charles. 'An Address on Micro-Organisms and Disease', *BMJ*, 1084 (8 October 1881), 583–6.

Cameron-Smith, Alexander. 'Australian Imperialism and International Health in the Pacific Islands', *Australian Historical Studies*, 41 (2002), 57–74.

Carothers, J.C. *The Psychology of Mau Mau* (Nairobi, 1954).

Carter, Vandyke H. 'Notes on the Spirillum Fever of Bombay, 1877', *Medical and Chirurgical Transactions*, 61 (1878), 273–300.

Chadwick, Edwin. *Report to Her Majesty's Principal Secretary of State for the Home Department, from the Poor Law Commissioners, on an Inquiry into the Sanitary Condition of the Labouring Population of Great Britain* (London, 1842).

Chakrabarti, Pratik. *Western Science in Modern India: Metropolitan Methods, Colonial Practices* (New Delhi, 2004).

——. '"Neither of Meate nor Drinke, but what the Doctor Alloweth": Medicine amidst War and Commerce in Eighteenth Century Madras', *Bulletin of the History of Medicine*, 80 (2006), 1–38.

——. 'Medical Marketplaces beyond the West: Bazaar Medicine, Trade and the English Establishment in Eighteenth Century India', in Wallis and Mark Jenner (eds) *Medicine and the Market*, pp. 196–215.

——. 'Empire and Alternatives: *Swietenia Febrifuga* and the Cinchona Substitutes', *Medical History*, 54 (2010), 75–94.

——. *Materials and Medicine; Trade, Conquest and Therapeutics in the Eighteenth Century* (Manchester, 2010).

——. *Bacteriology in British India; Laboratory Medicine and the Tropics* (Rochester, NY, 2012).

Chakraborty, Pratik. 'Science, Nationalism, and Colonial Contestations: P. C. Ray and his *Hindu Chemistry*', *Indian Economic and Social History Review*, 37 (2000), 185–213.

Charters, Erica. 'Disease, War, and the Imperialist State: The Health of the British Armed Forces during the Seven Years War, 1756–63', unpublished DPhil thesis, Faculty of Modern History, University of Oxford, 2006.

Chirol, V.I. *India Old and New* (London, 1921).

Chopra, R.N. *Pharmacopoeia of India* (Delhi, 1955).

Christopher, Emma. *Slave Ship Sailors and their Captive Cargoes, 1730–1807* (New York, 2006).

Churchill, Wendy D. 'Bodily Differences? Gender, Race, and Class in Hans Sloane's Jamaican Medical Practice, 1687–1688', *JHMAS*, 60 (2005), 391–444.

Cipolla, Carlo M. *Fighting the Plague in Seventeenth-Century Italy* (Madison, 1981).

Clarke, Edwin (ed.), *Modern Methods in the History of Medicine* (London, 1971).

Clément, Alain. 'The Influence of Medicine on Political Economy in the Seventeenth Century', *History of Economic Review*, 38 (2003), 1–22.

Clericuzio, Antonio. 'From van Helmont to Boyle; A Study of the Transmission of Helmontian Chemical and Medical Theories in Seventeenth-Century England', *The British Journal for the History of Science*, 26 (1993), 303–34.

Cohen, William B. 'Malaria and French Imperialism', *Journal of African History*, 24 (1983), 23–36.

Cohn, Samuel K. '4 Epidemiology of the Black Death and Successive Waves of Plague', *Medical History Supplement*, 27 (2008), 74–100.

Coley, Nigel G. '"Cures without Care" "Chymical Physicians" and Mineral Waters in Seventeenth-Century English Medicine', *Medical History*, 23 (1979), 191–213.

Columbus, Christopher (edited and translated with an Introduction and notes by B. W Ife). *Journal of the First Voyage (diario Del Premier Viaje) 1492* (Warminster, 1990).

Conklin, A.L. *A Mission to Civilize: The Republican Idea of an Empire in France and Africa* (Stanford, 1997).

Conway, Stephen. 'The Mobilization of Manpower for Britain's Mid-Eighteenth-Century Wars', *Historical Research*, 2004 (77), 377–404.

Cook, Harold J. *The Decline of the Old Medical Regime in Stuart London* (Ithaca, 1986).

——. 'Physicians and Natural History', in N. Jardine, J.A. Secord and E.C. Spary (eds) *Cultures of Natural History* (Cambridge, 1996), pp. 91–105.

——. *Trials of an Ordinary Doctor: Joannes Groenevelt in Seventeenth-Century London* (Baltimore, 1994).

——. *Matters of Exchange; Commerce, Medicine, and Science in the Dutch Golden Age* (New Haven & London, 2007).

——. 'Victories for Empiricism, Failures for Theory: Medicine and Science in the Seventeenth Century', in Charles T. Wolfe and Ofer Gal (eds), *The Body as Object and Instrument of Knowledge. Embodied Empiricism in Early Modern Science* (Dordrecht, 2010), pp. 9–32.

——. 'Markets and Cultures: Medical Specifics and the Reconfiguration of the Body in Early Modern Europe', *Transactions of the Royal Historical Society*, 21 (2011), 123–45.

Cook, Noble David. *'Born to Die': Disease and the New World Conquest, 1492–1650* (Cambridge, 1998).

——. 'Sickness, Starvation, and Death in Early Hispaniola', *Journal of Interdisciplinary History*, 32 (2002), 349–86.

Cooper, Alix. *Inventing the Indigenous: Local Knowledge and Natural History in Early Modern Europe* (Cambridge, 2007).

Cooper, Randolf G.S. *The Anglo-Maratha Campaigns and the Contest for India: The Struggle for Control of the South Asian Military Economy* (Cambridge, 2003).

Corbin, Alain. *The Foul and the Fragrant: Odor and the French Social* Imagination (Cambridge, MA., 1986).

Crellin, J.K. 'Pharmaceutical History and its Sources in the Wellcome Collections. I. The Growth of Professionalism in Nineteenth-Century British Pharmacy', *Medical History*, 11 (1967), 215–27.

Crimmin, P.K. 'British Naval Health, 1700–1800: Improvement over Time?', in Geoffrey L. Hudson (ed.), *British Military and Naval Medicine, 1600–1830* (Amsterdam & New York, 2007), pp. 183–200.

——. 'The Sick and Hurt Board and the Health of Seamen c. 1700–1806', *Journal for Maritime Research*, 1 (1999), 48–65.

Crosby, Alfred W. *The Columbian Exchange: Biological and Cultural Consequences of 1492* (Westport, 1972).

——. *Ecological Imperialism: The Biological Expansion of Europe (900–1900)* (Cambridge, 1986).

——. *The Columbian Voyages, the Columbian Exchange, and their Historians* (Washington, DC, 1987).

——. *Germs, Seeds & Animals: Studies in Ecological History* (New York, 1994).

Crowfoot, W.M. 'An Address on the Germ-Theory of Disease', *BMJ*, 1134 (23 September 1882), 551–4.

Crozier, Anna. *Practising Colonial Medicine: The Colonial Medical Service in British East Africa* (London & New York, 2007).

——. 'What Was Tropical about Tropical Neurasthenia? The Utility of the Diagnosis in the Management of British East Africa', *JHMAS*, 64 (2009), 518–48.

Cunningham, Andrew and Bridie Andrews (eds). *Western Medicine as Contested Knowledge* (Manchester, 1997).

Cueto, Marcos. *The Value of Health: A History of the Pan American Health Organization* (Washington, DC, 2007).

Curtin, Philip D. *The Image of Africa: British Ideas and Action, 1780–1850*, vol. 2 (Madison & London, 1973).

——. *Death by Migration: Europe's Encounter with the Tropical World in the Nineteenth Century* (Cambridge, 1989).

——. 'Disease and Imperialism', in Arnold (ed.), *Warm Climates and Western Medicine*: pp. 99–107.

Dancer, Thomas. *The Medical Assistant; Or Jamaica Practice of Physic: Designed Chiefly for the Use of Families and Plantations* (Kingston, 1801).

Das Gupta, Ashin. *Indian Merchants and the Decline of Surat: 1700–1750* (Wiesbaden, 1979).

——. *Merchants of Maritime India, 1500–1800* (Aldershot, 1994).

Datta, Partho. *Planning the City, Urbanization and Reform in Calcutta, c. 1800–1940* (New Delhi, 2012).

Davidovitch, Nadav and Rakefet Zalashik. 'Pasteur in Palestine: The Politics of the Laboratory', *Science in Context*, 23 (2010), 401–25.

Davis, Kingsley. 'Amazing Decline of Mortality in Underdeveloped Areas', *The American Economic Review*, 46 (1956), 305–31.

Dawson, Marc. 'Disease and Population Decline of the Kikuyu of Kenya, 1890–1925', in Christopher Fyfe and David McMaster (eds) *African Historical Demography: Proceedings of a Seminar Held in the Centre of African Studies, University of Edinburgh*, vol. 2 (Edinburgh, 1981) pp. 121–38.

De Vos, Paula. 'Natural History and the Pursuit of Empire in Eighteenth-Century Spain', *Eighteenth-Century Studies*, 40 (2007), 209–39.

De, Shambhu Nath. 'An Experimental Study of the Mechanism of Action of *Vibrio cholerœ* on the Intestinal Mucous Membrane', *Journal of Pathology and Bacteriology*, 66 (1953), 559–62.

Desmond, Ray. *The European Discovery of the Indian Flora* (Oxford, 1992).

Dewhurst, Kenneth. *The Quicksilver Doctor, the Life and Times of Thomas Dover, Physician and Adventurer* (Bristol, 1957).

Dias, Jill R. 'Famine and Disease in the History of Angola, 1830–1930', *Journal of African History*, 22 (1981), 349–78.

Digby, Anne and Helen Sweet. 'Social Medicine and Medical Pluralism: the Valley Trust and Botha's Hill Health Centre, South Africa, 1940s to 2000s', *Social History of Medicine* (2011) doi:10.1093/shm/hkr114.

Drayton, Richard. 'Science and the European Empires', *The Journal of Imperial and Commonwealth History*, 23 (1995), 503–10.

——. *Nature's Government: Science, Imperial Britain, and the 'Improvement' of the World* (New Haven & London, 2000).

Dritsas, Lawrence. 'Civilising Missions, Natural History and British Industry', *Endeavour*, 30 (2006), 50–4.

Driver, Felix. 'Moral Geographies: Social Science and the Urban Environment in Mid-Nineteenth Century England', *Transactions of the Institute of British Geographers*, 13 (1988), 275–87.

——. 'Geography's Empire: Histories of Geographical Knowledge', *Society and Space*, 10 (1992), 23–40.

——. *Geography Militant: Cultures of Exploration and Empire* (Oxford, 2001).

Duffy, John. 'Smallpox and the Indians in the American Colonies', *Bulletin of the History of Medicine*, 25 (1951), 324–41.

Dumett, Raymond E. 'The Campaign against Malaria and the Expansion of Scientific Medical and Sanitary Services in British West Africa, 1898–1910', *African Historical Studies*, 2 (1968), 153–97.

Duncan, Andrew. *Supplement to the Edinburgh New Dispensatory* (Edinburgh, 1829).

Dunn, Richard S. *Sugar and Slaves: The Rise of the Planter Class in the English West Indies, 1624–1713* (Chappell Hill, 1972).

——. *Moravian Missionaries at Work in a Jamaican Slave Community, 1754–1835* (Minneapolis, 1994).

Duran-Reynals, Marie Louise de Ayala. *The Fever Bark Tree: The Pageant of Quinine* (New York, 1946).

Durbach, Nadja. ''They Might as Well Brand us': Working-Class Resistance to Compulsory Vaccination in Victorian England', *Social History of Medicine*, 13 (2000), 45–63.

Echenberg, Myron. *Africa in the Time of Cholera: A History of Pandemics from 1817 to the Present* (Cambridge, 2011).

Eco, Umberto. 'In Praise of St. Thomas', *Travels in Hyperreality: Essays* (San Diego, 1987) pp. 257–68.

Eden, Trudy. *The Early American Table: Food and Society in the New World* (Dekalb, IL, 2010/2008).

Edmond, Rod. 'Returning Fears: Tropical Disease and the Metropolis', in Driver and Luciana Martins (eds). *Tropical Visions in an Age of Empire* (Chicago, 2005), pp. 175–94.

Eliot, Charles. *The East Africa Protectorate* (London, 1966/1905).

Eltis, David. *The Rise of African Slavery in the Americas* (Cambridge, 2000).

Emmer, P.C. *The Dutch Slave Trade 1500–1850* (Oxford, 2006).

Espinosa, Mariola. 'The Threat from Havana: Southern Public Health, Yellow Fever, and the U.S. Intervention in the Cuban Struggle for Independence, 1878–1898', *The Journal of Southern History*, 77 (2006), 541–68.

——. *Epidemic Invasions: Yellow Fever and the Limits of Cuban Independence, 1878–1930* (Chicago, 2009).

Esteban, Javier Cuenca. 'The British Balance of Payments, 1772–1820: India Transfers and War Finance', *The Economic History Review*, 54 (2001), 58–86.

Fabian, Johannes. *Out of Our Minds: Reason and Madness in the Exploration of Central Africa* (Berkeley, 2000).

Falconbridge, Alexander. *An Account of the Slave Trade on the Coast of Africa* (London, 1788).

Fan, Ruiping. 'Modem Western Science as a Standard for Traditional Chinese Medicine: A Critical Appraisal', *The Journal of Law, Medicine & Ethics*, 31 (2003), 213–21.

Fang, Xiaoping. *Barefoot Doctors and Western Medicine in China* (Rochester, NY, 2012).

Farley, John. *To Cast out Disease: A History of the International Health Division of the Rockefeller Foundation (1913–1951)* (New York, 2004).

Farmer, Paul, *Infections and Inequalities: The Modern Plagues* (Berkeley, 2001).

Feierman, Steven. 'Struggles for Control: The Social Roots of Health and Healing in Modern Africa', *African Studies Review*, 28 (1985), 73–147.

Fenger, Johan Ferdinand. *History of the Tranquebar Mission: Worked out from Original Papers, Published in Danish and translated in English from the German of Emil Francke* (Tranquebar, 1863).

Fett, Sharla M. *Working Cures: Health, Healing and Power on the Southern Slave Plantations* (Chapel Hill, 2002).

Findlen, Paula, and Pamela H. Smith (eds). *Merchants & Marvels; Commerce, Science, and Art in Early Modern Europe* (New York & London, 2002).

Fisher, Michael H. 'Indirect Rule in the British Empire: The Foundations of the Residency System in India (1764–1858)', *Modern Asian Studies*, 18 (1984), 393–428.

Flint, Karen. 'Competition, Race, and Professionalization: African Healers and White Medical Practitioners in Natal, South Africa in the Early Twentieth Century', *Social History of Medicine*, 14 (2001), 199–221.

——. *Healing Traditions: African Medicine, Cultural Exchange, and Competition in South Africa, 1820–1948* (Ohio, 2008).

Ford, John. *The Role of Trypanosomiases in African Ecology: A Study of the Tsetse Fly Problem* (Oxford, 1971).

Foucault, Michel. *Madness and Civilization: A History of Insanity in the Age of Reason* (London, 1967).

——. *Order of Things; An Archaeology of the Human Sciences* (New York, 1994/1970).

Fruehauf, F. 'Chinese Medicine in Crisis: Science, Politics and the Making of "TCM"', *Journal of Chinese medicine – HOVE*, 61 (1999), 6–14.

Furber, Holden. 'Asia and the West as Partners before "Empire" and after', *Journal of Asian Studies* 28 (1969), 711–21.

———. *Rival Empires of Trade in the Orient, 1600–1800* (Minneapolis, 1976).

Gallagher, J. and R. Robinson, 'The Imperialism of Free Trade', *The Economic History Review*, 6 (1953), 1–15.

García, Mónica. 'Producing Knowledge about Tropical Fevers in the Andes: Preventive Inoculations and Yellow Fever in Colombia, 1880–1890', *Social History of Medicine*, 25 (2012), 830–47.

Gascoigne, John. *Science in the Service of Empire: Joseph Banks, the British State and the Uses of Science in the Age of Revolution* (Cambridge, 1998).

Geggus, David. 'Yellow Fever in the 1790s: The British Army in Occupied Saint Dominique', *Medical History*, 23 (1979), 38–58.

Giblin, James. 'Trypanosomiasis Control in African History: An Evaded Issue?', *The Journal of African History* , 31 (1990), 59–80.

Gibson, Charles. *The Aztecs Under Spanish Rule: A History of the Indians of the Valley of Mexico, 1519–1810* (Stanford, 1964).

Goldman, Alvin. 'Social Epistemology', *The Stanford Encyclopedia of Philosophy* (Summer 2010 edition), Edward N. Zalta (ed.), http://plato.stanford.edu/archives/sum2010/entries/epistemology-social/.

Good, Charles. *The Steamer Parish: The Rise and Fall of Missionary Medicine on an African Frontier* (Chicago & London, 2004).

Gradmann, Christoph (translated by Elborg Forster). *Laboratory Disease: Robert Koch's Medical Bacteriology* (Baltimore, 2009).

———. 'Robert Koch and the Invention of the Carrier State: Tropical Medicine, Veterinary Infections and Epidemiology around 1900', *Studies in History and Philosophy of Biological and Biomedical Sciences*, 41 (2010), 232–40.

Griffiths, Leuan. 'The Scramble for Africa: Inherited Political Boundaries', *The Geographical Journal*, 152 (1986), 204–16.

Griffiths, Nicholas and Fernando Cervantes (eds). *Spiritual Encounters: Interactions between Christianity and Native religions in Colonial America* (Birmingham, 1999).

Grove, Richard. *Green Imperialism: Colonial Expansion, Tropical Island Edens and the Origins of Environmentalism, 1660–1800* (Cambridge, 1995).

Guenel, Annick. 'The Creation of the First Overseas Pasteur Institute, or the Beginning of Albert-Calmette's Pastorian Career', *Medical History*, 43 (1999), 1–25.

Guha, Ranajit. 'On Some Aspects of the Historiography of Colonial India', Guha (ed.), *Subaltern Studies; Writings on South Asian History and Society*, vol.1 (Delhi, 1982), pp. 1–9.

Guy, Alan J. *Oeconomy and Discipline, Officership and Administration in the British Army 1714–63* (Manchester, 1985).

Habib, Irfan. *The Agrarian System of Mughal India, 1556–1707* (New Delhi, 1963).

Haffkine, W.M. 'Le cholera asiatique chez la cobbaye', *Comptes Rendus des Séances et Mémoires de la Société de Biologie*, 44 (1892), 635–7.

Haines, Robin and Ralph Shlomowitz, 'Explaining the Modern Mortality Decline: What can we Learn from Sea Voyages?', *Social History of Medicine*, 11 (1998), 15–48.

Hajeebu, S. 'Emporia and Bazaars', in J. Mokyr (ed.), *Oxford Encyclopaedia of Economic History*, vol. 2 (Oxford, 2003), p. 258.

Haller, J.S. Jr., 'The Negro and the Southern Physician: A Study of Medical and Racial Attitudes 1800–1860', *Medical History*, 16 (1972), 238–53.

Hamilton, Douglas. 'Private Enterprise and Public Service: Naval Contracting in the Caribbean, 1720–50', Journal of Maritime Research, 6 (2004), 37–64.

Hamlin, Christopher. 'Providence and Putrefaction; Victorian Sanitarians and the Natural Theology of Health and Disease', *Victorian Studies*, 28 (1985), 381–411.

——. *Cholera: The Biography* (Oxford, 2009).

Handler, Jerome S. 'Slave Medicine and Obeah in Barbados, Circa 1650 to 1834', *New West Indian Guide*, 74 (2000), 57–90.

Hannaford, Ivan. *Race: The History of an Idea in the West* (Washington, DC, 1996).

Hardiman, David. *The Coming of the Devi: Adivasi Assertion in Western India* (Delhi, 1995).

Harding, R. *Amphibious Warfare in the Eighteenth-Century: The British Expedition to the West Indies 1740–1742* (Suffolk, 1991).

Harries, Lyndon. 'The Arabs and Swahili Culture', *Africa: Journal of the International African Institute*, 34 (1964), 224–9.

Harris, B. 'War, Empire, and the "National Interest" in Mid-Eighteenth-Century Britain', in J. Flavell and S. Conway (eds) *Britain and America Go to War: The Impact of War and Warfare in Anglo-America, 1754–1815* (Gainesville, 2004), pp. 13–40.

Harris, Barbara. 'Agricultural Merchants' Capital and Class Formation in India', *Sociologia Ruralis*, 29 (1989), 166–79.

Harris, Steven J. 'Jesuit Scientific Activity in the Overseas Missions, 1540–1773', *Isis*, 96 (2005), 71–9.

Harrison, Mark, 'Tropical Medicine in Nineteenth-Century India', *The British Journal for the History of Science*, 25 (1992), 299–318.

——. *Public Health in British India: Anglo-Indian Preventive Medicine 1859–1914* (Cambridge, 1994).

——. '"The Tender Frame of Man": Disease, Climate, and Racial Difference in India and the West Indies, 1760–1860', *Bulletin of the History of Medicine*, 70 (1996), 68–93.

——. 'A Question of Locality: The Identification of Cholera in British India, 1860–1890', in Arnold (ed.), *Warm Climates and Western Medicine*, pp. 133–59.

——. 'Medicine and the Management of Modern Warfare: An Introduction', in Harrison, Roger Cooter and Steve Sturdy (eds). *Medicine and Modern Warfare* (Amsterdam, 1999), pp. 1–22.

——. *Climates and Constitutions: Health, Race, Environment and British Imperialism in India 1600–1850* (Delhi, 1999).

——. *Disease and the Modern World: 1500 to the Present Day* (Cambridge, 2004).

——. 'Science and the British Empire, *Isis*, 96 (2005), 56–63.

——. *Medicine in an Age of Commerce and Empire: Britain and its Tropical Colonies, 1660–1830* (Oxford, 2010).

——. *Contagion: How Commerce has Spread Disease* (New Haven & London, 2012).

Harrison, Mark and Worboys. '"A Disease of Civilization": Tuberculosis in Africa and India', in Lara Marks and Worboys (eds) *Migrants, Minorities and Health; Historical and Contemporary Studies* (London, 1997) pp. 93–124.

Hart, Ernest A. 'Cholera: Where it Comes from and how it is Propagated', *BMJ*, 1696 (1 July 1893), 1–4.

——. 'The West Indies as a Health Resort: Medical Notes of a Short Cruise among the Islands', *BMJ*, 920 (16 October 1897), 1097–9.

Hasan, Farhat. 'Indigenous Cooperation and the Birth of a Colonial City: Calcutta, c. 1698–1750', *Modern Asian Studies*, 26 (1992), 65–82.

Haynes, Douglas M. *Imperial Medicine: Patrick Manson and the Conquest of Tropical Disease, 1844–1923* (Philadelphia, 2001).

Headrick, Daniel. *Tools of Empire; Technology and European Imperialism in the Nineteenth Century* (Oxford, 1981).

Heniger, J. *Hendrik Adriaan van Reede tot Drakenstein (1636–1691) and Hortus Malabaricus: A Contribution to the History of Dutch Colonial Botany* (Rotterdam, 1986).

Henze, Charlotte E. *Disease, Health Care and Government in Late Imperial Russia; Life and Death on the Volga* (Abingdon & New York, 2011).

Heyne, Benjamin. *Tracts, Historical and Statistical, on India with Journals of Several Tours. Also an Account of Sumatra in a Series of Letters* (London, 1814).

Hobsbawm, E. J. *Industry and Empire* (London, 1968).

Hobsbawm, E.J. and Ranger (eds). *The Invention of Tradition* (Cambridge, 1983).

Hokkanen, Markku. 'Imperial Networks, Colonial Bioprospecting and Burroughs Wellcome & Co.: The Case of *Strophanthus Kombe* from Malawi (1859–1915)', *Social History of Medicine* (2012) doi: 10.1093/shm/hkr167.

Holmes, Timothy (ed.), *David Livingstone: Letters and Documents 1841–1872* (London, 1990).

Hong, Francis F. 'History of Medicine in China; When Medicine Took an Alternative Path', *McGill Journal of Medicine*, 8 (2004), 79–84.

Hsü, Elisabeth. 'The Reception of Western Medicine in China: Examples from Yunnan', in Patrick Petitjean, Cathérine Jami (eds), *Science and Empires: Historical Studies about Scientific Development and European Expansion* (Dordrecht, 1992), pp. 89–102.

Huber, Valesca. 'The Unification of the Globe by Disease? The International Sanitary Conferences on Cholera, 1851–1894', *The Historical Journal* 49 (2006), 453–76.

Hudson, Geoffrey L. (ed.), *British Military and Naval Medicine, 1600–1830* (Amsterdam & New York, 2007).

Huguet-Termes, Teresa. 'New World Materia Medica in Spanish Renaissance Medicine: From Scholarly Reception to Practical Impact', *Medical History*, 45 (2001), 359–76.

Hulme, Peter. *Colonial Encounters; Europe and the Native Caribbean, 1492–1797* (London & New York, 1986).

Hunt, Nancy Rose. *A Colonial Lexicon: Of Birth Ritual, Medicalization, and Mobility in the Congo* (Durham, NC, 1997).

Hunter, Michael. *Establishing the New Science: The Experience of the Early Royal Society* (Woodbridge, 1989).

Hyam, R. *Britain's Imperial Century, 1815–1914: A Study of Empire and Expansion* (Batsford, 1976).

Iliffe, John. 'The Organization of the Maji Maji Rebellion', *The Journal of African History*, 8 (1967), 495–512.

——. *East African Doctors: A History of the Modern Profession* (Cambridge, 1998).

Inkster, Ian. 'Scientific Enterprise and the Colonial "Model": Observations on Australian Experience in Historical Context', *Social Studies of Science*, 15 (1985), 677–704.

Isaacs. Jeremy D. 'D D Cunningham and the Aetiology of Cholera in British India, 1889–97', *Medical History*, 42 (1998), 279–305.

Jardine, Lisa. *Ingenious Pursuits: Building the Scientific Revolution* (London, 1999).

Jeffery, Roger. ''Recognizing India's Doctors: The Institutionalization of Medical Dependency, 1918–1939', *Modern Asian Studies*, 13 (1979), 301–26.

——. 'Doctors and Congress: The Role of Medical Men and Medical Politics in Indian Nationalism', in Mike Shepperdson and Colin Simmons (eds) *The Indian National Congress and the Political Economy of India, 1885–1985* (Avebury, 1988), pp. 160–73.

Jennings, Eric T. *Curing the Colonizers; Hydrotherapy, Climatology and French Colonial Spas* (Durham, NC, 2006).

Jensen, Niklas Thode. 'The Medical Skills of the Malabar Doctors in Tranquebar, India, as Recorded by Surgeon T L F Folly, 1798', *Medical History*, 49 (2005), 489–515.

John, T.J. 'Polio Eradication and Ethical Issues', *Indian Journal of Medical Ethics*, 2 (2005), 1–4.

Johnson, Ryan. '"An All-White Institution": Defending Private Practice and the Formation of the West African Medical Staff', *Medical History*, 54 (2010), 237–54.

Johnson, Walter. 'On Agency', *Journal of Social History*, 37 (2003), 113–24.

Jones, Colin. *The Charitable Imperative: Hospitals and Nursing in Ancien Regime and revolutionary France* (London, 1989).

Jones, Geoffrey. *Merchants to Multinationals: British Trading Companies in the Nineteenth and Twentieth Centuries* (Oxford, 2002).

Jones, Margaret. *Health Policy in Britain's Model Colony: Ceylon (1900–1948)* (Hyderabad, 2004).

Joyce, Patrick. 'What is the Social in Social History', *Past & Present*, 206 (2010), 213–48.

Kalusa, Walima T. 'Language, Medical Auxiliaries, and the Re-interpretation of Missionary Medicine in Colonial Mwinilunga, Zambia, 1922–51', *Journal of Eastern African Studies*, 1 (2007), 57–78.

Kavadi, Shirish N. *The Rockefeller Foundation and Public Health in Colonial India, 1916–1945; A Narrative History* (Pune & Mumbai, 1999).

——. '"Parasites Lost and Parasites Regained" Rockefeller Foundation's Anti-Hookworm Campaign in Madras Presidency', *Economic and Political Weekly*, 42 (2007), 130–7.

Keller, Richard. 'Madness and Colonization: Psychiatry in the British and French Empires, 1800–1962', *Journal of Social History*, 35 (2001), 295–326.

——. *Colonial Madness: Psychiatry in French North Africa* (Chicago, 2007).

Kelly, James William. 'Wafer, Lionel (*d.* 1705)', *Oxford Dictionary of National Biography* [Hereafter *Oxford DNB*], www.oxforddnb.com/view/article/28392, accessed 9 Sept 2011.

Kennedy, Dane. 'The Perils of the Midday Sun: Climatic Anxieties in the Colonial Tropics', in John M. MacKenzie (ed.), *Imperialism and the Natural World* (Manchester & New York, 1990), pp. 118–40.

Kidambi, Prashant. 'An Infection of Locality: Plague, Pythogenesis and the Poor in Bombay, *c.* 1896–1905', *Urban History*, 31 (2004), 249–67.

Kim, Elizabeth. 'Race Sells: Racialized Trade Cards in 18th-Century Britain', *Journal Material Culture*, 7 (2002), 137–65.

Kim, Jeong-Ran. 'The Borderline of 'Empire': Japanese Maritime Quarantine in Busan c.1876–1910', *Medical History*, 57 (2013), 226–48.

Kiple, Kenneth F. *The Caribbean Slave: A Biological History* (Cambridge, 1984).

——. 'Response to Sheldon Watts, "Yellow Fever Immunities in West Africa and the Americas in the Age of Slavery and beyond: A Reappraisal"', *Journal of Social History*, 34 (2001), 969–74.

Kiple, K. and Virginia H. Kiple, 'Deficiency Diseases in the Caribbean', *Journal of Interdisciplinary History*, 11 (1980), 197–215.

Kjekshus, Helge. 'The Villagization Policy: Implementational Lessons and Ecological Dimension, *Canadian Journal of African Studies*, 11 (1977), 262–82.

——. *Ecology, Control and Economic Development in East African History* (London, 1977).

Klein, Herbert S. and Stanley L. Engerman, 'Long Term Trends in African Mortality in the Transatlantic Slave Trade', *A Journal of Slave and Post-Slave Studies*, 18 (1997), 36–48.

Klein, Ira. 'Death in India: 1871–1921', *Journal of Asian Studies*, 32 (1973), 639–59.

——. 'Plague, Policy and Popular Unrest in British India', *Modern Asian Studies*, 22 (1988), 723–55.

Koch, Robert. 'An Address on Cholera and its Bacillus, delivered before the Imperial German Board of Health, at Berlin', *BMJ*, 1236 (6 September 1884), 453–9.

Kohn, Margaret. 'Colonialism', Edward N. Zalta (ed.), *The Stanford Encyclopedia of Philosophy* (Summer 2012 Edition), http://plato.stanford.edu/archives/sum2011/entries/colonialism/.

Koponen, Juhani. *People and Production in Late Precolonial Tanzania: History and Structures* (Helsinki, 1988).

Kopperman, Paul E. 'Medical Services in the British Army, 1742–1783', *JHMAS*, 34 (1979), 428–55.

——. 'The British Army in North America and the West Indies, 1755–83: A Medical Perspective', in Geoffrey L. Hudson (ed.), *British Military and Naval Medicine 1600–1830* (Amsterdam, 2007), pp. 51–86.

Kriz, Kay Dian. 'Curiosities, Commodities, and Transplanted Bodies in Hans Sloane's 'Natural History of Jamaica', *The William and Mary Quarterly*, 57 (2000), 35–78.

Kuhn, Thomas. *The Structure of Scientific Revolution* (Chicago, 1962).

Kumar, Deepak. *Science and the Raj, 1857–1905* (Delhi, 1995).

——. 'Unequal Contenders, Uneven Ground: Medical Encounters in British India, 1820–1920' in Andrew Cunningham & Bridie Andrews (eds), *Western Medicine as Contested Knowledge* (Manchester & New York, 1997), pp. 172–90.

Kunitz, Stephen J. *Disease and Social Diversity: The European Impact on the Health of Non-Europeans* (Oxford, 1994).

Kupperman, Karen Ordahl. 'Fear of Hot Climates in the Anglo-American Colonial Experience', *The William and Mary Quarterly*, 41 (1984), 213–40.

Ladurie, Emmanuel Le Roy. 'A Concept: The Unification of the Globe by Disease', in Ladurie, *The Mind and Method of the Historian* (Brighton, 1981), pp. 28–83.

Land, Isaac. 'Customs of the Sea: Flogging, Empire, and the "True British Seaman" 1770 to 1870', *Interventions: International Journal of Postcolonial Studies*, 3 (2001), 169–85.

Landers, Jane. 'Gracia Real de Santa Teresa de Mose: A Free Black Town in Spanish Colonial Florida', *The American Historical Review*, 95 (1990), 9–30.

Latour, Bruno. *Science in Action: How to Follow Scientists and Engineers through Society* (Milton Keynes, 1987).

——. *Pasteurization of France* (Cambridge, MA, 1988).

Laudan, Larry. *Progress and its Problems, Towards a Theory of Scientific Growth* (London, 1977).

Lawrence, C. 'Disciplining Disease: Scurvy, the Navy, and Imperial Expansion, 1750–1825', in D.P. Miller and P.H. Reill (eds), *Visions of Empire: Voyages, Botany, and Representations of Nature* (Cambridge, 1996), pp. 80–106.

Lee, Cristopher J. 'Subaltern Studies and African Studies', *History Compass* 3 (2005) doi: 10.1111/j.1478-0542.2005.00162.x.

Levine, Philippa. 'Modernity, Medicine, and Colonialism: The Contagious Diseases Ordinances in Hong Kong and the Straits Settlements', *Positions*, 6 (1998), 675–705.

Lind, James. *An Essay on the Most Effectual Means of Preserving the Health of Seamen, in the Royal Navy* (2nd edition, London, 1762).

——. *An Essay on Diseases Incidental to Europeans in Hot Climates with the Method of Preventing their Fatal Consequences* (6th edition, London, 1808).

Livi-Bacci, M. 'The Depopulation of Hispanic America after the Conquest', *Population and Development Review*, 32 (2004), 199–232.

Livingstone, David N., 'Tropical Climate and Moral Hygiene: The Anatomy of a Victorian Debate' *The British Journal for the History Science*, 32 (1999), 93–110.

Livingstone, David *Missionary Travels and Researches in South Africa* (London, 1899).

Long, Edward. *The History of Jamaica; Or, General Survey of the Antient and Modern State of that Island: With Reflection on its Situations, Settlements, Inhabitants; In Three Volumes*, vol. 2 (London, 1774).

Longfield-Jones, G.M. 'Buccaneering Doctors', *Medical History*, 36 (1992), 187–206.

Lonie, Iain M. 'Fever Pathology in the Sixteenth Century: Tradition and Innovation', *Medical History*, Supplement (1981), 19–44.

Love, Henry Davison. *Vestiges of Old Madras 1640–1800, Traced from the East India Company's Records Preserved at Fort St. George and the India Office, and from other Sources* (London, 1913).

Lovell, W. George. '"Heavy Shadows and Black Night": Disease and Depopulation in Colonial Spanish America', *Annals of the Association of American Geographers*, 82 (1992), 426–43.

Low, Gordon. 'Thomas Sydenham: The English Hippocrates', *Australian and New Zealand Journal of Surgery* (1999), 258–62.

Löwy, Ilana. 'Yellow Fever in Rio de Janeiro and the Pasteur Institute Mission (1901–1905), the Transfer of Science to the Periphery', *Medical History*, 34 (1990), 144–63.

——. 'From Guinea Pigs to Man: The Development of Haffkine's Anticholera Vaccine', *JHMAS*, 47 (1992), 270–309.

Lucas, Charles Prestwood. *A Historical Geography of the British Colonies*, parts 2 and 4 (Oxford, 1888–1901).

Lyons, Maryinez. 'Sleeping Sickness in the History of Northwest Congo (Zaire), *Canadian Journal of African Studies*, 19 (1985), 627–33.

——. *The Colonial Disease; A Social History of Sleeping Sickness in Northern Zaire, 1900–1940* (Cambridge, 1992).

MacKenzie, John M. (ed.), *Imperialism and the Natural World* (Manchester, 1990).

——. 'Empire and the Ecological Apocalypse: The Historiography of the Imperial Environment', in Tom Griffiths and Libby Robin (eds) *Ecology and Empire: Environmental History of Settler Societies* (Melbourne, 1997), pp. 215–28.

Mackie, Eric S. 'Welcome the Outlaw: Pirates, Maroons, and Caribbean Countercultures', *Cultural Critiques*, 59 (2005), 24–62.

Maehle, Andreas-Holger. *Drugs on Trial: Experimental Pharmacology and Therapeutic Innovation in the Eighteenth Century* (Amsterdam, 1999).

Mahal, Ajay. et al (eds), *India Health Report 2010* (New Delhi, 2010).

Majeed, Javed. *Ungoverned Imaginings, James Mill's `The History of British India' and Orientalism* (Oxford, 1992).

Manson, Patrick. 'The Life-History of the Malaria Germ outside the Human Body', *BMJ*, 1838 (21 March 1896), 712–17.

——. 'On the Development of *Filaria sanguinis hominis*, and on the Mosquito Considered as a Nurse', *Journal of the Linnean Society of London, Zoology*, 14 (1878), 304–11.

Manson-Bahr, P. *Patrick Manson: The Father of Tropical Medicine* (London, 1962).

Markham, Clements R. *Peruvian Bark: A Popular Account of the Introduction of Cinchona Cultivation into British India, 1860–1880* (London, 1880).

Marks, Shula. 'What is Colonial about Colonial Medicine?; And What has Happened to Imperialism and Health?', *Social History of Medicine*, 10 (1997), 205–19.

Marshall, P.J. 'Britain and the World in the Eighteenth Century: I, Reshaping the Empire', *Transactions of the Royal Historical Society*, 8 (1998), 1–18.

——. 'Eighteenth-Century Calcutta' in Raymond F. Betts, Robert J. Ross and Gerard J. Telkamp (eds). *Colonial Cities: Essays on Urbanism in a Colonial Context* (Lancaster, 1984), pp. 87–104.

——. 'II, Britons and Americans', *Transactions of the Royal Historical Society*, 9 (1999), 1–16.

——. 'III, Britain and India', *Transactions of the Royal Historical Society* 10 (2000), 1–16.

——. 'IV the Turning Outwards of Britain', *Transactions of the Royal Historical Society*, 11 (2001), 1–15.

——. *The Making and Unmaking of Empire: Britain, India, and America c. 1750–1783* (Oxford, 2005).

Masson, Francis. 'An Account of Three Journeys from the Cape Town into the Southern Parts of Africa; Undertaken for the Discovery of New Plants, towards the Improvement of the Royal Botanical Gardens at Kew', *Philosophical Transactions*, 66 (1776), 268–317.

Mathias, Peter. 'Swords into Ploughshares: the Armed Forces, Medicine and Public Health in the Late Eighteenth Century', in Jay Winter (ed.), *War and Economic Development: Essays in Memory of David Joslin* (Cambridge, 1975), pp. 73–90.

Maynard, Kent. 'European Preoccupations and Indigenous Culture in Cameroon: British Rule and the Transformation of Kedjom Medicine', *Canadian Journal of African Studies*, 36 (2002), 79–117.

Mazrul, Ali A. 'Black Africa and the Arabs', *Foreign Affairs*, 53 (1975), 725–42.

McCandless, Peter. *Slavery, Disease, and Suffering in the Southern Lowcountry* (Cambridge, 2011).

McDonald, Dedra S. 'Intimacy and Empire; Indian-African Interaction in Spanish Colonial New Mexico, 1500–1800', *American Indian Quarterly*, 22 (1998), 134–56.

McKeown, T. *The Modern Rise of Population* (New York, 1976).

McNeill, J.R. 'The Ecological Basis of Warfare in the Caribbean, 1700–1804', in M. Ultee (ed.), *Adapting to Conditions: War and Society in the 18th Century* (Alabama, 1986), pp. 26–42.

——. 'Observations on the Nature and Culture of Environmental History', *History and Theory*, 42 (2003), 5–43.

——. *Mosquito Empires: Ecology and War in the Greater Caribbean, 1620–1914* (Cambridge, 2010).

Mead, Teresa A. *'Civilizing' Rio: Reform and Resistance in a Brazilian City, 1889–1930* (Philadelphia, 1997).

——. '"Civilizing Rio de Janeiro": The Public Health Campaign and the Riot of 1904', *Journal of Social History*, 20 (1986), 301–22.

Mishra, Saurabh. *Pilgrimage, Politics, and Pestilence; The Haj from the Indian Subcontinent, 1860–1920* (Delhi, 2011).

Mohanavelu, C.S. *German Tamilology: German Contribution to Tamil Language, Literature and Culture during the Period 1706–1945* (Madras, 1993).

Monnais, Laurence. 'Preventive Medicine and "Mission Civilisatrice"; Uses of the BCG Vaccine in French Colonial Vietnam between the Two World Wars', *The International Journal of Asia Pacific Studies*, 2 (2006), 40–66.

Montgomery, Scott L. 'Naming the Heavens: A Brief History of Earthly Projection', Part II: Nativising Arab Science', *Science as Culture*, 6 (1996), 73–129.

Moore, Henrietta L. and Todd Sanders (eds). *Magical Interpretations, Material Realities: Modernity, Witchcraft and the Occult in Postcolonial Africa* (London & New York, 2002).

Moore, W.J. 'The Causes of Cholera', *Indian Medical Gazette*, 20 (1885), 270–3.

Morton, Julia F. 'Medicinal Plants – Old and New', *Bulletin of the Medical Library Association*, 56 (1968), 161–7.

Moseley, Benjamin. *A Treatise on Tropical Diseases*, Second edition (London, 1789).

Moulin, Anne Marie. 'Patriarchal Science: The Network of the Overseas Pasteur Institutes', in Patrick Petitjean, Catherine Jami and Moulin (eds), *Science and Empires; Historical Studies about Scientific Development and European Expansion* (Dordrecht, 1992), pp. 307–22.

——. 'Bacteriological Research and Medical Practice in and out of the Pasteurian School', in Ann La Berge, and Mordechai Feingold (eds). *French Medical Culture in the Nineteenth Century* (Amsterdam & Atlanta, 1994), pp. 327–49.

——. 'Tropical without Tropics: The Turning Point of Pastorian Medicine in North Africa', in D. Arnold (ed.), *Warm Climates and Western Medicine*, pp. 160–80.

Mtika, Mike Mathambo. 'Political Economy, Labor Migration, and the AIDS Epidemic in Rural Malawi', *Social Science & Medicine*, 64 (2007), 2454–63.

Münch, R. 'Robert Koch', *Microbes and Infection*, 5 (2003), 69–74.

Nair, G. Balakrish and Jai P Narain, 'From Endotoxin to Exotoxin: De's Rich Legacy to Cholera', *Bulletin of the WHO*, 88 (2010), 237–40.

Nandy, A. *Alternative Sciences: Creativity and Authenticity in Two Indian Scientists* (Delhi, 1995).

National Planning Committee, Subcommittee on National Health Report (Bombay, 1948).

Needham, J. *The Grand Titration: Science and Society in East and West* (London, 1969)

Neild-Basu, Susan M. 'Colonial Urbanism: The Development of Madras City in the Eighteenth and Nineteenth Centuries', *Modern Asian Studies*, 13 (1979), 217–46.

Nunn, N. and N. Qian, 'The Columbian Exchange: A History of Disease, Food, and Ideas', *Journal of Economic Perspectives*, 24 (2010), 163–88.

Ogawa, Mariko. 'Uneasy Bedfellows: Science and Politics in the Refutation of Koch's Bacterial Theory of Cholera', *Bulletin of the History of Medicine* 74 (2000), 671–707.

Osborne, Michael A. *Nature, the Exotic and the Science of French Colonialism* (Bloomington, 1994).

——. 'Science and the French Empire', *Isis*, 96 (2005), 80–7.

Osseo-Asare, Abena. 'Bioprospecting and Resistance: Transforming Poisoned Arrows into Strophantin Pills in Colonial Gold Coast, 1885–1922', *Social History of Medicine*, 21 (2008), 269–90.

Packard, Randall M. 'Maize, Cattle and Mosquitoes: the Political Economy of Malaria epidemics in colonial Swaziland', *The Journal of African History*, 25 (1984), 189–212.

——. *White Plague, Black Labor; Tuberculosis and the Political Economy of Health and Disease in South Africa* (Berkeley & London, 1989).

——. '"Malaria Blocks Development" Revisited: The Role of Disease in the History of Agricultural Development in the Eastern and Northern Transvaal Lowveld, 1890–1960', *Journal of Southern African Studies*, 27 (2001), 591–612.

——. *The Making of a Tropical Disease: A Short History of Malaria* (Baltimore, 2007).

Palmer, Steven. 'Beginnings of Cuban Bacteriology: Juan Santos Fernandez, Medical Research, and the Search for Scientific Sovereignty, 1880–1920', *Hispanic American Historical Review*, 91 (2011), 445–68.

Panikkar, K.N. 'Indigenous Medicine and Cultural Hegemony: A Study of the Revitalization Movement in Keralam', *Studies in History*, 8 (1992), 287–308.

Papin, Father. 'A Letter from Father Papin, to Father Le Gobïen, Containing some Observations upon the Mechanic Arts and Physick of the Indians', *Philosophical Transactions*, 28 (1713), 225–30.

Pardo, Osvaldo. 'Contesting the Power to Heal: Angels, Demons and Plants in Colonial Mexico', in Griffiths and Cervantes (eds) *Spiritual Encounters*, pp. 163–84.

Pares, R. *War and Trade in the West Indies, 1739–1763* (London, 1963).

Parle, Julie. *States of Mind: Searching for Mental Health in Natal and Zululand, 1868–1918* (Scottsville, 2007).

Peard, Julyan G. *Race, Place, and Medicine: The Idea of the Tropics in Nineteenth-Century Brazilian Medicine* (Durham, NC, 1999).

Pearson, M.N. 'The Thin End of the Wedge. Medical Relativities as a Paradigm of Early Modern Indian-European Relations', *Modern Asian Studies*, 29 (1995), 141–70.

Peck, Linda L. *Consuming Splendor: Society and Culture in Seventeenth-Century England* (Cambridge, 2005).

Pelis, Kim. 'Prophet for Profit in French North Africa: Charles Nicolle and Pasteur Institute of Tunis, 1903–1936', *Bulletin of the History of Medicine*, 71 (1997), 583–622.

——. *Charles Nicolle, Pasteur's Imperial Missionary: Typhus and Tunisia* (Rochester, NY, 2006).

Petiver, James. 'An Account of Mr Sam. Brown, his Third Book of East India Plants, with their Names, Vertues, Description', *Philosophical Transactions*, 22 (1700–1), 843–64.

——. 'Some Attempts Made to Prove that Herbs of the Same Make or Class for the Generallity, have the Like Vertue and Tendency to Work the Same Effects', *Philosophical Transactions*, 21 (1699), 289–94.

——. 'The Eighth Book of East India Plants, Sent from Fort St George to Mr James Petiver Apothecary, and F. R. S. with His Remarks on Them', *Philosophical Transactions*, 23 (1702–3), 1450–60.

Petiver, James and Samuel Brown. 'Mr Sam. Brown His Seventh Book of East India Plants, with an Account of Their Names, Vertues, Description, etc', *Philosophical Transactions*, 23 (1702–3), 1252–3.

Pitman, Frank Wesley. 'Fetishism, Witchcraft, Christianity among the Slaves', *The Journal of Negro History*, 11 (1926), 650–68.

Playfair, G. *Taleef Shareef or the Indian Materia Medica* (Calcutta, 1833).

Pocock, Tom 'Pocock, Sir George (1706–1792)', *Oxford DNB*, www.oxforddnb.com/view/article/22421, accessed 12 Nov 2005

——. *Battle for Empire: the Very First World War, 1756–63* (London, 1998).

Pollitzer, R. 'Cholera studies: 1. History of the Disease', *Bulletin of the WHO*, 10 (1954), 421–61.

Porter, Dorothy. 'The Mission of Social History of Medicine: An Historical View', *Social History of Medicine*, 8 (1995), 345–59.

Porter, Roy. 'The Historiography of Medicine in the United Kingdom' in F. Huisman and J. H. Warner (eds), *Locating Medical History: The Stories and their Meanings* (Baltimore & London, 2004), pp. 194–208.

——. 'The Imperial Slaughterhouse', review of *Romanticism and Colonial Disease* by Alan Bewell, *Nature*, 404 (2000), 331–2.

Porter, Roy and Andrew Wear (eds). *Problems and Methods in the History of Medicine* (New York, 1987), pp. 1–12.

Porter, Roy and Dorothy Porter. 'The Rise of the English Drugs Industry: The Role of Thomas Corbyn', *Medical History*, 33 (1989), 277–95.

Portilla, Miguel León. *The Broken Spears: The Aztec Account of the Conquest of Mexico* (Boston, 1992/1962).

Power, Helen J. 'The Calcutta School of Tropical Medicine: Institutionalizing Medical Research in the Periphery', *Medical History*, 40 (1996), 197–214.

——.'Sir Leonard Rogers FRS (1868–1962), Tropical Medicine in the Indian Medical Service', thesis submitted to the University of London, for the degree of doctor of philosophy, 1993.

Prakash, Gyan. 'Science between the Lines', in Shahid Amin and Dipesh Chakrabarty (eds). *Subaltern Studies*, vol. 9 (Delhi, 1996), pp. 59–82.

Prakash, Om. 'Bullion for Goods: International Trade and the Economy of Early Eighteenth Century Bengal', *Indian Economic and Social History Review*, 13 (1976), 159–86.

Price, Jacob M. 'What did Merchants Do? Reflections on British Overseas Trade, 1660–1790', *Journal of Economic History*, 49 (1989), 267–84.

Priestley, Eliza. 'The Realm of the Microbe', *The Nineteenth Century*, 29 (1891), 811–31.

Pringle, John. *Observations on the Diseases of the Army, in Camp and Garrison. In Three Parts. With an Appendix, Containing some Papers of Experiments* (London, 1753).

Quah, Stella R. 'The Social Position and Internal Organization of the Medical Profession in the Third World: The Case of Singapore', *Journal of Health and Social Behavior*, 30 (1989), 450–66.

Quevedo, Emilio, et al. 'Knowledge and Power: The Asymmetry of Interests of Colombian and Rockefeller Doctors in the Construction of the Concept of Jungle Yellow Fever, 1907–1938', *Canadian Bulletin of Medical History*, 25 (2008), 71–109.

Quinlan, Sean. 'Colonial Encounters; Colonial Bodies, Hygiene and Abolitionist Politics in Eighteenth-Century France', *History Workshop*, 42 (1996), 107–26.

Raghunath, Anita. 'The Corrupting Isles: Writing the Caribbean as the Locus of Transgression in British Literature of The 18th Century' in Vartan P. Messier and Nandita Batra (eds) *Transgression and Taboo: Critical Essays* (Puerto Rico, 2005), pp. 139–52.

Raina, Dhruv and Irfan S. Habib, 'Bhadralok Perceptions of Science, Technology and Cultural Nationalism', *Indian Economic and Social History Review*, 32 (1995), 95–117.

Ramanna, Mridula. *Western Medicine and Public Health in Colonial Bombay, 1845–1895* (Hyderabad, 2002).

Ranger, Terence. 'Connexions between "Primary Resistance" Movements and Modern Mass Nationalism in East and Central Africa', *The Journal of African History*, 9 (1968), 631–41.

——. 'Godly Medicine: The Ambiguities of Medical Mission in Southeast Tanzania'. *Social Science and Medicine*, 15b (1981), 261–77.

——. 'Power, Religion, and Community: The Matobo Case', in Partha Chatterjee and Gyanadra Pandey (eds) *Subaltern Studies¸* vol. 7 (Delhi, 1993), pp. 221–46.

Report of the Health Survey and Development Committee (Delhi, Manager of Publications, 1946).

Report of the Plague Commission of India, vol. 5 (London, 1901).

Report, The Infant and Child Mortality India; Levels, Trends and Determinants, New Delhi, India, 2012, http://www.unicef.org/india/Report.pdf.

Representation of the Bombay Medical Union to the Royal Commission on the Public Services in India, (Bombay, 1 May 1913).

Reveal, J.L. and J.S. Pringle, 'Taxonomic Botany and Floristics', in: *Flora of North America North of Mexico*, 1 (1993), 157–92.

Riley, James C. 'Mortality on Long-Distance Voyages in the Eighteenth Century', *Journal of Economic History*, 41 (1981), 651–6.

Roberts, William. 'Address in Medicine', *BMJ*, 867 (11 August 1877), 168–73.

Robertson, William. *The History of America*, vol. 3 (London, 1800–1)

Robinson, David. *Muslim Societies in African History* (Cambridge, 2004)

Rodger, N.A.M. *The Command of the Ocean: A Naval History of Britain, 1649–1815* (London, 2006).

Roff, William R. 'Sanitation and Security: The Imperial Powers and the Nineteenth-Century Hajj', *Arabian Studies*, 6 (1982), 143–60.

Rogers, Leonard. 'The Conditions Influencing the Incidence and Spread of Cholera in India', *Proceedings of the Royal Society of Medicine*, 19 (1926), 59–93.

——. *Happy Toil: Fifty-Five Years of Tropical Medicine* (London, 1950).

Rosenberg, Charles E. 'Framing Disease: Illness, Society and History', Rosenberg, *Explaining Epidemics and Other Studies in the History of Medicine* (Cambridge, 1992), pp. 305–18.

Ross, Ronald. 'Observations on a Condition Necessary to the Transformation of the Malaria Crescent', *BMJ*, 1897 (30 January 1883), 251–5.

——. 'The Progress of Tropical Medicine', *Journal of the Royal African Society*, 4 (1905), 271–89.

——. 'Tropical Medicine—A Crisis', *BMJ*, 2771 (7 February 1914), 319–21.

——. 'On some Peculiar Pigmented Cells Found in Two Mosquitos Fed on Malarial Blood', *BMJ*, 1929 (18 December 1897), 1786–1788.

Rowley, Henry. *The Story of the Universities' Mission to Central Africa*, 2nd edition (London, 1867).

Royle, J.F. *An Essay on the Antiquity of the Hindoo Medicine* (London, 1837).

Rutten, A.M.G. *Dutch Transatlantic Medicine Trade in the Eighteenth Century Under the Cover of the West India Company* (Rotterdam, 2000).

Said, Edward. *Orientalism* (New York, 1978).

Savitt, Todd L. *Medicine and Slavery: The Disease and Healthcare of Black in Antebellum Virginia* (Urbana & London, 1978).

Scharfe, Hartmut. 'The Doctrine of the Three Humors in Traditional Indian Medicine and the Alleged Antiquity of Tamil Siddha Medicine', *Journal of the American Oriental Society*, 119 (1999), 609–29.

Schawb, Raymond. *The Oriental Renaissance: Europe's Rediscovery of India and the East, 1680–1880* (New York, 1984).

Scheid, Volker. *Chinese Medicine in Contemporary China: Plurality and Synthesis* (London & Durham, 2002).

Schiebinger, Londa. 'The Anatomy of Difference: Race and Sex in Eighteenth-Century Science', *Eighteenth-Century Studies*, 23 (1990), 387–405.

——. *Plants and Empire: Colonial Bioprospecting in the Atlantic World* (Cambridge, MA & London, 2004).

Schoepf, Brooke Grundfest. 'AIDS, Sex and Condoms: African Healers and the Reinvention of Tradition in Zaire' *Medical Anthropology*, 14 (1992), 225–42.

Schwartz, Stuart (ed.), *Victors and Vanquished: Spanish and Nahua Views of the Conquest of Mexico* (Bedford, 2000).

Scott, James C. *Seeing Like a State: How Certain Schemes to Improve the Human Condition Have Failed* (New Haven& London, 1998).

Seidi, Jacob. 'The Relationship of Garcia de Orta's and Cristobal Acosta's Botanical Works', *Actes du VIIe Congress International d'Histoire des Sciences* (Paris, 1955).

Sergent, Edmond. 'Address Delivered by Dr. Sergent on the Occasion of the Award of the Darling Medal to Dr. Swellengrebel', Geneva (17 September 1938) http://whqlibdoc.who.int/malaria/CH_Malaria_266.pdf

Shannon, R. *Practical Observations on the Operation and Effects of Certain Medicines, in the Prevention and Cure of Diseases to which Europeans are Subject are Subject in Hot Climates, and in these Kingdoms* (London, 1793).

Shapin, Steven. 'The History of Science and its Sociological Reconstructions', *History of Science*, 20 (1982), 157–211.

——. *A Social History of Truth; Civility and Science in Seventeenth Century England* (Chicago & London, 1994).

Shapter, Thomas. *The History of the Cholera in Exeter in 1832* (London, 1849).

Sharp, Lesley A. 'The Commodification of the Body and Its Parts', *Annual Review of Anthropology*, 29 (2000), 287–328.

Sheridan, Richard B. 'The Slave Trade to Jamaica, 1702–1808' in B.W. Higman (ed.), *Trade, Government and Society: Caribbean History 1700–1920* (Kingston, 1983).

——. 'The Doctor and Buccaneer: Sir Hans Sloane's Case History of Sir Henry Morgan, Jamaica, 1688', *JHMAS*, 41 (1986), 76–87.

Sigerist, Henry E. *Socialised Medicine in the Soviet Union* (London, 1937).

Simms, F.W. *Report on the Establishment of Water-Works to Supply the City of Calcutta, with other Papers on Watering and Draining the City* [1847–52] (Calcutta, 1853).

Singer, Charles and E. Ashworth Underwood. *A Short History of Medicine* (New York & Oxford, 1962).

Sivasundaram, Sujit. 'Natural History Spiritualized: Civilizing Islanders, Cultivating Breadfruit, and Collecting Souls', *History of Science*, 39 (2001), 417–43.

Sloane, Hans. 'A Description of the Pimienta or Jamaica Pepper-Tree, and of the Tree That Bears the Cortex Winteranus', *Philosophical Transactions*, 16 (1686), 462–8.

——. 'An Account of a China Cabinet, Filled with Several Instruments, Fruits, &c. Used in China: Sent to the Royal Society by Mr. Buckly, Chief Surgeon at Fort St George', *Philosophical Transactions*, 20 (1698), 390–2.

——. *A Voyage to the Islands of Madera, Barbados, Nieves, S. Christophers and Jamaica, with the Natural History*, vol. 1 (London, 1707).

Smallwood, Stephanie. *Saltwater Slavery: A Middle Passage from Africa to American Diaspora* (Cambridge, MA, 2007).

Smith, David B. *Report on the Drainage and Conservancy of Calcutta*, (Calcutta, 1869).

Smith, Pamela H. and Paula Findlen (eds). *Merchants & Marvels; Commerce, Science, and Art in Early Modern Europe* (New York & London, 2002).

Smith, Woodruff D. 'The Function of Commercial Centers in the Modernization of European Capitalism: Amsterdam as an Information Exchange in the Seventeenth Century', *The Journal of Economic History*, 44 (1984), 985–1005.

Sokhey, Sahib S. *The Indian Drug Industry and its Future* (New Delhi, 1959).

Spangenberg, August Gottlieb. *An Account of the Manner in which the Protestant Church of the Unitas Fratrum, or United Brethren, Preach the Gospel, and Carry on their Missions among the Heathen*. Transl, H. Trapp (London, 1788).

Spary, Emma C. '"Peaches Which the Patriarchs Lacked": Natural History, Natural Resources, and the Natural Economy in France', *History of Political Economy*, 35 (2003), 14–41.

Stanley, Henry Morton. *How I Found Livingstone; Adventures, and Discoveries in Central Africa; Including Four Months' Residence with Dr. Livingstone* (London, 1874).

——. *Through the Dark Continent* (London, 1880).

Steel, Henry Draper. *Portable Instructions for Purchasing the Drugs and Spices of Asia and the East-Indies: Pointing out the Distinguishing Characteristics of those that are genuine, and the Arts Practised in their Adulteration* (London, 1779).

Stepan, Nancy. *Picturing Tropical Nature* (Ithaca, 2001).

Stephenson, Marcia. 'From Marvelous Antidote to the Poison of Idolatry: The Transatlantic Role of Andean Bezoar Stones during the Late Sixteenth and Early Seventeenth Centuries', *Hispanic American Historical Review*, 90 (2010), 3–39.

Stewart, John. *'The Battle for Health': A Political History of the Socialist Medical Association, 1930–51* (Aldershot, 1999).

Stokes, Eric. 'The Road to Chandrapore', *London Review of Books*, 2 (17 April 1980), 17–18.

——. (ed C.A. Bayly). *The Peasant Armed: Indian Revolt of 1857* (New Delhi, 1986).

Strachan, John 'The Pasteurization of Algeria?', *French History*, 20 (2006), 260–75.

Stringer, Chris. *The Origin of Our Species* (London & New York, 2011).

Subrahmanyam, Sanjay. 'Asian Trade and European Affluence? Coromandel, 1650–1740', *Modern Asian Studies*, 22 (1988), 179–188.
——. *Penumbral Visions: Making Polities in Early Modern South India* (Ann Arbor, 2001).
Subramanian, Lakshmi. *Indigenous Credit and Imperial Expansion: Bombay, Surat and the West Coast* (Delhi, 1996).
Szreter, Simon. 'The Importance of Social Intervention in Britain's Mortality Decline c. 1850–1914: A Reinterpretation of the role of Public Health', *Social History of Medicine*, 1 (1988), 1–38.
——. 'Economic Growth, Disruption, Deprivation, Disease, and Death: On the Importance of the Politics of Public Health for Development', *Population and Development Review*, 23 (1997), 693–728.
Taylor, Kim. 'Divergent Interests and Cultivated Misunderstandings: The Influence of the West on Modern Chinese Medicine', *Social History of Medicine* 17 (2004), 93–111
Taylor, Norman. *Cinchona in Java: The Story of Quinine* (New York, 1945).
Terry, P.T. 'African Agriculture in Nyasaland 1858 to 1894', *The Nyasaland Journal*, 14 (1961), 27–35.
The Gazetteer of Bombay City and Island, vol. 3 (1909).
The Trade Granted to the South-Sea-Company: Considered with Relation to Jamaica. In a Letter to One of the Directors of the South-Sea-Company by a Gentleman who has Resided several Years in Jamaica (London, 1714).
Thomas, Jennifer. 'Compiling 'God's Great Book [of] Universal Nature', The Royal Society's Collecting Strategies', *Journal of the History of Collections*, 23 (2011), 1–13.
Thomson, James. *Treatise on the Diseases of the Negroes, as they Occur in the Island of Jamaica with Observations on the Country Remedies, Aikman Junior, Jamaica* (Kingston, 1820).
Thomson, T.R.H. 'On the Value of Quinine in African Remittent Fever', *Lancet* (28 February 1846), 244–5.
Thoral, Marie-Cecile. 'Colonial Medical Encounters in the Nineteenth Century: The French Campaigns in Egypt, Saint Domingue and Algeria', *Social History of Medicine* (2012), hks020v1-hks020.
Thornton, John. *Africa and Africans in the Making of the Atlantic World 1400–1800* (Cambridge, 1998).
Ticktin, M. 'Medical Humanitarianism in and beyond France: Breaking Down or Patrolling Borders?' in Alison Bashford (ed.), *Medicine at the Border: Disease, Globalization and Security, 1950 to the Present* (Basingstoke, 2006) pp. 116–35.
Tilley, Helen. 'Ecologies of Complexity: Tropical Environments, African Trypanosomiasis, and the Science of Disease Control Strategies in British Colonial Africa, 1900–1940', *Osiris*, 19 (2004), 21–38.
'Traditional Medicine', fact sheet N°134, December 2008, WHO, http://www.who.int/mediacentre/factsheets/fs134/en/#.

Trevelyan, G.O. *The Competition Wallah* (London & New York, 1863).

Turshen, Meredeth. *The Political Ecology of Disease in Tanzania* (Rutgers, 1984).

van Heyningen, W.E. and John R. Seal, *Cholera: The American Scientific Experience, 1947–1980* (Boulder, Colorado, 1983).

van Rheede, Henry. *Hortus Indicus Malabaricus* (Amsterdam, 1678).

Vaughan, Megan. *Curing their Ills; Colonial Power and African Illness* (Stanford, 1991).

——. 'Healing and Curing: Issues in the Social History and Anthropology of Medicine in Africa', *Social History of Medicine*, 7 (1994), 283–95.

Viswanathan, Gauri. *Masks of Conquest; Literary Study and British Rule in India* (London, 1989).

Voeks, Robert. 'African Medicine and Magic in the Americas', *Geographical Review*, 83 (1993), 66–78.

Wagenitz, G. 'The "Plantae Malabaricae" of the Herbarium at Göttingen Collected near Tranquebar', *Taxon*, 27 (1978), 493–4.

Wallerstein, Immanuel. *The Modern World-System: Capitalist Agriculture and the Origins of the European World-Economy in the Sixteenth Century* (New York, 1976).

Wallis, Patrick. 'Consumption, Retailing, and Medicine in Early-Modern London', *The Economic History Review*, 61 (2008), 26–53.

——. 'Exotic Drugs and English Medicine: England's Drug Trade, *c.* 1550–*c.* 1800', *Social History of Medicine*, 25 (2012), 20–46.

Wallis, Patrick and Mark Jenner (eds). *Medicine and the Market in Early Modern England and its Colonies, c. 1450–c. 1850* (Basingstoke, 2007).

Watts, Sheldon J. *Epidemics and History, Disease, Power and Imperialism* (New Haven, 1997).

——. 'British Development Policies and Malaria in India 1897–*c*.1929', *Past & Present* 165 (1999), 141–81.

Webster, Jane. 'Looking for the Material Culture of the Middle Passage', *Journal for Maritime Research*, 7 (2005), 245–58.

Weindling, Paul. *Epidemics and Genocide in Eastern Europe, 1890–1945* (Oxford, 2000).

White, Owen and and J.P. Daughton (eds). *In God's Empire: French Missionaries in the Modern World* (Oxford, 2012).

White, Luise. *Speaking with Vampires; Rumor and History in Colonial Africa* (Berkeley & London, 2000).

William, M.C. and James Ormiston. *Medical History of the Expedition to the Niger, during the Years 1841, 2, Comprising an Account of the Fever which Led to its Abrupt Termination* (London, 1843).

Wilson, Renate. *Pious Traders in Medicine: A German Pharmaceutical Network in Eighteenth-Century North America* (Philadelphia, 2000).

Wilson, W.J. *History of the Madras Army* (Madras, 1882).

Worboys, Michael. 'Colonial Medicine', in Cooter and John V. Pickstone (eds) *Companion to Medicine in the Twentieth Century* (London, 2003), pp. 67–80.

——. 'Almroth Wright at Netley: Modern Medicine and the Military in Britain, 1892–1902', in Cooter, Harrison and Sturdy (eds) *Medicine and Modern Warfare*, pp. 77–97.

——. 'The Emergence of Tropical Medicine: A Study in the Establishment of a Scientific Speciality', in G. Lemaine et. al. (eds) *Perspectives on the Emergence of Scientific Disciplines* (The Hague & Paris, 1976) pp. 76–98.

Worboys, Michael and Paolo Palladino. 'Science and Imperialism', *Isis*, 84 (1993), 84–102.

Wujastyk, Dagmar (ed.), *Modern and Global Ayurveda: Pluralism and Paradigms* (Albany, 2008).

Wujastyk, Dominik. '"A Pious Fraud": The Indian Claims for Pre-Jennerian Smallpox Vaccination', in Jan Meulenbeld and Dominik Wujastyk (eds) *Studies on Indian Medical History* (New Delhi, 2001), pp. 131–67.

Young, Robert. *Postcolonialism: A Very Short Introduction* (New York, 2003).

Zahedieh, Nuala. 'London and the Colonial Consumer in the Late Seventeenth Century', *The Economic History Review*, 47 (1994), 239–61.

Zhan, Mei. *Other-Worldly: Making Chinese Medicine through Transnational Frames* (Hastings, 2009).

Zunino, Francesca. 'A Marvellous and Useful New World: Constructions of American Ecology in Christopher Columbus' Indies', *Language and Ecology*, 2 (2008), 1–26.

Županov, Ines G. *Missionary Tropics: The Catholic Frontier in India, 16th–17th Centuries* (Ann Arbor, 2005).

Index

Academie de sciences (Paris) xiv
Acclimatization 43, 61, 64, 65, 66, 67,
  68–9, 70, 144
Acosta, Cristobal 21–2
Africa 66
  constructive imperialism in 150, 152
  Christianity in 138. *See also*
    missionaries; medical missionaries
    below
  civilizing mission in xi, 123, 136,
    137, 164, 190
  colonial medicine in 133–4, 135,
    154–6, 190
  colonization of x, 122–3, 124,
    127, 128, 129, 151, 152. *See also*
    protectorate system *below*
  European expeditions into 124–6
  European mortality in 124–5
  healing traditions in 7–8, 190–1,
    192, 193
  history of medicine in xxv–xxvi
  imperialism, early histories
    of xviii–xix
  madness in 134–6, 137
  Niger Expedition (of 1841) 125–6
  protectorate system 150–1, 152
  medical missionaries, role in 28,
    129–33, 190. *See also* missionaries
    below; Livingstone, David
  missionaries, role in xiv, 123,
    127, 129–33. *See also* medical
    missionaries *above*
  Western medicine in 129, 133
  witchcraft in 135, 136, 137. *See also*
    Voodoo; Obeah

African agency xv, xxii, xxviii
  and African history writing xxi,
    xxiii
  and African madness 136–7
  and history of medicine xxv
African Inland Company 124
African Lakes Company 124
African otherness 133–4, 135, 136.
    See also Africa, madness in
African slaves 60
  agency of. *See* African agency *above*
  and Amerindians, interaction
    between 7, 15
  and dirt eating 67, 80–1
  mortality in transatlantic slave
    trade 77, 78–81
  See also Slave trade
Age of Commerce 1–6, 15, 16, 73
  imperialism and ix, x
  markets and medicines in. *See*
    Markets and medicines
  medical exchanges in xi, 2, 6–9
  slave trade in 15
Age of Discovery xi, 5, 31, 131
Age of Empire 122
  and constructive imperialism 152.
    See also Constructive
    imperialism
  imperialism and ix, x
  laboratory medicine in xii. *See also*
    Laboratories
  tropical medicine in xv. *See also*
    Tropical medicine
Age of New Imperialism ix, x, xix.
    See also New Imperialism

Agency xxvii–xxviii, xxix
  African. *See* African agency
  and colonial medicine xxiv–xxv
  loss of, and colonialism xxi, xxii,
    xxiii
  romanticism and xxvii–xviii, xxix
  and subaltern studies xxii
Ainslie, Whitelaw 14, 31, 187
Ajmal Khan, Hakim 187, 189
Albuquerque, Alphonse de 2
Algeria 165, 176–7
Algiers School of Psychiatry 136
Ali, M. Athar 102
All India Institute of Medical
    Sciences 117
Allen, William 125
America, intervention in Cuba 95, 96
Americas
  colonization of xx, 1–2, 4, 7
  depopulation of 15, 73, 75–7
  early histories of imperialism xvii
  histories of colonization, and
    agency xxii
  Spanish civilizing mission xi,
    xviii, xx
  See also Amerindians *below*
American Society of Tropical
    Medicine 146
Amerindians xviii, 24, 76, 97
  and Africans, interaction
    between xxv, 6–8
  depopulation of 75–7
  healing traditions, and colonial
    medicine 27–8
  Spanish colonization, historical
    studies on xxiv
  See also Americas *above*
Anatomy of Plants 32
Anderson, Warwick 68
Anglo Maratha wars 105
Annales 12
Anthrax 166
Antiquity of Hindoo Medicine 186
Antiseptics 45, 46

Apothecaries, importance of 2, 13,
    31, 42
Arabs 1, 2, 122, 123, 136, 177
Army Medical Department 48–9, 52
Army Medical School, Netley 149
Arnold, David xxiv, 111, 113
Asian markets (bazaars) 14
Asiatic Cholera. *See* Cholera
Asiatic Society 186
Australia 68, 69, 168–9
Ayurveda xii, xxvii, 184–5, 186,
    187–8, 189–90
Aztecs 2, 76–7

Bacteriological laboratories 112, 172
Bacteriological research 112, 170–1
Bacteriology 142, 143, 147, 164, 168,
    169, 170, 175–6, 178
Baikie, Dr W.B. 127
Barber surgeons 42
Barham, Joseph Foster 28
Barrére, Pierre 67
Basalla, George xiii
Bazaar medicines 14
BCG vaccine 174
Bégué, Jean-Michel 136
Bengal 40, 102, 104, 105, 109
  as home of cholera 148, 170
Berlin Conference (1884–5) 151–2, 171
Bertin, Antoine 67
Bezoar stones 27–8
Bhore, Joseph 117
Black doctors 107
Black towns 61, 110
Blantyre Mission, in Africa 131
Blumenbach, Johann 59
Böhm, Dr Leo 169
Bombay 103, 109
Bombay hospital 104, 105, 108
Bombay Improvement Trust 110
Bombay Medical Union 116
Bombay plague 110–11, 113
Bonaparte, Napoleon, invasion of
    Egypt 165

Botanical gardens 20, 21, 22, 28, 29, 35
Brantlinger, Patrick 64
Braudel, Fernand 12
British Institute of Preventive Medicine 149
Broken Spears, The xxv
Browne, Samuel 11
Bruce, David 149
Bryson, Alexander 126
Buccaneer surgeons 10
Buccaneers of America, The 10
Buckley, Edward 11
Buffon, Comte de 59, 67
Bullionism, growth of 3
Burmanni, Nicolai Laurentii 23
Buxton, Thomas Foxwell 123

Cabanis, Pierre-Jean-Georges 65
Calcutta 4, 89, 90, 109, 148, 186
  public health measures in 110, 113–14
Calcutta hospital 52, 104
Calcutta Medical Club 116
Calcutta School of Tropical Medicine 146
Calmette, Albert 173, 174
Cape of Good Hope 24
Caribbean islands 2–3, 6, 25–6, 60, 61, 76
Carothers, J.C. 135
Carrier state, Robert Koch's theory of 171
Cassia fistula 24, 26
Castellani, George 156
Caventou, Joseph-Bienaimé 34, 126
Chadwick, Edwin 87
Chamberlain, Joseph 152–3, 154
Charaka Samhita 185
Chemistry, and modern medicine 32, 33–4, 36
China 193–4
  Traditional Chinese Medicine 195–6
Chirol, Valentine Ignatius xviii

Cholera 58, 82, 83–5, 113, 144–5, 148, 167, 170, 200
  as airborne disease 87
  contagionist theory of 85–6, 89–90
  as disease of locality 144–5, 148
  in Egypt 90
  in Europe 65, 83–4, 86
  international sanitary conferences and 86, 88, 89, 90, 91–2
  in Jamaica 87
  and mortality 83–4
  non-contagionist theories of 86, 87, 89, 148. *See also* Koch, Robert
  public health and sanitation movements xii, 87
  research, in India 148
  transmission, debates regarding 85–6, 89. *See also* Quarantine
  vaccine 170. *See also* Haffkine, Waldemar
Chopra, R.N. 146, 188
Christianity
  in Americas xi, 27, 28, 30
  and civilizing mission in Africa xi, 164. *See also* Civilizing mission; Missionaries
  and native religions 7, 8, 27, 192
Christophers, Samuel Rickard 148–9
Church Missionary Society 131
Cinchona ix, 20, 21, 24–5, 33, 126–7
Civilization and germs 167–9
Civilizing mission xi
  and Western medicine xxiv
  African colonization and xi, 122–3, 136–7, 164–5, 174, 190
  Pasteur institutes and xii, 173–8
Climate
  and race 58–62, 66
  and colonial settlement patterns 68, 69
  and disease 62–4, 69, 125, 145, 168
  and putrefaction. *See* Putrefaction
  See also Tropics
Climatic determinism 66, 169

Coffee 4, 7, 20, 21, 25, 31, 33, 103
Cohen, William 127–8
Collegium Medicum (Amsterdam) xiv
Colonial Administrative Service, in
    Africa 154
Colonialarmy, reforms and mortality
    revolution 48–51, 53
Colonial bioprospection 7, 20, 34,
    126–7, 190, 191
  and medical botany 20–6
Colonial gardens 6, 20–1
Colonial hospitals xxv, 42, 43, 51–2,
    107, 117, 185
Colonial Medical Service, in
    Africa 154
Colonial medicine xii, xiii–xvi, xxiv,
    xxix, 6, 63
  in Africa 133–4, 135, 154–6, 190
  and Amerindian healing
    traditions 27–8
  assimilation and divergence in 200
  and 'colonization of the body' 111, 113
  definition of xiii, 51
  and germ theory 167. See also Germ
    theory
  historiography of xvi–xvii
  and history of agency and
    resistance xxiv–xxix
  and ideas of difference xvi. See also
    African otherness
  in India xxvi, 101, 104–8, 115–17
  and indigenous medicines,
    engagement between xxiv–xxv
  and modern medicine xv–xvi
  and Pasteurism 170. See also Pasteur
    institutes; Pasteurian science
  and postcolonial history
    writing xxi–xxvi
  and public health 112. See also
    Public health
  and Western medicine 189. See also
    Western medicine
Colonial migration and
    mortality 78–82

Colonial science xiii
Colonial warfare 40
  and European military crisis 41–3
Colonialism 53
  and acclimatization. See
    Acclimatization
  and civilizing mission. See Civilizing
    mission
  and commerce 3–6
  critique of xix–xxi
  and disease xv, xx, xxiii, 62–4, 97,
    157–8
  and ecological destruction xxiii,
    xxvi
  economic imperialism and xix
  expansion, and army reforms 49–50
  European mortality decline and ix,
    81–2
  and loss of agency xxi, xxiii. See also
    Agency
  medicine and ix, x, xvi, xx, 35–6
  and power and anxiety 65, 67–8, 70
  and race 64–9
'Colonization of the body', concept
    of 111, 113
Columbus, Christopher ix, 1, 2, 6
Commission for Sick and Hurt
    Seamen 48
Company of Apothecaries 47
Competition Wallah, The xviii
Congo 156, 157–8
Conrad, Joseph xviii
Conseil, Ernest 176
Constructive imperialism xii, 150,
    152, 158
  tropical medicine and 152–6
Contagionist theories of disease 85–6,
    89–90, 96
Contingent contagionism,
    Pettenkofer's theory 144–5
Cook, Harold J. 13
Cortés, Hernán 76
Creolization 65
Crosby, Alfred xviii

Cruz, Oswaldo 168
Cuba, US intervention in 94–6
Cultural imperialism xxi–xxii
Cummins, Stevenson Lyle 147
*Curing their Ills: Colonial Power and
     African Illness* xxv
Curtin, Philip D. 50, 51, 81, 127
Cuvier, Georges 65

D'Orta, Garcia 21
Da Gama, Vasco ix, 102
Danish Halle mission,
     Tranquebar 29–30
Dazille, J.B. 67
*Decline of the Old Medical Regime,
     The* 13
Decolonization ix, x–xi, 202
   and alternative medicine xiii, 182,
     189, 197. *See also* Traditional
     medicine
Depopulation 96–7
   in colonial Spanish America 73, 75–8
   in Polynesia 78
   and settler capitalism 78
Descourtilz, Michel Etienne 26
Dialogues on Samples and Drugs 21
Difference, ideas of, and colonial
     medicine xvi. *See also* African
     otherness
Dirt eating, among African slaves 67,
     80–1
Disease
   colonialism and xv, xx, xxiii, 97, 157–8
   commercial expansion and 15
   contagionist theories of 85–6,
     89–90, 96
   and European mortality 40–1
   geographical determinism and 146
   global migration and 73–5
   miasmatic theories of 44, 62, 63,
     69, 125, 144
   and modernity, link between 156–7
   non-contagionist theories of 66,
     85–6, 89–90, 96, 148

poverty and 205
   and putrefaction, link
     between 44–5, 46, 62
   tropics and 145. *See also* Tropics;
     Tropical disease
Dover, Thomas 11
Duncan Jr, Andrew 34–5
Dutch x, 1, 3, 4, 6, 24, 40, 57, 73
   and medicinal plants 21, 22–3
   and spice trade 3, 21, 22
Dutch East India Company 4, 22, 23
Dutton, Joseph E. 153

East African Medical Staff 154
East African School of Psychiatry 134
Eco, Umberto 184
Ecole Coloniale 172
Ecological Imperialism xxiii
*Ecology, Control, and Economic
     Development in East African
     History* xxiii, 158
Economic imperialism xix, xxiii
Edinburgh New Dispensatory 34–5
Egypt, French invasion of 165
Elcano, Juan Sebastian 2
Eliot, Charles xviii–xix
'Empire and the Ecological
     Apocalypse' xxvi
English East India Company 4,
     102, 103
Esquemeling, Alexander Olivier 10, 24
*Essay on Diseases Incidental to
     Europeans in Hot Climates* 63–4
Etienne, Eugéne 172
European drug markets, growth of xv.
     See also Markets and medicines
European/Western medicine ix, xix,
     10, 15–16, 32, 35, 182
   and colonialism x, xx, xxiii–xxiv.
     See also Tropical medicine
   and exotic drugs 14, 15–16, 20–21,
     24–5, 32. *See also* Medical botany
   histories of xx–xxi, xxiii
   and imperialism, and ix, xi

European/Western medicine – *continued*
in India 101–2, 104, 107
and laboratory medicine xii.
missionaries and 30. *See also*
Missionaries; Medical missionaries
Surgeons and. *See* Surgeons
in Africa 129–33, 138
and African otherness 133–7
and civilizing mission xxiv. *See also*
Civilizing mission
and colonialism 152–3
See also Modern medicine
European mortality 40–1, 57
decline in 42–3, 80, 81, 127. *See also*
Preventive medicine
decline in, and colonial
expansion ix, 81–2, 128–9
Evans, Thomas 107

Fabian, Johannes 136
Feierman, Steven 190–1
Fernel, Jean 62
Fevers xi, 45, 125
and climate, link between 62–3,
69, 125
Field hospitals 49, 105
Finlay, Carlos J. 95
Fitzpatrick, Jeremiah 49
Flora Indica (Burmanni) 23
Flora Indica (Roxburgh) 23
Flora Zeylanica 23
Ford, John 158
Foucault, Michel 134
Fraser, Thomas 192
Freire, Domingos 94
French civilizing mission 136, 164–5,
174, 177
Pasteur institutes and xii, 173–8
French Colonial Health Service 172–3
French imperialism x, 4, 57, 102, 165
and civilizing mission 136, 164–5,
174, 177
Pasteur institutes and 171, 173,
175, 176

French missionaries, role of 165, 176

Galenic medicine xiv
Gallagher, John 151, 152
Germ theory xii, xv, xxiv, 89–90, 94–5,
144–5, 146, 166, 167, 169, 170
See also Germs
German Cholera Mission 148
Germs 143, 144, 161n.21, 169–71
and civilization 167–9
theory of human carrier 168
and tropics, link between 170. *See
also* Tropics
universality of 145
Global health initiatives xiii, xv, 203
Gordon, Dr H.L. 135
Gorgas, William 95, 146
Government of India Act (1905) 115
Grant Medical College 107
Grew, Nehemiah 32
Grundler, Johann Ernst 29
Guinea Coast and slave trade 124
Günther, August 23

Habib, Irfan 102
Haffkine, Waldemar 112, 148, 170
Haines, Robin 80
Haitian Revolution 93, 96
Haj pilgrims, quarantine of 87–8
Hakims 186, 187
Hamdard laboratories 188
Hardiman, David xxv
Harrison, Mark 113
Harvey, W.F. 147
Havana, yellow fever in 94
Headrick, Daniel 123
Health Survey and Development
Committee 117
Heart of Darkness xviii
Hermann, Paul 23
Hippocrates xiv
Hippocratic medicine 63, 69, 145
Hippocratic Oath, invention of 184
Historia Plantarum 32

History of America 9
HIV/AIDS 45, 97, 157, 204, 205
Hobsbawm, Eric xix, 183
Hobson, J.A. xix
Hortus Indicus Malabaricus 23, 28
Hunter, William G. 89

Ibn Sina(Avicenna) 185
Imperial British East Africa
  Company 154
Imperialism ix, x, xi, xvii, xxix, 74
  in Africa, early histories xviii–xix
  in America, early histories xviii
  British 57, 141. *See also* Colonialism
  and civilizing mission. *See* Civilizing
    mission
  constructive xii, 150, 158
  commerce and 2–5
  critique of xix–xxi
  cultural xxi–xxii, xxv–xxvi
  ecological xxiii
  economic xix, xxiii
  and European anxiety 64–5, 70
  French. *See* French imperialism
  histories of xvii–xix
  impact of, debates
    regarding xvi–xxii
  Marxist theories of xix
  and medicine ix, x–xiii, xv, xix
  and mortality, link between
    127–8, 129
  New, Age of ix, x, xix
  See also Colonialism
*Imperialism, the Highest Stage of
  Capitalism* xix
Imperialism: A Study xix
Incas 77
India
  Anglo-French rivalry 102, 105, 106
  anti-imperial movements
    103–4
  colonial medicine in xvi, 101,
    104–8. *See also* Colonial medicine
  colonization of 4, 101, 102–4

European medicine in 101–2,
  104, 107
European traders in 103, 117
medical education in India 116
medical research in 147–9
public health in xxvi, 108–15, 117
traditional medicine in. *See*
  Traditional medicine, in India
See also individual entries
India, Old and New xviii
Indian Medical Degrees Act 185–6
Indian Medical Service 108–109,
  116–17, 154
Indian nationalism 115
  and medicine 115–7
  and traditional medicine 187, 188
Indian pharmacology 188
Indian Plague Commission 110, 111
Indian plants, study of 21, 28–30
Indian Research Fund Association 112
Indigenous medicines, European
  interaction with 182–3, 186, 187
See also Traditional medicine
Indigenous physicians and modern
  medicine xii–xiii
Indo-China, French civilizing mission
  in 174
Industry and Empire xix
*Influence of Tropical Climates on
  European Constitution* 65
Informal empire 93–4, 95, 150–1, 168
Institute for Infectious Diseases,
  Berlin 149
International Health Board 202
International Health Commission 202
International Health Division 202
International Sanitary
  Conferences 88, 89, 90, 91–2, 97
International Sanitary Convention
  (1892) 91
Invention of Tradition 183
Ipecacuanhaix, 20, 24, 33

Jacobsz, Jan 13

Jalap ix, 7, 31
Jamaica 4, 8, 11, 14, 26, 51–2, 60, 87
Jardin du Roi 21
Jesuits 27, 28, 101, 194. *See also*
    Missionaries
Johnson, James 58, 65, 67
Johnson, Walter xxvii–xxviii
Jones, William 186
Joyce, Patrick xxix
Joyfull News out of the New Founde
    Worlde 23–4

Kala-azar, research in 149
Kalusa, Walima T. xxv
Kennedy, Dane 69
King Leopold, of Belgium 156
Kingston Naval Hospital 51–2
Kitasato, Shibasaburo 170
Kjekshus, Helge xiii, 158
Klein, Ira xxiii
Koch, Robert 50, 85, 89–90, 112, 143,
    144, 145, 148, 166, 169, 170
    theory of carrier state 171
    in Egypt 89
    in Calcutta 89–90
Kunitz, Stephen 78

Laboratories xii, 112, 172
    and public health 167
    and medicine xiv, 178
Ladurie, Emanuel Le Roy 74
Latour, Bruno 171, 178
Laveran, Charles Louis Alphonse 145
Lavoisier, Antoine 32, 34, 186
League of Nations Health
    Organization xiii, 115, 149,
    201–2
Leishman, William 149
Leishmania donovani 149
Lenin, Vladimir I. xix
Leprosy 44, 45, 147–8
Lind, James xi, 43, 46–7, 63–4, 125, 144
Linnaeus, Carl 23, 186
    taxonomic codification of plants 32

Liverpool School of Tropical
    Medicine 143, 146, 153
Livi-Bacci, Massimo 75–6
Livingstone, David xvi, xviii, 123,
    133, 191
    Zambezi Expedition xi, 130–1
Livingstonia Mission, in Africa 131
London Missionary Society 130, 131
London School of Tropical
    Medicine 143, 146, 153
Long, Edward 60
Lugard, Fredrick 150
Lustig, A. 112, 170
Lyons, Mariynez 157

MacKenzie, John M. xxvi
Madness 134, 135–6, 137, 157
Madness and Civilization: A History of
    Insanity in the Age of Reason 134
Madras 103, 104, 106, 109
Madras General Hospital 106
Madras hospital 104, 105, 106
Madras Medical School 107
Madras Medical College 107
Magellan, Ferdinand 2
Malabar Medicus 29
Malaria xx, xxiii, 125, 126, 138n.7,
    142, 143, 154, 200, 201. *See also*
    Quinine
    in Americas 25, 37n.25, 77
    colonialism and 156–7
    discovery of malaria parasite 145–6
    eradication programmes 149, 177,
    202, 203
    in Europe 62
    research in 148–9
Manson, Patrick 126, 142, 143, 145,
    148, 153, 154
Markets and medicines 12–15
Markham, Clements R. 25, 127
Marks, Shula xvi, xxviii
Marshall, Henry, and army
    reforms 49–50
Marshall, P.J. 57

Materia Indica 187
Materia Medica of Hindoostan 31–2
Materia medica, as a medical
   discipline 35
Mau Mau Rebellion (Kenya) 135
McGrigor, James 48–9, 50
McKeown, Thomas 204
McWilliam, Dr James O. 125
Medical botany
   and chemistry 32, 33, 34–5, 36
   classification and codification 30,
      31–3, 183
   and colonial bioprospection 6–7,
      20–6, 34, 126–7, 190, 191
   missionaries and 26–30
   modern medicine and 30–5
   Spanish engagement with 23–4, 25
Medical College of Bengal 107
Medical Commentaries 12
Medical Department, establishment
   of 106
Medical marketplace 13, 15
Medical missionaries, role in
   Africa 28, 129–33, 190. *See also*
   Livingstone, David
Medical traditions, invention
   of 183–4
Medicinal plants. *See* Medical botany
Medicine xiii–xv
   in Africa, history of xxv–xxvi
   colonial. *See* Colonial medicine
   and colonialism ix, x, xv–xvi,
      xx, 35–6, 123, 128–9. *See also*
      imperialism *below*
   and commerce 2, 5–9, 15
   European. *See* European/Western
      medicine
   history of xx–xxi
   and imperialism ix, x–xiii, xv, xix.
      See also colonialism *above*
   invented traditions of 197. *See also*
      Traditional medicine
   military 48, 49, 51–3
   modern. *See* Modern medicine

minerals and 9
   and natural history xiv, xv. *See also*
      Medical botany
   naval, changes in 42, 43–8
   negative impact of xx–xxi, xxiii
   preventive. *See* Preventive medicine
   as tool of empire. *See* colonialism *above*
   traditional. *See* Traditional medicine
   tropical. *See* Tropical medicine
   and warfare 48
Mediterranean trade 1, 2
Mehta, Dr Jivraj N. 116–7
Mercantilism, rise of 3
Miasmatic theories of disease 44, 62,
      63, 69, 125, 144
Migration, and disease 73–5, 96–7
Military medicine 48–9, 51–3
Minerals and medicine 9
Missionaries ix, 129–30
   in Africa xiv, 123, 127, 132–3, 192.
      See also Medical missionaries
   and Amerindian healing
      traditions 27–8
   botanical gardens of 28
   and hybrid medical
      traditions 129–130, 192–3
   in India 28–30
   medical botany and 26–30
   and medicine xxiv, 27, 30
   plantations, role in 26, 28
Missionary Travels and Researches 131
Modern medicine xiv–xv
   and chemistry 32, 33–4, 36
   and colonial medical botany 30–5
   and colonial medicine xv–xvi
   and imperialism ix, x–xiii, xv
   laboratory research and 178
   and indigenous medicine,
      interaction xii–xiii, 182, 183
   See also European/Western medicine
Monardes, Nicolas 23, 62
Moravian missionaries 28
Morgan, Henry 41
Morphine 33

Morrison, Robert 32
Mortality
  colonial migration and 78–82
  decline, and preventive medicine.
    See Preventive medicine
  European. *See* European mortality
  global decline in 203–5
  imperialism and 127–8, 129
  patterns, in sea voyages 79t, 80, 81
  slave trade and 77, 78–81
Mozambique 124
Musaeum Zeylanicum 23
Mysore wars 105, 106

Natal Government Asylum 137
National Health Service, in
  Britain 203
National Planning Commission 117
Native Medical Institution 187
Natural history 30, 31
  and chemistry 34–5
  See also Medical botany
Navy
  healthcare, changes in 42, 43–8
  hospitals 47–8
New Imperialism, Age of ix, x, xii,
  xix, 164
  and bacteriology 171, 176. *See also*
    Bacteriology
  and germ theory xii. *See also* Germ
    theory
  and tropical medicine 150–2. *See*
    also Tropical medicine
Nicolle, Charles 174, 175
Niger River 124, 125, 127
Nigeria 150
Non-contagionist theories of
  disease 66, 85, 86, 89–90, 96, 148
Nyerere, Julius K. xxvii

Obeah 8, 137, 182
Observation on the Nature and Cure of
  Hospital and Jayl Fevers 45
Oldenburg, Henry 32

Onslow Commission 52
Opium 20, 33
Opium Wars 193–4
Orientalism 186
Orientalism xxi. *See also* Said, Edward
  below
Osborn, Michael 68
Oudh Akhbar 187
Out of Our Minds: Reason and Madness
  in the Exploration of Central
  Africa 136

Packard, Randall xxiii, 157
Palestine, germs and civilization in 169
Panama Canal, and yellow fever
  epidemic 95
Paracelsus 8
Parasites 143
Parasitology 147
  and British tropical medicine 141,
    142, 143
Parle, Julie 136
Pasteur, Louis xii, 50, 145, 166, 169,
  176, 178. *See also* Pasteurian
  science; Pasteur Institutes
Pasteur Institute, Paris 166, 173
Pasteur Institutes xii, 147, 166–7,
  172–3
  economic influence of 175
  and French civilizing mission xii,
    173–8
  French imperialism and 171, 173
  in India 147–8
  vaccination programmes 173, 174
  See also Pasteurian science
Pasteurian population 171
Pasteurian science 164, 170, 171, 173,
  176, 178
Pasteurization 166, 168, 175, 178
Pelletier, Pierre-Joseph 34, 126
Petiver, James 11, 13, 21, 32
Pettenkofer, Max Von 144–5
Pharmaceutical industry, growth
  of xii, 34, 35, 36

Pharmacology 35
Philadelphia, yellow fever in 93
Philosophical Transactions of the Royal Society 11
Physicians 42
Pietist missionaries 28–9
Plague 58, 170
  in Bombay 110–11, 113
  in Europe 61, 74
  in Italy, and quarantine system 82–3
  vaccine for 170
Plague Research Laboratory, Bombay 111
Plantarum Historiae Universalis Oxoniensis 32
Plantations 7, 20, 21, 25, 35, 60, 96
  missionaries' role in 26, 28
  and yellow fever 93
Plants, and colonialism 20, 35. *See also* Medical botany
Plütschau, Heinrich 29
Pocock, George 52
Polynesia, depopulation in 78
Polygenism 58, 65–6, 69
Porot, Antoine 136
Porter, Roy xxi
Portuguese x, 2, 21–2, 101, 102
Preventivemedicine 9, 43, 50, 53, 201
  and European mortality decline 48, 49, 50–1, 53, 203
Pringle, John xi, 41, 45
Protectorate system, in Africa 150–1, 152, 176, 190
Psychiatry 134, 135, 136, 137
Public health xii, xv, 87, 159, 201, 204
  in Britain 87, 96
  and colonialism in India xxvi, 108–15, 117
  in China 194
  and decline in mortality 204
  international health initiatives in xiii, 95–6, 115, 117, 202
  laboratories and 166–7

preventive medicine and. *See* Preventive medicine
  socialist ideas of 203
Public Health Act, in Britain 87
Putrefaction 167, 169
  climate and 63–4, 144
  and disease 44–5, 46, 62

Qanun (Canon of Medicine) 185
Quarantine, system of xv, 86, 87–8, 89–90, 91–3, 96
  in Australia 169
  British opposition to 86, 87, 89
  in Italy 82–3
  and political sovereignty 91
Quinine, 33–4, 51, 81, 125–6, 128. *See also* Cinchona
  and African colonization 123, 127–8

Rabies 112
Rabies vaccine xii, 166
Race ix, 57, 59–61, 68, 69
  and colonial settlement patterns 58, 61, 68
  French ideas of 67
  and labour, link between 60
  modern ideas of 66
  monogenic theories of 59, 61, 69–70
  Polygenic theories of 58, 65–6, 69
  and power 66–7
  theories of, and colonialism 64–9
  and tropical climate 58–62, 66
Racial determinism 66
Racial pathology, and theory of carrier state 171
Racism 66, 70
Ranger, Terence xxi, 130, 132, 183
Ray, John 32
Reed, Walter 95
Rentier class, and colonial state, link 114
Report of the Sanitary Conditions of the Labouring Population 87

Revolt of 1857 103, 109
Rhazi, Muhammad bin Zakaria
　(Rhazes) 185
Rinderpest xxiii, 158
Rio de Janeiro 93, 114, 168
Robertson, William 9
Rockefeller Foundation, and public
　health 96, 115, 202
Rogers, Leonard 142, 147, 149, 154
Ross, Ronald 126, 142, 143, 148, 156,
　201, 202
　discovery of malarial parasite
　145–6
Rottler, John Peter 29–30
Roux, Émile 145, 176
Roxburgh, William 23
Royal College of Physicians
　(London) xiv
Royal Society of London xiv, 32, 35
Royle, J.F. 186

Said, Edward xxi–xxii, 136, 165
Saint-Dominique 25–6, 92
Sanitary conferences, and cholera 85,
　86–7, 88, 89, 90–2
Sanitary reforms, in colonial
　army 48–50
Sanitation and hygiene xi–xii, xv,
　43, 49, 94, 95, 201, 204. See also
　Sanitary reforms; Public health
Sarsaparilla 20
Scheid, Volker 195
Schiebinger, Londa xvi, 31
Scurvy 40, 43–4, 45, 46, 47, 102
Semple, David 147
Sergent, Edmond 177
Sergent, Emile 176
Sergent, Etienne 177
Seven Years War 40, 57, 165
Sheridan, Richard 10
Shlomowitz, Ralph 80
Sick and Hurt Board 47
Sigerist, Henry 115, 201
Simms, F.W. 110

Slave trade 15, 122–3, 124
　and African mortality 78–81
　See also African slaves
Sleeping Sickness
　(trypanosomiasis) xx, 149, 153,
　154–6, 157–8, 171
Sloane, Hans 8, 11, 13, 26, 32, 61, 62
Smallpox xx, 194
　and Amerindian depopulation 75–7
　eradication of 111, 148, 203
　in India xxv, 111, 148
Snow, John 87
Society for the Extinction of the Slave
　Trade and for the Civilization of
　Africa 123
Sokhey, S.S. 117, 188
South Africa, insanity in 137, 157
South Sea Company 4
Spangenberg, Augustus Gottlieb 28
Spanish imperialism x, xx, xxiv,
　1–2, 59
　and Amerindian depopulation xx,
　73, 75–8
　and civilizing mission xi, xviii, xx,
　1–2
　and medical botany 23–4, 25
Spanish flu xiii, 201, 204
Sphere of influence, system of 151,
　152
Spice trade 3, 22, 33
Spices 1, 20
Stanley, Henry Morton xvi, xviii, 131
Steel, Henry Draper 21
Strophanthus 20, 191–2
Strychnine 33
Subaltern studies xvi, xvii, xxii, xxv
Suez Canal, quarantine on 85, 89, 90
Surgeons xv, 10, 42, 43, 48, 50,
　106–7
　and European medicine 2, 4, 10–12
Sushruta Samihita 185, 186
Swaziland, malaria in 157
Sydenham, Thomas 63, 69, 125
Szreter, Simon 204

Taleef Shareef or the Indian Materia Medica 187
Tanzania, villagization in xxvii, 158
Taylor, Kim 195
Thompson, Dr T.R.H. 125–6
Thomson, James 26
Through the Dark Continent xviii
Tobacco ix, 4, 20, 25, 33, 60, 73, 94, 175
Trading companies, establishment of 4
Traditional medicine xiii, xxiv, 197
    African, and colonialism 8, 190–3
    Chinese, modern invention
        of 193–6
    Indian 182, 184–90
    invention of 182–3, 184, 187, 197
    and modern medicine xii, 182, 183
Traditions, invention of 183–4, 197
Treatise of the Drugs and Medicines of the East Indies 22
Trevelyan, George Otto xviii
Tropical disease 159
    ecological change and 156–8
    climate and 62–4, 69, 125, 145, 168
    and climatic determinism 66, 169
    constructive imperialism and 158
    See also individual diseases
Tropical medicine x, xii, xv, 68, 81, 96, 141–4, 152, 159
    Asian *vs* African practice of 153–4
    and colonial burden 156–8
    and constructive imperialism 152–6
    field surveys in 153, 154–6
    hybrid traditions of 147–50
    institutional development of 146–7, 153
    and New Imperialism 150–2
    origins of 144–7
    research in colonies 147–9
Tropics 20, 58–9, 61, 66, 145, 159, 169
    climate of, and racial
        difference 58–62, 66
    and disease 62–4, 69, 125, 145, 168.
        See also Tropical disease

European acclimatization of. *See* Acclimatization
    and germs, link between 169, 170.
        See also Germ theory;
        Putrefaction
    putrefaction in. *See* Putrefaction
Trotter, H.D. 125
Trypanosomiasis. *See* Sleeping sickness
Tuberculosis, in South Africa 157
Typhoid 171

Uganda, sleeping sickness in 153, 156
Ujamaa xxvii, 158
Unani-tibb xii, 184, 185, 186, 187–8, 189–90
Universities Mission to Central Africa 1, 31, 132–3

Vaccinations xv, 112, 168, 169, 178
    global programmes xiii, 203
    Pasteur institutes and 173, 174
Vaccines xii, 144, 166, 167
Vaids 186, 187
Van Rheede, Henry 22–3, 28
Varier, P.S. 189
Vaughan, Megan xxv, 130, 134
Venereal disease 52, 96
Voeks, Robert xxv
Voodoo 8, 137, 182

Wallerstein, Immanuel xix
Warm Climates and Western Medicine 142
Watts, Sheldon xxiii
West African Medical Services 154
Western Medicine as Contested Knowledge xxiv
White Plague, Black Labor 157
White towns 61, 110
World Health Organization xiii, 149, 159, 203
Worboys, Michael 141

Wright, Almroth 147

Yellow fever 40, 77, 92–6, 146
Yellow Fever Commission 96
Yersin, Alexandre 112, 170, 173, 174

Zambezi Expedition 130–1. *See also* Livingstone, David
Ziegenbalg, Bartholomäus 29
Zomba Lunatic Asylum 134
Zymotic theory 167